T0396686

Arthropod-Plant Interactions

Progress in Biological Control

Volume 14

Published:

Volume 4
J. Gould, K. Hoelmer and J. Goolsby (eds.):
Classical Biological Control of *Bemisia tabaci* in the United States. 2008
ISBN 978-1-4020-6739-6

Volume 5
J. Romeis, A.M. Shelton and G. Kennedy (eds.):
Integration of Insect-Resistant Genetically Modified Crops within IPM Programs. 2008
HB ISBN 978-1-4020-8372-3; PB ISBN 978-1-4020-8459-1

Volume 6
A.E. Hajek, T.R. Glare and M.O'Callaghan (eds.):
Use of Microbes for Control and Eradication of Invasive Arthropods. 2008
ISBN: 978-1-4020-8559-8

Volume 7
Jonathan G. Lundgren:
Relationships of Natural Enemies and Non-Prey Foods. 2008
ISBN: 978-1-4020-9234-3

Volume 8
S.S. Gnanamanickam:
Biological Control of Rice Diseases
ISBN: 978-90-481-2464-0

Volume 9
F.L. Cônsoli, J.R.P. Parra and R.A. Zucchi (eds.):
Egg Parasitoids in Agroecosystems with Emphasis on *Trichogramma*
ISBN: 978-1-4020-9109-4

Volume 10
W.J. Ravensberg:
A Roadmap to the Successful Development and Commercialization of Microbial Pest
Control Products for Control of Arthropods
ISBN: 978-94-007-0436-7

Volume 11
K. Davies and Y. Spiegel (eds.):
Biological Control of Plant-Parasitic Nematodes. 2011
ISBN 978-1-4020-9647-1

Volume 12
J.M. Mérillon and K.G. Ramawat (eds.):
Plant Defence: Biological Control
ISBN 978-94-007-1932-3

Volume 13
H. Roy, P. De Clercq, L.J. Lawson Handley, J.J. Sloggett, R.L. Poland and E. Wajnberg (eds.):
Invasive Alien Arthropod Predators and Parasitoids: An Ecological Approach
ISBN 978-94-007-2708-3

For further volumes:
http://www.springer.com/series/6417

Guy Smagghe · Isabel Diaz
Editors

Arthropod-Plant Interactions

Novel Insights and Approaches for IPM

 Springer

Editors

Guy Smagghe
Department of Crop Protection
Faculty of Bioscience Engineering
Ghent University
Coupure Links 653
Ghent, Belgium

Isabel Diaz
Centro de Biotecnología y
Genómica de Plantas
Universidad Politécnica de Madrid
Autovía M40 (km 38)
Pozuelo de Alarcón, Madrid, Spain

ISBN 978-94-007-3872-0   ISBN 978-94-007-3873-7 (eBook)
DOI 10.1007/978-94-007-3873-7
Springer Dordrecht Heidelberg New York London

Library of Congress Control Number: 2012936129

Printed on acid-free paper

Springer is part of Springer Science+Business Media (www.springer.com)

Foreword

Insect pests constitute a very important constraint in global agriculture. The potential yield of all agricultural crops and/or farmer income are reduced substantially due to direct or indirect effects of insect attack. In order to control damage to crops by insects hundreds of millions of US$ need to be spent annually on crop protection chemicals such as insecticides. In addition to the financial cost of insecticides, we are becoming increasingly aware of the environmental and health hazards associated with conventional pesticides. In more recent years biotechnology has provided additional tools to limit the damage caused by insects while at the same time minimizing or perhaps eliminating some or all of the environmental and health risks associated with chemical insecticides. This is even more critical as pressure mounts on regulators to put sustainability, health and environmental protection high on their list of criteria for evaluating the fate of existing or new chemical pesticides.

In order to harness the power of biotechnology in the context of crop protection in its broadest senses it is crucial that we develop an in depth understanding of biological, developmental and evolutionary aspects of the complex interactions between crop plants and their insect pests.

One mechanism through which plants develop resistance against their insect pests is through constant evolution of endogenous defence pathways stimulated by insect attack. The power of next generation sequencing technologies for plants and also their insect pests has the potential to provide invaluable information through comparative genomics that can be used to design better strategies for IPM combining conventional as well as molecular/biotechnological approaches. To that effect bioinformatic tools and resources as well as sophisticated algorithms for data analysis and interpretation are already making an impact in terms of new potential strategies for insect pest control.

Little is known about the mechanism responsible for the physiological adaptation of the insect gut when insects feed on living plant material. However more information is emerging from studies on the digestion of plant tissues by insects feeding through different means (e.g. leaf chewing or phloem sucking) and elucidation of factors which determine the physiological processes leading to adaptation and subsequent resistance of feeding insects to deal with plant secondary metabolites and insecticidal proteins. More recently it has been established that insect feeding on plants are adapted to circumvent the effects of insecticidal proteins of plant origin that specifically target their digestive enzymes. Thus transcriptome analysis of phytophagous insects is revealing new aspects of the physiology of the insect gut that may be involved in the adaptation to plant defences. However, investigations aiming towards elucidating both components of the plant-insect interaction at the whole genome level are needed in order to develop a better understanding of this interaction.

In order to remain one step ahead of the pest and move away from reliance on single resistance genes, inevitably resulting in a never ending co-evolution of plants and their insect pests, multi-mechanistic resistance must be breed/ engineered into our crops. Transgenic crops are an essential and inevitable component towards sustainable food security as part of IPM where molecular breeding strategies are also used to develop pest resistant crops. In depth knowledge of how plants respond to insect attack, at the molecular level is very important in order to identify appropriate genes, gene families or markers (QTLs) for subsequent breeding programmes.

Biological control has rapidly become an essential component of modern horticultural. However most success stories come from greenhouse applications with open field biological control presenting a big challenge. More recently encouraging results are emerging from studies involving the development of targeted precision biocontrol and pollination enhancement using different species of bees. The "entomovector" technology utilizes insects as vectors of biological control agents targeting plant pests and diseases. The technology depends on bee management, manipulation of bee behavior, components of the cropping system, and on the plant-pathogen-vector-antagonist-system.

When insects commence feeding on a plant the amount of volatile compounds emitted by the plant under attack increases dramatically. Concomitantly, the composition of the volatiles changes often in such a way to attract natural enemies of the pest or on occasion to increase repellency to herbivores. Herbivore-induced volatiles promote a natural form of biological pest control referred to as "indirect plant defense". It has been suggested that this phenomenon could be exploited to enhance crop protection. Different approaches are being explored to manipulate indirect defenses through the application of synthetic volatiles or via transgene-mediated modulation of plant-volatile production.

The pace of the development of insect-resistant crops through genetic engineering has increased dramatically since the commercial introduction of the first crop plants (cotton and corn) expressing a single *Bacillus thuringiensis* (Bt) toxin, 15 years ago. Amongst the key benefits of these first generation insect resistant crops has been a huge reduction in pesticide application. In order to address issues of durability of resistance and also to extend the technology to pests outside the host range of Bt, substantial investments have been made in the search for alternative/ complementary strategies to protect crops from insect pests. Technological advances in plant genetic transformation provided tools for transferring multiple pest resistance traits into agronomically important crops while novel strategies based on a better understanding of endogenous plant resistant mechanisms are emerging and appear to constitute further elements in our efforts to achieve durable resistance against insect pests.

Transgene-encoded dsRNAs provide interesting alternative strategies for insect pest control which can either be exploited independently or complement *Bacillus thuringiensis* (*Bt*) toxins in insect management programs. Transgenic crops with such "pyramided" insect-protection traits are most likely superior to single-trait crops with respect to development of resistance by insect pests.

Transgenic (GM or GE) crops are subjected to safety and environmental risk assessment which is far more stringent than for conventionally bred varieties. In the European Union, regulation is influenced by political expediency, and implementation of regulatory assessments by scientifically competent bodies are almost invariably overruled. This anti-science approach to regulation in combination with further complex, onerous and hardly justified policies on traceability, labeling, coexistence, socio-economic issues and liability constitute de facto trait barriers to legitimize the stance of the EU towards GM crops.

Paul Christou

Paul Christou

Currently Dr. Christou works at Universitat de Lleida as an ICREA Research Professor and head of the Applied Plant Biotechnology Laboratory.

He received BS degree in Chemistry and PhD in plant biochemistry from University College London, worked as senior scientist at Agracetus Inc. Madison WI USA. Subsequently head of Molecular Biotechnology Unit, John Innes Centre, Norwich UK and then was head of Crop Genetics and Biotechnology Department at the Fraunhofer Institute for Molecular Biotechnology and Applied Ecology, Aachen/Schmallenberg, Germany.

For the past several years he has investigated the organization of foreign genes in important crops such as maize, rice and wheat and its impact on transgene expression levels and stability. Applied aspects of our research include production of high value recombinant pharmaceuticals (vaccines and antibodies) in plants for use in human health and veterinary medicine; engineering crop plants for enhanced nutrition and novel strategies of sustainable and environmentally friendly agriculture using transgenic approaches, all with emphasis on developing countries, poverty alleviation and food security. His group is heavily involved in training and capacity building in the area of plant biotechnology focusing on developing countries.

Dr. Christou is Chief Editor of Transgenic Research and Molecular Breeding and belongs to the Editorial Advisory board of several scientific journals. He is author of books, chapters in books and has published near 150 articles in the most prestigious peer reviewed journals. He is also author of several patents, has supervised PhD theses, has imparted lectures in international meetings and workshops and has been the coordinator or principal investigator of research projects financed by international public and private institutions.

Progress in Biological Control

Series Preface

Biological control of pests, weeds, and plant and animal diseases utilising their natural antagonists is a well-established and rapidly evolving field of science. Despite its stunning successes world-wide and a steadily growing number of applications, biological control has remained grossly underexploited. Its untapped potential, however, represents the best hope to providing lasting, environmentally sound, and socially acceptable pest management. Such techniques are urgently needed for the control of an increasing number of problem pests affecting agriculture and forestry, and to suppress invasive organisms which threaten natural habitats and global biodiversity.

Based on the positive features of biological control, such as its target specificity and the lack of negative impacts on humans, it is the prime candidate in the search for reducing dependency on chemical pesticides. Replacement of chemical control by biological control – even partially as in many IPM programs – has important positive but so far neglected socio-economic, humanitarian, environmental and ethical implications. Change from chemical to biological control substantially contributes to the conservation of natural resources, and results in a considerable reduction of environmental pollution. It eliminates human exposure to toxic pesticides, improves sustainability of production systems, and enhances biodiversity. Public demand for finding solutions based on biological control is the main driving force in the increasing utilisation of natural enemies for controlling noxious organisms.

This book series is intended to accelerate these developments through exploring the progress made within the various aspects of biological control, and via documenting these advances to the benefit of fellow scientists, students, public officials, policy-makers, and the public at large. Each of the books in this series is expected to provide a comprehensive, authoritative synthesis of the topic, likely to stand the test of time.

Heikki M.T. Hokkanen, Series Editor

Preface

The book consists of multiple chapters by leading experts on the different aspects in the unique relationship between arthropods and plants, the underlying mechanisms, realized successes and failures of interactions and application for IPM, and future lines of research and perspectives. Interesting is the availability of the current genomes of different insects, mites and nematodes and different important plants and agricultural crops to bring better insights in the cross talk mechanisms and interacting players. This book will be the first one that integrates all this fascinating and newest (from the last 5 years) information from different leading research laboratories in the world and with perspectives from academia, government and industry.

Contents

1 Co-evolution of Genes for Specification in Arthropod-Plant Interactions: A Bioinformatic Analysis in Plant and Arthropod Genomes 1
Manuel Martinez

2 The Impact of Induced Plant Volatiles on Plant-Arthropod Interactions 15
Juan M. Alba, Petra M. Bleeker, Joris J. Glas,
Bernardus C.J. Schimmel, Michiel van Wijk, Maurice W. Sabelis,
Robert C. Schuurink, and Merijn R. Kant

3 Physiological Adaptations of the Insect Gut to Herbivory 75
Félix Ortego

4 Successes and Failures in Plant-Insect Interactions: Is it Possible to Stay One Step Ahead of the Insects? 89
Angharad Gatehouse and Natalie Ferry

5 Multitrophic Interactions: The Entomovector Technology 127
Guy Smagghe, Veerle Mommaerts, Heikki Hokkanen,
and Ingeborg Menzler-Hokkanen

6 Biotechnological Approaches to Combat Phytophagous Arthropods 159
Isabel Diaz and M. Estrella Santamaria

7 Use of RNAi for Control of Insect Crop Pests 177
Luc Swevers and Guy Smagghe

8 Regulatory Approvals of GM Plants (Insect Resistant) in European Agriculture: Perspectives from Industry 199
Jaime Costa and Concepcion Novillo

Index 217

Chapter 1
Co-evolution of Genes for Specification in Arthropod-Plant Interactions: A Bioinformatic Analysis in Plant and Arthropod Genomes

Manuel Martinez

1.1 Plant-Arthropod Co-evolution

Plants and arthropods are organisms that continuously interact in natural ecosystems. These relations provide them with reciprocal benefits. Several arthropods species protect plants and take active part in plant pollination. Likewise, plants provide shelter, oviposition sites and food to arthropods. However, arthropods might also act as plant pests and be extremely harmful to plants. During evolution, plants have developed sophisticated defence mechanisms to avoid pest damage and arthropod pests have varied their mechanism of attack to overcome plant defences. Then, arthropod-plant interaction is a highly dynamic system, subjected to endless variation. A diagram of plant-herbivorous interactions is shown in Fig. 1.1. When arthropods attack a plant (1), the response of the plant is a consequence of the perception of herbivory associated molecular patterns (HAMPs) (Felton and Tumlinson 2008; Mithofer and Boland 2008; Wu and Baldwin 2009). HAMPs can be classified into two categories: (i) chemical elicitors derived from herbivore oral secretions and oviposition fluids; and (ii) those that are originated from the specific patterns of wounding. Although chemical and mechanical stimuli lead to tissue damage, plants respond differentially to both stresses. Attacked plants have several different defence levels (2). Defences include physical barriers such as cuticles, trichomes and thorns, a battery of compounds with toxic, repellent or anti-digestive effects on arthropod pests, and the emission of volatile compounds to attract the predators of arthropod pests (Heil 2008; Unsicker et al. 2009).

These biochemical and physiological changes can be costly to plants. Therefore, plants have acquired complex regulatory networks to maintain a balance between

M. Martinez (✉)
Centro de Biotecnología y Genómica de Plantas (UPM-INIA), Campus Montegancedo,
Universidad Politécnica de Madrid, Autovía M40 (Km 38), 28223 Pozuelo de Alarcón,
Madrid, Spain
e-mail: m.martinez@upm.es

G. Smagghe and I. Diaz (eds.), *Arthropod-Plant Interactions: Novel Insights and Approaches for IPM*, Progress in Biological Control 14, DOI 10.1007/978-94-007-3873-7_1, © Springer Science+Business Media B.V. 2012

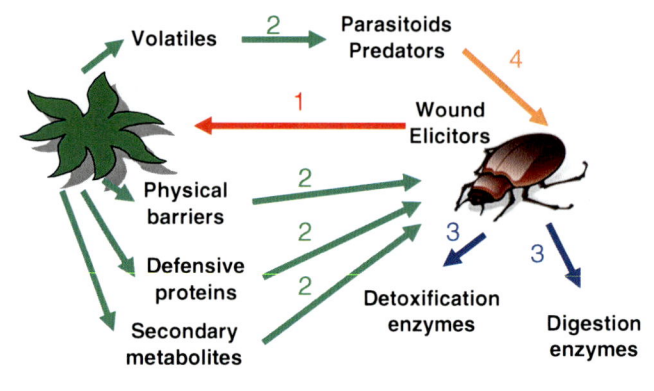

Fig. 1.1 Plant-herbivorous interactions

development and defence response when attacked by herbivores (Tian et al. 2003). In parallel, insects developed strategies to overcome plant barriers (3). Detoxification and sequestration of toxic compounds and alteration of gene expression pattern leading to variations of digestive enzymes are some examples of arthropod adaptation to the plant (Carrillo et al. 2011; Li et al. 2002; Zeng et al. 2009). In addition, arthropod pests might develop alternative mechanisms to fight against of parasitoids and predators attracted by plant volatiles (4). Plant secondary metabolites can be sequestered and used in pest defence against predators (Opitz et al. 2010).

This plant-herbivore co-evolution was formerly explained by Ehrlich and Raven (Ehrlich and Raven 1964), who suggested that in response to herbivory a plant species may evolve a novel and highly effective defense that enables escape from its associated herbivores. Moreover, after some time one herbivorous species will adapt to it and will colonize this plant. Over time, such stepwise adaptive mechanisms will be continuously repeated. However, most recent research leads to a more complex explanation of plant-herbivory co-evolution (Futuyma and Agrawal 2009) which require integrating macroevolutionary patterns with functional, genetic, and ecological evidences.

Thus, changes during evolution have determined the actual battery of genes involved in the arms-race between plants and pests. In this context, the importance of knowing the gene content of both systems becomes evident. There are an extremely large number of amino acid sequences in the public databases belonging to most known species that can be used to search for key genes in the plant-pest interaction. Of these, the genomic sequences obtained from genome projects are crucial tools to establish the repertoire of genes putatively involved in the plant-pest interaction and to develop plant-herbivore model systems (Whiteman and Jander 2010).

1.2 Plant Genomes

Recent advances in genomic studies have led to a deep increase in our knowledge of plant genomes (Feuillet et al. 2011). Table 1.1 summarizes the current state-of-art of main plant genomic sequencing projects with genome sequences available in the web. According to the Genomes On-Line Database (GOLD, http://www.genomesonline.org/), more than 20 plant genomes have been already completed and there are more than 200 ongoing plant genomic projects. The first genome completely sequenced was that of the eudicot model plant for plant biology, *Arabidopsis thaliana*. Then, the monocot crop plant rice was sequenced. Nowadays, several other plant species from both, eudicots and monocots clades have been completely sequenced. Among eudicot species, there are examples of species from the most important orders, whereas all monocots species, with the exception of the date palm, belongs to the Poales order. Main genome centers as JGI (Joint Genome Institute), BGI (Beijing Genomics Institute), JCVI (J. Craig Venter Institute) or MSU (Michigan State University) support most complete plant genomic projects. Plant genomes have been usually sequenced by the Sanger sequencing technology and have relatively small genomes. The recent development of next generation sequencing technologies that offer improvements in throughput and cost efficiency by using massively parallel sequencing systems has lead to their use in plant genomic sequencing. Then, the genomes of the cucumber, the cassava, the cacao and the strawberry have been mostly based on Illumina, Roche 454 or SOLiD reads. This technology will make possible the sequencing of large and complex genomes of most crop plant species.

Recently, a group of internationally recognized sequencing centers (Chain et al. 2009) proposed six community-defined categories of standards that better reflect the quality of a genome: (i) standard draft: minimally or unfiltered data that are assembled into contigs. This is the minimum standard for a submission to the public databases. Sequence of this quality will likely harbour many regions of poor quality and can be relatively incomplete; (ii) high-quality draft: overall coverage representing at least 90% of the genome. This is still a draft assembly with little or no manual review of the product. This is appropriate for general assessment of gene content; (iii) improved high-quality draft: additional work has been performed beyond the initial sequencing and high-quality draft assembly, by using either manual or automated methods. This standard is normally adequate for comparison with other genomes; (iv) annotation-directed improvement: may overlap with previous standards, but the term emphasizes the verification and correction of anomalies within coding regions; (v) non-contiguous finished: describes high-quality assemblies that have been subject to automated and manual improvement, and where closure approaches have been successful for almost all gaps; (vi) finished: refers to the current gold standard. All sequences are complete and have been reviewed and edited.

Based on this classification, as new updates from every genome are released, the quality of the genomic sequences is step-by-step upgrading from the standard

Table 1.1 Higher plant genome sequencing projects with publicly available sequences

Clade/Order	Species	Genome size (Mb)	Current status	Database/Reference	Protein coding transcripts
Eudicots					
Brassicales	*Arabidopsis thaliana* (thale cress)	135	Finished	TAIR Release 10/ AGI (2000)	33,410
Brassicales	*Arabidopsis lyrata* (rock cress)	207	Standard draft	JGI v1.0/ unpublished	32,670
Brassicales	*Carica papaya* (papaya)	135	High-quality draft	University of Hawaii v0.4/ Ming et al. (2008)	27,796
Cucurbitales	*Cucumis sativus* (cucumber)	203	Improved high-quality draft	BGI-JGI v1/ Huang et al. (2009)	32,528
Fabales	*Lotus japonicus* (trefoil)	201	High-quality draft	Kazusa DNA Institute v2.5/ unpublished	43,146
Fabales	*Medicago truncatula* (barrel medic)	257	Standard draft	JCVI Mt3.5/ unpublished	53,423
Fabales	*Glycine max* (soybean)	975	Improved high-quality draft	JGI Glyma1.0/ Schmutz et al. (2010)	66,153
Lamiales	*Mimulus guttatus* (spotted monkeyflower)	321	Standard draft	JGI v1.1/ unpublished	28,282
Malpighiales	*Manihot esculenta* (cassava)	533	Standard draft	JGI v4.1/ unpublished	34,151
Malpighiales	*Populus trichocarpa* (poplar)	403	Improved high-quality draft	JGI v2.2/ Tuskan et al. (2006)	45,033
Malpighiales	*Ricinus communis* (castor bean)	400	Standard draft	JCVI v0.1/ Chan et al. (2010)	31,221
Malvales	*Theobroma cacao* (cacao)	430	Standard draft	CIRAD/ Argout et al. (2011)	28,798
Myrtales	*Eucalyptus grandis* (rose gum)	691	Standard draft	JGI v1.0/ unpublished	44,974
Ranunculales	*Aquilegia coerulea* (colorado blue columbine)	302	Standard draft	JGI/ unpublished	27,583
Rosales	*Fragaria vesca* (woodland strawberry)	240	Standard draft	University of North Texas/ Shulaev et al. (2011)	34,809
Rosales	*Malus x domestica* Borkh (apple)	742	Standard draft	IASMA Research Center v1.0/ Velasco et al. (2010)	84,380

(continued)

Table 1.1 (continued)

Clade/Order	Species	Genome size (Mb)	Current status	Database/Reference	Protein coding transcripts
Rosales	*Prunus persica* (peach)	227	High-quality draft	JGI v1.0/ unpublished	28,702
Sapindales	*Citrus clementina* (Clementine mandarin)	296	Standard draft	JGI v0.9/ unpublished	25,385
Sapindales	*Citrus sinensis* (sweet orange)	319	Standard draft	JGI v1.1/ unpublished	25,376
Solanales	*Solanum lycopersicum* (tomato)	950	Standard draft	MIPS v2.40/ unpublished	n/d
Solanales	*Solanum tuberosum* (potato)	720	Standard draft	MSU/ unpublished	n/d
Vitales	*Vitis vinifera* (grapevine)	487	Improved high-quality draft	Genoscope/ Jaillon et al. (2007)	26,346
Monocots					
Arecales	*Phoenix dactylifera* (date palm)	550	Standard draft	Weill Cornell Medical College/ unpublished	n/d
Poales	*Brachypodium distachyon* (purple false brome)	272	Improved high-quality draft	JGI, v1.0/ TIBI (2010)	32,255
Poales	*Oryza sativa* ssp. indica (rice)	466	High-quality draft	BGI/ Yu et al. (2002)	39,578
Poales	*Oryza sativa* ssp. japonica (rice)	372	Finished	MSU Release 6.1/ Goff et al. (2002)	50,939
Poales	*Setaria italica* (foxtail millet)	406	Standard draft	JGI, v1.1/ unpublished	38,038
Poales	*Sorghum bicolor* (sorghum)	760	Improved high-quality draft	JGI, v1.0/ Paterson et al. (2009)	36,338
Poales	*Zea mays* (popcorn)	2,100	Standard draft	Arizona Genome Institute 5b.60/ Schnable et al. (2009)	110,028

Abbreviation: *n/d* no data

draft to the finished categories. As shown in Table 1.1, and based in Feuillet et al. (2011) current plant genomic sequences are in different quality categories. Rice and Arabidopsis are the only plant species that have finished genome sequences to date. For the establishment of the accuracy of the different protein-coding genes

Table 1.2 Several in progress crop plant genome projects

Clade/Order	Species	Genome size (Mb)	Main sequencing center
Eudicots/Brassicales	*Brassica napus* (rapeseed)	1,100	BGI
Eudicots/Brassicales	*Brassica rapa* (Chinese cabbage)	500	JGI
Eudicots/Brassicales	*Capsella rubella* (pink Shepherd's purse)	250	JGI
Eudicots/Caryophyllales	*Beta vulgaris* (sugar beet)	758	GABI
Eudicots/Fabales	*Phaseolus vulgaris* (common bean)	630	JGI
Eudicots/Fabales	*Vigna radiata* (mungbean)	579	BIOTEC Genome Institute
Eudicots/Fagales	*Castanea mollissima* (Chinese chestnut)	800	CUGI
Eudicots/Gentianales	*Asclepias syriaca* (milkweed)	800	Oregon State University
Eudicots/Malvales	*Gossypium raimondii* (cotton)	880	JGI
Eudicots/Rosales	*Prunus avium* (sweet cherry)	338	WSU
Eudicots/Sapindales	*Citrus sinensis* (sweet orange)	380	JGI
Monocots/Poales	*Hordeum vulgare* (barley)	5,000	GABI
Monocots/Poales	*Oryza glaberrima* (African Rice)	357	Arizona Genomics Institute
Monocots/Poales	*Pennisetum glaucum* (pearl millet)	2,450	University of Groningen
Monocots/Poales	*Triticum aestivum* (bread wheat)	16,000	Genoscope
Monocots/Zingiberales	*Musa acuminata* (banana)	600	Genoscope

annotated in every genome, a systematic use of these standards in each of the published genome sequences is strongly advisable.

Besides complete genomic sequence projects, a great number of genomic projects for crop species are now in progress. Some of them with available on-line information are compiled in Table 1.2. Most of these projects combine Sanger and next generation sequencing technologies and are being developed by international consortiums.

1.3 Pest Genomes

Although the number of pest genomic projects is scarce comparing with plant genomic projects, some insects closely related to plants have also been sequenced during the last years (Table 1.3).

The Human Genomic Sequencing Center located at the Baylor College of Medicine (BCM) has been the institution involved in all these projects. The insect species sequenced are: the silkworm *Bombyx mori*; the honey bee *Apis mellifera*;

Table 1.3 Arthropod plant-related genomic projects

Clade/Order	Species and genotype	Genome size (Mb)	Current status	Database / Reference	Protein coding transcripts
Insects/Diptera	*Mayetiola destructor* (hessyan fly)	158	Standard draft	BCM Mdes v1.0/ unpublished	n/d
Insects/ Hemiptera	*Acyrthosiphon pisum* (pea aphid)	525	Standard draft	BCM Acyr 1.0/ TIAGC (2010)	34,604
Insects/ Coleoptera	*Tribolium castaneum* (red flour beetle)	975	Improved high-quality draft	BCM Tcas 3.0/ Richards et al. (2008)	16,561
Insects/ Lepidoptera	*Bombyx mori* (silkworm)	432	Improved high-quality draft	BCM SilkDB v2.0/ Xia et al. (2004)	14,623
Insects/ Hymenoptera	*Apis mellifera* (honey bee)	231	Improved high-quality draft	BCM Amel v4.5/ HGSC (2006)	10,157
Insects/ Lepidoptera	*Helicoverpa armigera* (cotton bollworm)	400	In progress	BCM	n/d
Acari/ Prostigmata	*Tetranychus urticae* (two-spotted spider mite)	135	In progress	JGI	n/d
Insects/Diptera	*Ceratitis capitata* (mediterranean fruit fly)	500	In progress	BCM	n/d
Insects/ Hemiptera	*Myzus persicae* (green peach aphid)	313	In progress	BCM	n/d
Insects/ Hemiptera	*Bemisia tabaci* (sweet potato whitefly)	1,020	In progress	Zhejiang University	n/d

Abbreviation: *n/d* no data

the red flour beetle *Tribolium castaneum*; the pea aphid *Acyrthosiphon pisum*; and the hessyan fly *Mayetiola destructor*. These insect species are agriculturally important and belong to different orders. Some of them are beneficial insects, such as the hymenoptera honey bee *A. mellifera*, which is a major pollinator of food plants and producer of honey, and the lepidoptera silkworm *B. mori*, which produces silk. In contrast to them, the coleopteran red flour beetle *T. castaneum*, which destroys stored grain and many other dried and stored commodities for human consumption, and the hemiptera pea aphid *A. pisum* and the diptera hessyan fly *M. destructor*, which causes severe damage to green food plants, are serious agricultural pests. Besides, several other plant pests, such as the insects *Helicoverpa armigera*, *Ceratitis capitata* and *Myzus persicae* and the acari *Tetranychus urticae* are currently being sequenced.

1.4   Gene Family Bioinformatics Tools

The annotation of sequenced genomes provides us a great number of putative protein-coding sequences. In plants, they rank from about 25,000 genes in *Citrus* species to more than 100,000 putative genes in maize (Table 1.1). In arthropods, the number of genes is lower but all species have more than 10,000 protein-coding genes (Table 1.3). Several of these proteins will be crucial in the interaction between pests and plants. To date, some protein families important in these interactions have been identified and characterized, but some others are still unknown. Thus, it is crucial to establish the arsenal of proteins present in pests to attack plants and in plants to defend against pests. For that, several databases have developed different ways to cluster proteins in gene families. These bioinformatics databases are available on-line and can be used to perform analyses of gene families, and to know the presence and the number of members of each gene family in a specific species. Over last few years, some of these databases have rapidly become obsolete or have not been properly updated while new databases are continuously being created. Table 1.4 compiles the main databases currently available for gene family classification. Traditional gene family databases based on signatures have been extensively used to classify proteins. Sequence signatures are amino acid motifs conserved among a set of homologous proteins and typically derived from multiple sequence alignments. Searches in signature databases allow genes to be grouped according to similarities with known sequence signatures. These methods have different limitations, such as missing of gene families with yet uncharacterized motifs or domains, or deficient updating. In addition to single databases, integrative databases provide a powerful resource to classify proteins on multiple levels: from protein families to structural superfamilies and function-ally close subfamilies. The best known integrative databases are InterPro (Hunter et al. 2009) and CDD (Marchler-Bauer et al. 2011), which integrates signatures from several signature databases.

All aforementioned signature databases use sequences obtained from prokary-otic and eukaryotic species, without focussing on pests or plants. The continuously increasing number of plant genomic projects has led to the creation of comparative genomic databases specifically from plants. These databases can be used to perform evolutionary and comparative analyses, and to study gene families and genomes organization. Based on orthologous genes (genes sharing common ancestry and that have diverged by speciation) comparative genomics provides a powerful approach to translate functional information from model species to crops. Likewise, the analysis of genes in a phylogenetic context has shown to provide us with useful information about the processes that have contributed to the evolutionary divergence in gene content (Paterson et al. 2010; Van de Peer et al. 2009). The most

Table 1.4 Bioinformatics tools for plant and pest gene family analyses

Bioinformatic tool	URL	Protein families or signatures
Signature Databases		
ProtClustDB Dec 2 2010	http://www.ncbi.nlm.nih.gov/proteinclusters	10,885
Pfam 24.0	http://pfam.sanger.ac.uk/	11,912
PROSITE 20.68	http://expasy.org/prosite/	1,598
PRINTS 41.1	http://www.bioinf.manchester.ac.uk/dbbrowser/PRINTS/index.php	2,050
ProDom 2006.1/ CG267	http://prodom.prabi.fr/prodom/current/html/home.php	574,656/301,126
SMART 6.1	http://smart.embl-heidelberg.de/	895
TIGRFAMs 10.0	http://www.jcvi.org/cms/research/projects/tigrfams/overview/	4,025
PIRSF 2.73	http://pir.georgetown.edu/pirwww/dbinfo/pirsf.shtml	3,233
SUPERFAMILY 1.75	http://supfam.cs.bris.ac.uk/SUPERFAMILY/	2,019
GENE3D 10.0.0	http://gene3d.biochem.ucl.ac.uk/Gene3D/	2,549
PANTHER 7.0	http://www.pantherdb.org/	6,594
Integrative Signature Databases		
InterPro 30.0	http://www.ebi.ac.uk/interpro/	21,178
CDD 2.26	http://www.ncbi.nlm.nih.gov/Structure/cdd/cdd.shtml	41,593
Plant Comparative Genomic Databases		
PLAZA 2.0	http://bioinformatics.psb.ugent.be/plaza/	32,332
Phytozome 6.0	http://www.phytozome.net/	69,664
GreenPhylDB 2.0	http://greenphyl.cirad.fr/v2/cgi-bin/index.cgi	8,227

comprehensive comparative genomic databases that focus on plant gene families are PLAZA (Proost et al. 2009), GreenPhylDB (Rouard et al. 2011) and Phytozome. These databases differ in the genomes they include and are based in methods that typically involve new clustering techniques, which allow to group genes into families not covered by signature methods. For pests, there are no specific databases focused on arthropod gene families yet.

1.5   Gene Families Involved in Plant-Pest Interactions

As previously mentioned, molecular biology is strongly influencing the research on the ecology of plant-pest interactions. To find out how herbivores are able to feed on their host plants, and to understand how plants became resistant to pests, many of the most important genes involved in plant-herbivore interactions have been identified through molecular mapping of mutations, mainly in plants (Whiteman and Jander 2010). Besides, comparative transcriptome analyses of plant defenses have been used to detect differences in global expression when different herbivores are attacking the plant, or to address questions about the responses of a particular gene of interest. This approach has been used to identify candidate genes for further analysis or to suggest novel unknown defense traits (Ehlting et al. 2008; Rasmann and Agrawal 2009).

From these approaches, several plant and pest gene families involved in their reciprocal interaction have been identified. In the plant side, they include (Howe and Jander 2008): (i) genes implicated in the early events such as pest recognition and signalling cascades (e.g., NBS-LRR receptors, MAPK kinases); (ii) genes that lead to the production of secondary metabolites (e.g., terpenoids, alkaloids, furanocoumarins, cardenolites, tannins, saponins, glucosinolates); (iii) genes that encode proteins that directly serves as direct defences (e.g., proteases, protease inhibitors, lectins, chitinases, lipoxygenases, arginases); and (iv) genes that encode volatiles (e.g., terpenes, amines) . In the pest side, it is remarkable the importance of: (i) genes encoding elicitors that are present in oral secretions such as peptides or fatty acids; (ii) genes involved in the detoxification of plant compounds; and (iii) genes directly involved in feeding that can change their expression pattern in response to host.

Searches in gene family databases permit to discover the presence and the number of members of each gene family involved in the plant-pest interaction in different plant and pest species. Thus, as an example, searches in the comparative genomic database GreenPhylDB for lipoxygenase encoding genes show a great variation for this defensive gene family, ranking from 83 putative genes in *Glycine max* to only six genes in *A. thaliana*. These searches, with the caution required when working with automatically annotated sequences from genomes, allow us to predict how these putative gene families from both plant and pest systems have evolved and specialized in different species.

1.6 Future Directions

Next generation sequencing technologies are leading to the completion of many plant genome projects and promises to be involved in many others, mainly in crop species. Furthermore, the number of pest sequencing projects is increasing and in next years, many pest genomes will be completed. In this context, comparative analyses will determine the extent to which plant defense pathways are truly conserved and how microevolutionary forces have shaped both plant defense against arthropod herbivores and herbivore responses to these plant defenses. For that, recent comparative genomic databases have been developed for plant species. Developments of these databases and the creation of new comparative genomic databases will increase the number and accuracy of tools to deal with plant gene families in the near future. From pests, it will be desirable the existence of similar comparative genomic tools to establish pest protein-coding gene families. In this way, comparative analyses will come useful to know: (i) the extent of protein-coding families involved in the plant-pest interaction in each plant or arthropod species; (ii) the existence of species-specific protein-coding families or family members implicated in the plant mechanisms to reduce pest attack or in the pest strategies to overcome plant defense barriers.

Appendix: URLs of the Complete Plant and Pest Genome Projects

Clade/Species	URL
Eudicots	
Aquilegia coerulea (colorado blue columbine)	http://www.phytozome.net/aquilegia.php
Arabidopsis lyrata (rock cress)	http://www.phytozome.net/alyrata.php
Arabidopsis thaliana (thale cress)	http://www.arabidopsis.org/
Carica papaya (papaya)	http://asgpb.mhpcc.hawaii.edu/papaya/
Citrus clementina (Clementine mandarin)	http://www.phytozome.net/clementine.php
Citrus sinensis (sweet orange)	http://www.phytozome.net/citrus.php
Cucumis sativus (cucumber)	http://www.icugi.org/cgi-bin/ICuGI/genome/cuke.cgi
Eucalyptus grandis (rose gum)	http://www.phytozome.net/eucalyptus.php
Fragaria vesca (woodland strawberry)	http://www.strawberrygenome.org/
Glycine max (soybean)	http://www.phytozome.net/soybean.php
Lotus japonicus (trefoil)	http://www.kazusa.or.jp/lotus/summary2.5.html
Malus x domestica Borkh (apple)	http://www.rosaceae.org/projects/apple_genome
Manihot esculenta (cassava)	http://www.phytozome.net/cassava.php
Medicago truncatula (barrel medic)	http://www.jcvi.org/cgi-bin/medicago/overview.cgi
Mimulus guttatus (spotted monkeyflower)	http://www.phytozome.net/mimulus.php
Populus trichocarpa (poplar)	http://www.phytozome.net/poplar.php
Prunus persica (peach)	http://www.phytozome.net/peach.php
Ricinus communis (castor bean)	http://castorbean.jcvi.org/index.php
Solanum lycopersicum (tomato)	http://mips.helmholtz-muenchen.de/plant/tomato/index.jsp
Solanum tuberosum (potato)	http://potatogenomics.plantbiology.msu.edu/
Theobroma cacao (cacao)	http://cocoagendb.cirad.fr/
Vitis vinifera (grapevine)	http://www.genoscope.cns.fr/spip/Vitis-vinifera-e.html
Monocots	
Brachypodium distachyon (purple false brome)	http://www.phytozome.net/brachy.php
Oryza sativa ssp. indica (rice)	http://rice.genomics.org.cn/rice/index2.jsp
Oryza sativa ssp. japonica (rice)	http://www.phytozome.net/mimulus.php
Phoenix dactylifera (date palm)	http://qatar-weill.cornell.edu/research/datepalmGenome/index.html
Setaria italica (foxtail millet)	http://www.phytozome.net/foxtailmillet.php
Sorghum bicolor (sorghum)	http://www.phytozome.net/sorghum.php
Zea mays (popcorn)	http://www.maizesequence.org/index.html
Insects	
Acyrthosiphon pisum (pea aphid)	http://www.aphidbase.com/aphidbase
Apis mellifera (honey bee)	http://hymenopteragenome.org/beebase/
Bombyx mori (silkworm)	http://silkworm.genomics.org.cn/
Mayetiola destructor (hessyan fly)	http://www.hgsc.bcm.tmc.edu/project-species-i-Hessian_fly.hgsc?pageLocation=Hessian_fly
Tribolium castaneum (red flour beetle)	http://www.beetlebase.org/

Acknowledgements I thank Dr. Santamaría for critical reading of the manuscript. The financial support from the Ministerio de Ciencia e Innovación (project BFU2008-01,166) and from the Universidad Politecnica de Madrid/Comunidad de Madrid (project CCG10-UPM/AGR-5,242) is gratefully acknowledged.

References

AGI, Arabidopsis Genome Initiative (2000) Analysis of the genome sequence of the flowering plant Arabidopsis thaliana. Nature 408:796–815

Argout X, Salse J, Aury JM, Guiltinan MJ, Droc G, Gouzy J, Allegre M, Chaparro C, Legavre T, Maximova SN et al (2011) The genome of *Theobroma cacao*. Nat Genet 43:101–108

Carrillo L, Martinez M, Ramessar K, Cambra I, Castanera P, Ortego F, Diaz I (2011) Expression of a barley cystatin gene in maize enhances resistance against phytophagous mites by altering their cysteine-proteases. Plant Cell Rep 30:101–112

Chain PS, Grafham DV, Fulton RS, Fitzgerald MG, Hostetler J, Muzny D, Ali J, Birren B, Bruce DC, Buhay C et al (2009) Genomics. Genome project standards in a new era of sequencing. Science 326:236–237

Chan AP, Crabtree J, Zhao Q, Lorenzi H, Orvis J, Puiu D, Melake-Berhan A, Jones KM, Redman J, Chen G et al (2010) Draft genome sequence of the oilseed species *Ricinus communis*. Nat Biotechnol 28:951–956

Ehlting J, Chowrira SG, Mattheus N, Aeschliman DS, Arimura G, Bohlmann J (2008) Comparative transcriptome analysis of *Arabidopsis thaliana* infested by diamond back moth (*Plutella xylostella*) larvae reveals signatures of stress response, secondary metabolism, and signalling. BMC Genomics 9:154

Ehrlich PR, Raven PH (1964) Butterflies and plants: a study in coevolution. Evolution 18:586–608

Felton GW, Tumlinson JH (2008) Plant-insect dialogs: complex interactions at the plant-insect interface. Curr Opin Plant Biol 11:457–463

Feuillet C, Leach JE, Rogers J, Schnable PS, Eversole K (2011) Crop genome sequencing: lessons and rationales. Trends Plant Sci 16:77–88

Futuyma DJ, Agrawal AA (2009) Macroevolution and the biological diversity of plants and herbivores. Proc Natl Acad Sci U S A 106:18054–18061

Goff SA, Ricke D, Lan TH, Presting G, Wang R, Dunn M, Glazebrook J, Sessions A, Oeller P, Varma H et al (2002) A draft sequence of the rice genome (*Oryza sativa* L. ssp. japonica). Science 296:92–100

Heil M (2008) Indirect defence via tritrophic interactions. New Phytol 178:41–61

HGSC, Honeybee Genome Sequencing Consortium (2006) Insights into social insects from the genome of the honeybee *Apis mellifera*. Nature 443:931–949

Howe GA, Jander G (2008) Plant immunity to insect herbivores. Annu Rev Plant Biol 59:41–66

Huang S, Li R, Zhang Z, Li L, Gu X, Fan W, Lucas WJ, Wang X, Xie B, Ni P et al (2009) The genome of the cucumber, *Cucumis sativus* L. Nat Genet 41:1275–1281

Hunter S, Apweiler R, Attwood TK, Bairoch A, Bateman A, Binns D, Bork P, Das U, Daugherty L, Duquenne L et al (2009) InterPro: the integrative protein signature database. Nucleic Acids Res 37:D211–D215

Jaillon O, Aury JM, Noel B, Policriti A, Clepet C, Casagrande A, Choisne N, Aubourg S, Vitulo N, Jubin C et al (2007) The grapevine genome sequence suggests ancestral hexaploidization in major angiosperm phyla. Nature 449:463–467

Li X, Schuler MA, Berenbaum MR (2002) Jasmonate and salicylate induce expression of herbivore cytochrome P450 genes. Nature 419:712–715

Marchler-Bauer A, Lu S, Anderson JB, Chitsaz F, Derbyshire MK, DeWeese-Scott C, Fong JH, Geer LY, Geer RC, Gonzales NR et al (2011) CDD: a Conserved Domain Database for the functional annotation of proteins. Nucleic Acids Res 39:D225–D229

Ming R, Hou S, Feng Y, Yu Q, Dionne-Laporte A, Saw JH, Senin P, Wang W, Ly BV, Lewis KL et al (2008) The draft genome of the transgenic tropical fruit tree papaya (*Carica papaya* Linnaeus). Nature 452:991–996

Mithofer A, Boland W (2008) Recognition of herbivory-associated molecular patterns. Plant Physiol 146:825–831

Opitz SE, Jensen SR, Muller C (2010) Sequestration of glucosinolates and iridoid glucosides in sawfly species of the genus *Athalia* and their role in defense against ants. J Chem Ecol 36:148–157

Paterson AH, Bowers JE, Bruggmann R, Dubchak I, Grimwood J, Gundlach H, Haberer G, Hellsten U, Mitros T, Poliakov A et al (2009) The *Sorghum bicolor* genome and the diversification of grasses. Nature 457:551–556

Paterson AH, Freeling M, Tang H, Wang X (2010) Insights from the comparison of plant genome sequences. Annu Rev Plant Biol 61:349–372

Proost S, Van Bel M, Sterck L, Billiau K, Van Parys T, Van de Peer Y, Vandepoele K (2009) PLAZA: a comparative genomics resource to study gene and genome evolution in plants. Plant Cell 21:3718–3731

Rasmann S, Agrawal AA (2009) Plant defense against herbivory: progress in identifying synergism, redundancy, and antagonism between resistance traits. Curr Opin Plant Biol 12:473–478

Richards S, Gibbs RA, Weinstock GM, Brown SJ, Denell R, Beeman RW, Gibbs R, Bucher G, Friedrich M, Grimmelikhuijzen CJ et al (2008) The genome of the model beetle and pest *Tribolium castaneum*. Nature 452:949–955

Rouard M, Guignon V, Aluome C, Laporte MA, Droc G, Walde C, Zmasek CM, Perin C, Conte MG (2011) GreenPhylDB v2.0: comparative and functional genomics in plants. Nucleic Acids Res 39:D1095–D1102

Schmutz J, Cannon SB, Schlueter J, Ma J, Mitros T, Nelson W, Hyten DL, Song Q, Thelen JJ, Cheng J et al (2010) Genome sequence of the palaeopolyploid soybean. Nature 463:178–183

Schnable PS, Ware D, Fulton RS, Stein JC, Wei F, Pasternak S, Liang C, Zhang J, Fulton L, Graves TA et al (2009) The B73 maize genome: complexity, diversity, and dynamics. Science 326:1112–1115

Shulaev V, Sargent DJ, Crowhurst RN, Mockler TC, Folkerts O, Delcher AL, Jaiswal P, Mockaitis K, Liston A, Mane SP et al (2011) The genome of woodland strawberry (*Fragaria vesca*). Nat Genet 43:109–116

The International Brachypodium Initiative (2010) Genome sequencing and analysis of the model grass *Brachypodium distachyon*. Nature 463:763–768

TIAGC, The International Aphid Genomics Consortium (2010) Genome sequence of the pea aphid Acyrthosiphon pisum. PLoS Biol 8

Tian D, Traw MB, Chen JQ, Kreitman M, Bergelson J (2003) Fitness costs of R-gene-mediated resistance in *Arabidopsis thaliana*. Nature 423:74–77

Tuskan GA, Difazio S, Jansson S, Bohlmann J, Grigoriev I, Hellsten U, Putnam N, Ralph S, Rombauts S, Salamov A et al (2006) The genome of black cottonwood, *Populus trichocarpa* (Torr. & Gray). Science 313:1596–1604

Unsicker SB, Kunert G, Gershenzon J (2009) Protective perfumes: the role of vegetative volatiles in plant defense against herbivores. Curr Opin Plant Biol 12:479–485

Van de Peer Y, Fawcett JA, Proost S, Sterck L, Vandepoele K (2009) The flowering world: a tale of duplications. Trends Plant Sci 14:680–688

Velasco R, Zharkikh A, Affourtit J, Dhingra A, Cestaro A, Kalyanaraman A, Fontana P, Bhatnagar SK, Troggio M, Pruss D et al (2010) The genome of the domesticated apple (*Malus x domestica* Borkh.). Nat Genet 42:833–839

Whiteman NK, Jander G (2010) Genome-enabled research on the ecology of plant-insect interactions. Plant Physiol 154:475–478

Wu J, Baldwin IT (2009) Herbivory-induced signalling in plants: perception and action. Plant Cell Environ 32:1161–1174

Xia Q, Zhou Z, Lu C, Cheng D, Dai F, Li B, Zhao P, Zha X, Cheng T, Chai C et al (2004) A draft sequence for the genome of the domesticated silkworm (*Bombyx mori*). Science 306:1937–1940

Yu J, Hu S, Wang J, Wong GK, Li S, Liu B, Deng Y, Dai L, Zhou Y, Zhang X et al (2002) A draft sequence of the rice genome (*Oryza sativa* L. ssp. indica). Science 296:79–92

Zeng RS, Wen Z, Niu G, Schuler MA, Berenbaum MR (2009) Enhanced toxicity and induction of cytochrome P450s suggest a cost of "eavesdropping" in a multitrophic interaction. J Chem Ecol 35:526–532

Chapter 2
The Impact of Induced Plant Volatiles on Plant-Arthropod Interactions

Juan M. Alba, Petra M. Bleeker, Joris J. Glas, Bernardus C. J. Schimmel, Michiel van Wijk, Maurice W. Sabelis, Robert C. Schuurink, and Merijn R. Kant

2.1 Introduction

Plants have very stressful lives since they are under continuous pressure by drought, heat, salinity, infection by pathogens and infestation by herbivores. Hence plants have adapted to resist such stresses and to defend themselves against attackers. Some of these defenses require the actions of other organisms, i.e. foraging predators and host-searching parasitoids, which use plant-derived herbivore-specific cues to track down plants with their prey. This process, referred to as indirect defense, will be the focus of this chapter. Plants certainly do not fully rely on this type of defenses. Plant organs are covered with structures adapted to resist abiotic stress or to make life more difficult to plant-eaters already in a much earlier stage. For example, they are equipped with a waxy cuticle that prevents dehydration and functions as a barrier against invaders (Eigenbrode and Espelie 1995); their cells are surrounded by cell walls which support and protect cellular integrity (Hematy et al. 2009) and the epidermis of several species is covered with, glandular and non-glandular, hairs that form structural barriers (Fahn 1988; Simmons and Gurr 2005) and secrete protective coatings (Shepherd et al. 2005). Often such structures have acquired multiple roles and functions (Hanley et al. 2007). For example, glandular trichomes are very efficient barriers that hinder small arthropods in their mobility and they produce, contain and secrete protective substances but also function to guide light properly to the leaf surface such that it can be used optimally for photosynthesis (Wagner 1991; Wagner et al. 2004). Hence the collective of these

J.M. Alba • J.J. Glas • B.C.J. Schimmel • M. van Wijk • M.W. Sabelis • M.R. Kant (✉)
IBED, Section Population Biology, University of Amsterdam, Science Park 904,
1098 XH Amsterdam, The Netherlands
e-mail: m.kant@uva.nl

P.M. Bleeker • R.C. Schuurink
SILS, Department of Plant Physiology, University of Amsterdam, Science Park 904,
1098 XH Amsterdam, The Netherlands

G. Smagghe and I. Diaz (eds.), *Arthropod-Plant Interactions: Novel Insights and Approaches for IPM*, Progress in Biological Control 14,
DOI 10.1007/978-94-007-3873-7_2, © Springer Science+Business Media B.V. 2012

structural barriers function as 'constitutive defenses' since they can hinder pests and are formed irrespective of the presence of these pests but often have also functions beyond defense.

2.1.1 Plant Defense Strategies

Upon arrival, herbivores start exploring the plant surface using their vision, olfaction, touch and taste and walk around to decide whether they will try feeding from the substrate or depart (Chapman and Bernays 1989) by flying or walking off or by taking position for ballooning (Reynolds et al. 2007). Resistant, or 'compatible', herbivores will in principle continue feeding. Continuous feeding by herbivores leads to rapid induced changes in host defense physiology which augment or increase the constitutive defenses. Such changes can include structural reinforcements, programmed local cell death, the production of toxins and the accumulation of proteins that interfere with the structural integrity of the attacker or inhibit their feeding activities and digestion (Howe and Jander 2008; Smith et al. 2009). The balance between constitutive and induced defenses seems determined by the energetic costs of each; by the relative auto-toxicity of each and by how much time it takes for each of them to become operational. At this stage plants can face two different challenges: while mobile pests can still be stimulated to depart, the immobile pests, like fungi or bacteria, do not have this choice and hence need to be killed, or to be put in protective isolation, on the spot. Hence, induced defenses are particularly efficient and successful when taking effect via an 'isolate-and-kill' strategy in defense against immobile pests. Especially the hyper-sensitive response (HR), the process of orchestrated local cell death, is a wide-spread and very successful type of plant resistance against pathogens (Dangl and Jones 2001). In contrast, there is not much evidence for the killing efficiency of induced defenses against mobile herbivores although it was shown that mobile herbivores tend to avoid such defenses (Paschold et al. 2007; Shroff et al. 2008) and hence herbivore-induced defenses may sometimes take effect as a 'go-away-or-die' type of strategy. However, most of these induced anti-herbivore defenses seem to come down to a 'slow-them-down' strategy characterized by an induced decrease in food quality and quantity and often is aimed to interfere with herbivore feeding and their digestive physiology to delay them developing into larger stages, which eat more, and to slow down their population growth. Two plant hormones play central roles in establishment en organization of the 'isolate-and-kill' strategy and the 'go-away-or-die/slow-them-down' strategies i.e. salicylic acid (SA) and jasmonic acid (JA), and these two hormones modulate each other's actions (Howe and Jander 2008; Pieterse et al. 2009). The 'slow-them-down' strategy can be initiated via accumulation of defensive products such as inhibitors of herbivore digestive proteases but, in order to be successful, the fitness penalty associated with the energy investment to establish these defenses must be lower than the net fitness gain and hence such defenses are often paralleled by resource allocation (Strauss et al. 2002).

During such 'rescue-the-resources' response, nutritious carbon- (Schwachtje et al. 2006; Babst et al. 2008) and nitrogen-containing substances (Newingham et al. 2007; Gomez et al. 2010) are re-allocated to either the reproductive tissues or to storage organs, such as roots (Anten and Pierik 2010). Hence the rescue-the-resources response of the plant can give rise to rapid flowering and seed set or to a period of dormancy (Stowe et al. 2000), for example to regrow from a rootstock, depending on the life-history characteristics of both plant and herbivore. However, this re-allocation of resources may not only serve to limit loosing resources but also augment the 'slow-them-down' response since depriving a herbivore from its food will yield similar effects as inhibition of its digestion. Accordingly, the 'slow-them-down' response was found to trigger compensatory feeding responses (Gomez et al. 2010) and plants put special efforts in controlling these (Steppuhn and Baldwin 2007). Herbivore-induced resource allocation can be induced, like defenses, by JA (Gomez et al. 2010) but does not depend on it (Schwachtje et al. 2006) and possibly plants start to prepare local resource-depletion of herbivore-feeding sites already early in the interaction since down-regulation of photosynthesis in the infested leaf is a commonly observed early JA-dependent phenomenon (Creelman and Mullet 1997). However, it is unknown if this implies a partial (Kahl et al. 2000) or complete induced shut-down of local metabolic activity for example to initiate senescence (Gross et al. 2004) or, and if so for how long, these tissues are still supported by photosynthates and defensive products from distal tissues (Kleiner et al. 1999; Delaney 2008; Nabity et al. 2009). It is worthwhile to note that even a 'scorched-earth'-like tactic, during which plants sacrifice tissues without rescuing resources, can help an individual plant to survive implying that events that may look like 'a waste of resources' could be indicative of targeted resource depletion to starve the feeding the herbivore or pathogen. Taken together, from the herbivore's point of view, especially when these are small and their movements are restricted by structural barriers like leaf hairs, resource depletion at the feeding site may represent a big problem possibly even more difficult to overcome than plant-derived toxins since they cannot become resistant to it. However, some herbivores have evolved abilities that enable them to manipulate plant resource-flows to their own benefit such as gall making organisms that cause their feeding site to become a sink for resources (Tooker et al. 2008) but also leaf cutters and trenchers (Dussourd and Denno 1991) that not only prevent plants from transporting defense compounds to the feeding site but, in principle, also from transporting resources away from it.

2.1.2 The Onset of Plant Defenses

Several herbivore-associated actions related to their feeding and movement contribute to the establishment of induced defenses as e.g. crawling can elicit the accumulation of proteinase inhibitors (Jongsma et al. 1995; Solomon et al. 1999) as well as the activation of genes involved in the production of secondary metabolites in tomato (Peiffer et al. 2009). Furthermore, on different host plants it was found that already the footsteps of *Heliothis virescens* gave rise to local accumulation of the γ-aminobutyrate (GABA) in the underlying leaf tissue within minutes (Bown et al. 2002) and it was suggested that this GABA could act as a toxin acting on neuromuscular junctions of invertebrates after ingestion (Bown et al. 2002). Substrate probing or biting (Chapman and Bernays 1989; Giordanengo et al. 2010; Tjallingii et al. 2010) lead to an array of changes that can make further penetration harder (Hao et al. 2008); that isolate the damaged tissues (Will et al. 2009); that block vascular tissue (Will et al. 2007); that inhibit digestion (Jongsma et al. 1995); that intoxicate (Duffey and Stout 1996); that may physically damage herbivore (gut) tissues (Carlini and Grossi-de-Sa 2002) and possibly make them more susceptible for diseases (Shikano et al. 2010) and most of these induced changes are elicited by a combination of mechanical damage and salivary elicitors (Howe and Jander 2008) but also egg-deposition can elicit a subset of very similar responses (Hilker and Meiners 2006) or suppress these (Bruessow et al. 2010). Metabolic alterations in plants induced by chewing herbivores are largely associated with the 'slow-them-down' strategy, and plant responses that interfere with growth and development are predominantly mediated by JA. In contrast, changes induced by pathogens, that cannot rely on brute force only to penetrate a cell but first have to loosen it up via secretion of cell wall degrading enzymes (Hematy et al. 2009), especially elicit the 'isolate-and-kill' strategy, i.e. local apoptosis, and such defenses are much more related to SA (Lam 2004) although it can differ between pathogens with different life styles (Spoel et al. 2007). The plant response to stylet feeding herbivores, however, is more ambiguous maybe because their feeding style has characteristics of both herbivore chewing and pathogen cell-wall penetration, and often comes down to a cocktail of JA and SA responses (Kaloshian and Walling 2005). For example, in tomato *Solanum lycopersicum* the *Mi* gene confers resistance to stylet feeding root-knot nematodes *Meloidogyne incognita* but also to the stylet feeding potato aphid *Macrosiphum euphorbiae* (Vos et al. 1998) and whitefly *Bemisia tabaci* (Nombela et al. 2000) while homologs of this gene confer resistance to pathogens (van der Vossen et al. 2005) and are associated with the recognition of pathogen elicitors (Friedman and Baker 2007) in a gene-for-a-gene like manner. Accordingly the *Mi*-homolog from melon is associated with the occurrence of a micro-HR during stylet penetration of plant cells by the aphid *Aphis gossypii* but not by *B. tabaci* or *Myzus persicae* (Villada et al. 2009). Taken together, the 'isolate-and-kill' response; the 'go-away-or-die' response; the 'slow-them-down' response and the 'relocate-resources' response, have regulatory, i.e. metabolic, interconnections converging in a highly complex manner on plant hormones such as JA and SA but are influenced also by ABA, ethylene and auxin (Pieterse et al. 2009). Despite this complexity several pathogens and some herbivores have found ways to interfere with these processes, i.e. to slow these down or stop accumulation of toxins and selective tissue death, and have acquired resistances to deal with some of these defenses.

2.2   Induced Plant Volatiles

Somewhere along the track of these offensive and defensive phases through which the plant passes, it starts with the increased production and emission of volatile organics compounds (VOCs), or 'odors', via its infested and systemic vegetative green (leaf) tissues, into its environment i.e. the air and soil. Volatile organic compounds are produced in, and released from, practically all plant tissues i.e. leaves, flowers and roots, and the majority is derived from biosynthesis pathways that take place in plastids or proplastids and although almost all tissues can produce these substances different plant species have evolved special structures where the production of volatiles is much higher than in the rest of the plant like glandular trichomes and resin ducts (Fahn 1988; Maffei 2010). These volatiles are of diverse metabolic origin and the majority belong to the terpenoids, C6-aldehydes derivatives (also called 'C_6-volatiles') or aromates (Dudareva and Pichersky 2000; Schuurink et al. 2006). The function of these volatiles has been investigated and debated for many years now and although clearly functions are diverse they seem often related to plant defenses and the key issue has not just been why they are produced as such but especially why they are so often after production released into the air. Several attempts were made to estimate the costs, i.e. the energy loss, of emitting volatiles but a clear consensus has not been reached (Lerdau and Gershenzon 1997; Hamilton et al. 2001; Hoballah et al. 2004). Most, if not all, induced plant volatiles are already produced under control conditions and released to the air in relatively small amounts (Kant et al. 2009). The release of volatiles was suggested to maybe have a 'safety valve' function to ensure the maintenance of metabolic flux (Penuelas and Llusia 2004). Moreover, constitutive and induced production of C_6-volatiles and of terpenoids may contribute to a plant's basal resistance against microbes (Raguso 2004) and it was also suggested they could function as anti-oxidants and that their emission plays a role in heat regulation (Penuelas et al. 2005). In resin, terpenoids serve as solvents to keep it soft when enclosed in ducts but after evaporation cause resins to harden, i.e. after the resin is exposed to air after wounding, thereby sealing the wound (Langenheim 1994; Abbott et al. 2010). Possibly, although even more speculative, terpenoids also determine the viscosity of the sticky-toxic glandular trichome contents (Wagner et al. 2004) and maybe plants upregulate the concentration of terpenoids in these glands when they are attacked (Kang et al. 2010) to make the contents more fluid to promote exudation. Some volatiles, like MeSA (Shulaev et al. 1997; Park et al. 2007), MeJA (Farmer and Ryan 1990; Weber 2002) and possibly *ent*-kaurene (Otsuka et al. 2004) could play a role in within-plant signaling since diffusion in the air is much faster than transport via the vascular tissues (Frost et al. 2007) reminiscent of ethylene signaling. Volatiles released into the air can not only be perceived by the plant's own distal tissues but can also influence the metabolism of neighboring plants (Baldwin et al. 2006). Moreover, plant volatiles can be used by parasitic plants to find a host (Runyon et al. 2006) and can change the behavior of vertebrates and invertebrates as, for example, the volatiles released from flowers mediate plant-pollinator interactions (Raguso et al. 2003). Some volatiles released from green leaf tissues, constitutive or after induction, are repelling to herbivores (De Moraes et al. 2001; Bleeker et al. 2009) while others are attractive and can take effect as feeding stimuli (Willmer et al. 1998).

2.2.1 Indirect Defenses

A well established effect of herbivore-induced plant volatiles (HIPV) is that foraging natural enemies often use these volatiles to track-down plants with potential prey, probably because the emission of these volatiles correlates so well with the absence or presence of herbivores (Kessler and Baldwin 2001; Sabelis et al. 2001). Subsequently, foraging natural enemies may liberate a plant from its attackers. This is called a 'tritrophic interaction' and the attraction of a herbivore's natural enemies via plant-derived cues is referred to as 'indirect defense' (Fig. 2.1). Hence, given its nature, such indirect defense may well be the only form of induced plant defense that is truly lethal for herbivores. While indirect defenses can be mediated by plant volatiles as prey-betraying signals, and although the overall presence of sufficient prey is a prerequisite, it can also be facilitated via other plant traits such as plant-produced alternative food, often in the form of extra floral nectar, and by providing shelter, such as via domatia, to foraging carnivores. The total sum of the direct and indirect physiological responses determine the survival chances of the plant after herbivory and natural selection will constantly re-adjust the balance between growth and defense for both plants and herbivores (Herms and Mattson 1992; Paré and Tumlinson 1999; Anten and Pierik 2010; Gomez et al. 2010). Finally, although plant survival clearly is significant since the terrestrial world still is largely green, we may not forget that in many cases individual plants may simply fall short in their attempts so survive.

2.2.2 The Onset of Indirect Defenses

In most plant species, mechanical damage will hardly cause sustained emission of volatiles while herbivore feeding is known to up-regulate a plant-species specific volatile blend from infested and distal leaves over longer periods of time, i.e. several days or longer (Turlings et al. 1990; Kessler and Baldwin 2001; Schmelz et al. 2003). It is not fully clear yet to which extent the fact that mechanical damage falls short in mimicking herbivore induced volatiles is due to the way such damage is experimentally applied. For example, continuous mechanical damage using a device, referred to as "MacWorm", that resembles the way and the rate of tissue removal by the caterpillar *Spodoptera littoralis,* induces the emission of a blend of volatiles similar in quality and quantity to the emission induced by natural herbivory on Lima bean *Phaseolus lunatus* (Mithöfer et al. 2005).

Fig. 2.1 The tritrophic interaction. A tritrophic system comprises of simultaneous interactions between members of three trophic levels i.e. the first trophic level (plants) which obtain energy from sunlight; the second throphic level (herbivores) which obtain energy by consuming plants and the third trophic level (predators and parasitoids) whose members obtain their energy by eating herbivores. The figure shows an herbivorous spider mite (**a**) eating from a tomato leaf i.e. emptying mesophyll cells with its 70–120 μm long stylets causing whitish chlorotic lesions. The leaf tissue (**b**) responds to the feeding damage by changing its physiology such that it becomes less palatable i.e. via production of feeding deterrents (toxins; inhibitors of digestive enzymes in the herbivore gut; thicker cell walls, etc.) and resource allocation. Simultaneously the production and release of a wide array of plant volatiles is upregulated. These volatiles are, in turn, used by foraging carnivores (**c**: the blind predatory mite *Iphiseius degenerans*) or host-searching parasitoids to track down plants with prey. Some of the local herbivore-induced defenses also occur in distal undamaged plant parts, albeit later in time, such that the herbivore cannot simply avoid induced defenses by walking to adjacent plant parts. Systemically induced volatiles are released in lower amounts than those locally released (Farag and Pare 2002). Molecules responsible for systemic signaling probably travel via the phloem and most likely are JA- and SA-derivatives (Photos by Merijn R. Kant and Jan van Arkel)

However, many studies have shown that the specificity in plant response is largely mediated by molecular recognition of components released by the herbivore during feeding i.e. constituents from saliva or regurgitant and oviposition-related wounding or fluids (Howe and Jander 2008). Following the terminology proposed by Felton and Tumlinson (2008), and in analogy to definitions used in phytopathology, herbivore-derived compounds that 'betray' the herbivore to the host plant and trigger induced defenses, are usually called *elicitors* while compounds that contribute to more efficient infection or infestation process, as for example molecules that interfere with plant defenses, are usually called *effectors*.

2.3   Elicitors That Induce Indirect Defenses

There is a rich body of literature on feeding-related insect secretions especially of insect saliva (Miles 1999; Thivierge et al. 2010). Many arthropods are stylet feeders meaning they have evolved needle-shaped mouthparts, usually consisting of a food channel and a salivary channel, which they insert into host plant tissue to withdraw its nutrients (Labandeira 1997; Rogers et al. 2002). The larger stylet feeding herbivores, like aphids and whiteflies, feed from vascular fluids i.e. predominantly from phloem which is very sugar rich and hence such herbivores, especially the juveniles, tend to secrete honeydew (Douglas 1993) which they use to cover their bodies for protection against predators but which also often gives rise to secondary fungal growth (Perez et al. 2009). Smaller stylet feeders, like nematodes and mites, feed from epidermal cells, parenchyma or mesophyll largely depending on their stylet length. While some species induce the formation of gall-like feeding cells (Hassan et al. 2010) many of them feed simply via piercing-and-sucking while walking from one feeding site to the other. Stylet feeders have in common that they cause only minor mechanical damage, in contrast to chewers, and that they secrete proteins that serve surface attachment or that coat the stylet track *in planta* (Miles 1972). For some aphid species it was found that salivary stylet secretions also contain substances that counter-act the formation of plant-borne sieve element occlusion off vascular bundles that have been probed by the insect and it was suggested aphids already inject substances into tissue surrounding the feeding site prior to feeding (Giordanengo et al. 2010). Hence the feeding activities of stylet feeders are in some aspects reminiscent of germ-tube formation by fungi (Markovich and Kononova 2003) and maybe for this reason plant responses to stylet feeders often are a mixture of the typical SA-dependent anti-pathogen responses on the one hand, and typical JA-dependent anti-herbivore responses on the other. Chewers use brute force to obtain plant material by taking bites. During chewing and ingestion such plant material gets mixed with gut enzymes to digest it and several insect species allocate part of this digestion to take place outside of their body referred to as regurgitation. Regurgitation allows insects to digest and detoxify relatively large volumes of food (Peiffer and Felton 2009) and generates plant-derived deterrents that decrease predation risk (Higginson et al. 2011).

However, this regurgitant also elicits strong and specific defense responses during feeding or when applied to mechanical wounds experimentally (McCloud and Baldwin 1997) and, accordingly, some species of caterpillars limit the amount of regurgitation possibly to avoid alerting the host plant (Peiffer and Felton 2009).

2.3.1 Fatty Acid Conjugates (FACs)

FACs are molecules found in the oral secretions of several caterpillar species but recently also in *Drosophila melanogaster* and two cricket species (Yoshinaga et al. 2007) and they are elicitors of plant defenses. The mixture of FACs is formed in the regurgitant of the insect where fatty acids derived from plant membranes are conjugated with other compounds, generally amino acids produced by the insect. The first FAC described was *N*-17- hydroxylinolenoyl-L-glutamine derived from *Spodoptera exigua* and it was named volicitin due to its capacity to induce the emission of a blend of terpenoids in *Zea mays* (Alborn et al. 1997) very similar to that induced by the caterpillar. Pare and co-workers (1998) demonstrated that the linolenic acid-part of volicitin was of plant origin and that it became conjugated with glutamine in the epithelial cells of the insect mid-gut during regurgitation. Hence it accumulates in the lumen of the gut where it is taken up by the insect possibly to facilitate nitrogen assimilation (Yoshinaga et al. 2008), while some of it is released back into the regurgitant mixture (Tumlinson and Lait 2005). Variation in FACs, e.g. their length can range from C_{16} to C_{18}, is probably mostly due to variation in the pool of fatty acids derived from the plant and their formation occurs in several Lepidopteran species reared on different diets. The predominant amino acid conjugated to these fatty acids is glutamate but also glutamine conjugates can be formed (Halitschke et al. 2001; De Moraes and Mescher 2004). Many of these FACs induce not only plant volatiles but elicit general defense responses very similar to true herbivory when applied after artificial wounding. Finally, also a not conjugated fatty acid, 2-hydroxyoctadecatrienoid acid (2-HOT) in the oral secretion of *M. sexta*, was found to induce the emission of a herbivore-induced terpene, trans-α-bergamotene, when applied to artificially wounded *N. attenuata* (Gaquerel et al. 2009).

2.3.2 Caeliferins

The elicitor family of the Caeliferins was originally obtained from the oral secretions of the American bird *Schistocerca americana* but these compounds are present in many groups of the suborder Caelifera to which the grasshoppers belong (Alborn et al. 2007). While also these elicitors are fatty acids, the structure is distinct from FACs. Caeliferins are compounds derived from the racemic dihydroxyl acids and although in grasshopper regurgitant fatty acids with different lengths can form Caeliferins, the most abundant and bioactive ones are C_{16} molecules and induce volatile emission in *Zea mays*. There are two families of caeliferins i.e. caeliferin A which has two sulphate hydroxyl esterified molecules conjugated at two different positions to the fatty acid backbone and which can have a double bond between carbon 6 and 7 or not and caeliferin B with one sulphate sulphite ester and one glycine conjugated to the fatty acid backbone and which can have also a double bond between carbon 6 and 7 or not. All four Caeliferin molecules can induce the emission of volatiles but (E)-2,16 disulfooxy-6-hexadecenoic acid (caeliferin A 16:1) has the highest abundance and the strongest effect on this induced emission (Alborn et al. 2007).

2.3.3   Inceptines

The inceptines were discovered because some plants that do not react to FACs, still emit a blend of volatiles when they are attacked by caterpillars. Inceptines are peptides of 10–11 amino acids long with a disulfide bound between the two cysteine residues and were isolated from the regurgitant of *Spodoptera frugiperda* reared on *Vigna unguiculata* (Schmelz et al. 2006). These peptides are of plant origin and are formed after the digestion of chloroplastic ATP synthase (Schmelz et al. 2006, 2007). Inceptines were found to induce increased levels of jasmonic acids (JA) and salicylic acid (SA) in cowpea leaves as well as the release of ethylene and terpenoid volatiles (Schmelz et al. 2006).

2.3.4   Orally Secreted Proteins

There is a lot of data published on the composition of herbivore saliva but we are aware of only two examples of enzymes that elicit the emission of herbivore-induced plant volatiles. Hopke and co-workers (1994) showed that the application of β-glucosidase to *Phaseolus lunatus* and *Z. mays* resembles the volatile emission induced by JA and by two-spotted spider mite *Tetranychus urticae* feeding, thereby demonstrating its volatile-inducing elicitor activity. The oral secretion of the caterpillar *Pieris brassicae* L. contains β-glucosidase, even if it has been reared on artificial diet, showing that the enzyme is purely insect-derived. Its applica-tion was found to induce volatiles as similar as those released during herbivory and host-searching parasitic wasps responded to the treated plants as if they were infested (Mattiacci et al. 1995). Unfortunately, the mode of action of this enzyme as elicitor of volatiles remains unclear. Since the enzyme removes sugars from sugar-conjugated metabolites it was suggested it may liberate volatiles that are stored as such conjugates (Felton and Tumlinson 2008). Furthermore, application of this enzyme on rice *Oryza sativa* was found to lead to the accumulation of SA, hydrogen peroxide (H_2O_2), and ethylene, but not of JA so it was suggested that maybe it induces emission of volatiles via up-regulation of hormonal signalling (Wang et al. 2008).

The second enzyme has not been identified but its activity was found in the oral secretions of *Manduca sexta* and it appeared a heat-labile isomerase-like component that markedly changed the Z/E isomer ratio of the C_6-volatiles released by *N. attenuata* during caterpillar feeding and this change was found to triple the predation response of *M. sexta*'s natural enemy *Geocoris spp* under natural conditions. It was shown that this unfortunate conversion occurs in the caterpillar's regurgitant and it was proposed that it may serve to increase the antimicrobial properties of the regurgitant (Allmann and Baldwin 2010).

2.3.5 Oviposition-Derived Cues

All the elicitors described until now have been found in the oral secretions of the herbivores and this makes sense since induced plant defenses are usually the result from feeding activities. Interestingly, also insect oviposition was found to induce volatile emission sufficient to attract natural enemies to these plants. This was first described to occur in *Pinus sylvestris* after oviposition of the pine sawfly *Diprion pini* since this induces a blend of volatiles that attract the egg parasitoid *Chrysonotomyia ruforum* (Hilker et al. 2002) and the elicitor that induces plant volatile production after ovipostion is likely a protein present in the oviduct secretion of the pine sawfly (Hilker et al. 2005). Other examples of egg-related elicitors are compounds called bruchins, a family of compounds present in the oviposition fluid of the elm leaf beetle *Bruchis pisorium*. However, bruchins induce the formation of neoplasms, i.e. callus on leaf surface, but do not induce the emission of volatiles (Doss et al. 1995; Oliver et al. 2000). In contrast, it has also been described that *Spodoptera littoralis* eggs can also inhibit local defenses of *Arabidopsis thaliana* by leaking substances into the plant tissues below that elicit SA responses which, in turn, inhibit the JA responses locally to which the young caterpillar is sensitive (Bruessow et al. 2010).

It was found that oviposition by mated females of *Pieri brassicae* induces the plant to become more attractive to the natural enemy *Trichogamma brassicae* while this effect was not observed using un-mated females. It appeared that the molecule that elicits the plant parasitic wasp attracting volatiles was the compound benzyl cyanide which originates from the male ejaculate of *P. brassicae* (Fatouros et al. 2008) where it serves as anti-aphrodisiac (Andersson et al. 2003).

Plant molecular mechanisms involved in elicitor recognition remain mostly unknown. Thus far, only a membrane receptor involved in volicitin recognition has been found representing the only complete receptor-ligand system described that triggers plant indirect defenses (Truitt et al. 2004).

2.4   Plant Volatiles and Their Upstream Signaling Pathways

Herbivory elicits accumulation of a wide range of secondary metabolites derived from an extremely complex network of biosynthetic pathways (Cowan 1999). The upstream organization, although still not fully understood, is far less complex and comes largely down to the action of a selected set of plant hormones. Similar to direct defenses, the induction of volatile secondary metabolites also is mainly depending on the action of JA, ethylene and, to some extent, to SA (Schmelz et al. 2009) and likely also other hormones like ABA, auxins and gibberellins modulate these processes up to a certain degree (Pieterse et al. 2009). The volatile blend induced by herbivory and emitted into the air surrounding the plant, i.e. the so-called 'headspace', predominantly is composed of terpenoids, C_6-volatiles and aromatics (Fig. 2.2). However, the qualitative and quantitative composition of the induced blends differs greatly across plant species and across herbivore species (Takabayashi et al. 1991).

2.4.1   Plant Hormone Signaling Cascades: Jasmonic Acid

Induction of volatiles in lima bean (Gols et al. 2003), tomato (Ament et al. 2004) and wild tobacco (Halitschke and Baldwin 2003) depends on JA. JA is an oxylipin-derived signaling molecule predominantly synthesized via the octadecanoid pathway from linolenic acid (C_{18}) but some plant species can also use hexadecatrienoic acid (C_{16}) isomers for its production (Gfeller et al. 2010). The polyunsaturated fatty-acid linolenic acid is released from the plasma membrane probably by one or more lipases (Hyun et al. 2008; Ellinger et al. 2010) and oxygenated in the plastid by 13-lipoxygenase (LOX) to form C_{13}-hydroperoxy linolenic acid (13-HP) (Wasternack 2007). 13-HP can then be further modified via different metabolic pathways. Consecutive dehydration by allene oxide synthase (AOS) and cyclisation by allene oxide cyclase (AOC) converts 13-HP into cis-oxophytodienoic acid (OPDA), the precursor of JA. OPDA is then converted to JA in the peroxisomes via reduction and ß-oxidation which can be derivatized into several components. MeJA is formed by the transfer of a methylgroup from S-adenosyl-L-methionine (SAM) to the carboxyl group of JA; it can be decarboxylated into the bioactive volatile cis-jasmone or be conjugated to amino-acids. Conjugation to isoleucine leads to the formation of the active derivative (+)-7-iso-jasmonoyl-L-isoleucine (JA-Ile) which can bind to the SCF^{COI1} complex which thereby undergoes a conformational change after which it starts to ubiquitinate JAZ (Jasmonate Zim Domain) transcriptional repressors. These repressors physically block the action of transcription factors, such as MYC2 under non-induced conditions. Hence, these 'liberated' transcription factors can then, in turn, assemble with the transcription machinery to give rise to expression of 'JA-responsive genes' (Chini et al. 2007; Browse 2009). Especially the induced emission of terpenoids, but also that of MeSA, is depending on JA (Ament et al. 2004; Li et al. 2004).

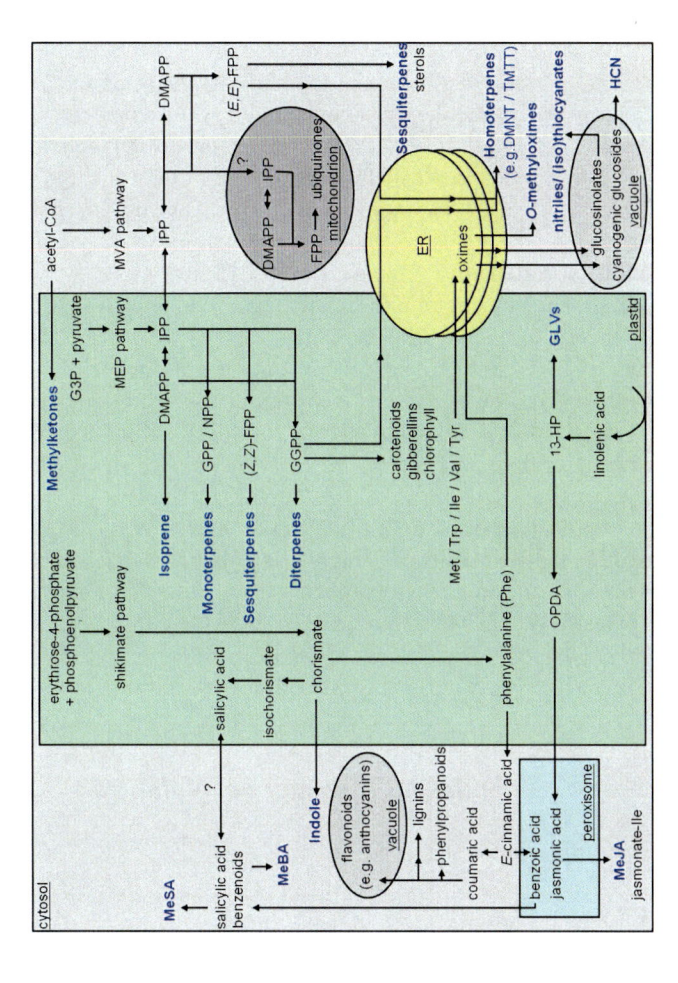

Fig. 2.2 Biosynthesis pathways of plant volatiles projected onto their cellular compartments. Plant volatiles are indicated in blue; a single arrow does not necessarily represent a single enzymatic conversion. Abbreviations used: *(E,E)-FPP* 2E,6E-farnesyl diphosphate, *(Z,Z)-FPP* 2Z,6Z-farnesyl diphosphate, *13-HP* C$_{13}$-hydroperoxy linolenic acid, *Acetyl-CoA* acetyl coenzyme A, *BA* benzoic acid, *DMAPP* dimethylallyl diphosphate, *DMNT* 4,8-dimethyl-1,3,7-nonatriene, *ER* endoplasmic reticulum, *FPP* farnesyl diphosphate, *G3P* glyceraldehyde 3-phosphate, *GGPP* geranylgeranyldiphosphate, *GLVs* green leaf volatiles, *GPP* geranyldiphosphate, *HCN* hydrogen cyanide, *Ile* isoleucine, *IPP* isopentenyl diphosphate, *JA-Ile* jasmonic acid isoleucine, *MeBA* methyl benzoate, *MeJA* methyl jasmonate, *MEP pathway* non-mevalonate pathway also known as the 2-C-methyl-D-erythritol 4-phosphate (DOXP) pathway, *MeSA* methyl salicylate, *Met* methionine, *MVA pathway* mevalonate pathway, *NPP* neryldiphosphate, *OPDA* cis-oxophytodienoic acid, *TMTT* 4,8,12-trimethyltrideca-1,3,7,11-tetraene, *Trp* tryptophan, *Tyr* tyrosine, *Val* valine

2.4.2   Plant Hormone Signaling Cascades: Salicylic Acid

Salicylic acid is a benzenoid derived from the shikimate pathway (Dudareva and Pichersky 2000; Schuurink et al. 2006) which is derived from chorismate (Catinot et al. 2008). Chorismate can be converted to tryptophane, tyrosine, phenylalanine (Phe) or isochorismate and the latter two can be subsequently converted into SA. Phe is first converted to trans-cinnamic acid which can be converted to SA via benzoic acid or via o-coumaric acid. However, the bulk of the induced SA in Arabidopsis, tomato and *Nicotiana benthamiana* is derived from isochorismate (Wildermuth et al. 2001). SA, in turn, can be converted into SA-glucoside (SAG) via the enzyme *SAGT* and, in a lesser extent, into salicyloyl glucose ester (SGE) but also into the volatile methyl salicylate (MeSA) via the enzyme *SAMT* and its glucosilated derivative MeSAG. In tobacco cell suspensions the ratio of SA, SAG and MeSAG is almost 1:1:1. Moreover, MeSA and SAG are biologically inactive while a hydroxylated form of SA, gentisic acid, was found to have specific PR-protein inducing activities in tomato not induced by SA (Belles et al. 2006). In tobacco, MeSA is probably the phloem mobile signal that accounts for systemic SA-responses and an SA-binding protein (SABP2), which displays MeSA-esterase activity and whose activity is SA-inhibitable locally but not systemically, is responsible for generating active SA from inactive MeSA. SA inhibits auxin responses and SA/JA as well as SA/ABA antagonize each others actions locally (Vlot et al. 2009). MeSA is a common component in the headspace of insect-infested plants (Arimura et al. 2002; Zhao et al. 2010) and is involved in the recruitment of beneficiary predators (van Wijk et al. 2008; Ament et al. 2010) and airborne MeSA can affect the metabolism of neighboring plants (Shulaev et al. 1997). It has been proposed that some phytopathogens hijack the SA-defense pathway by increasing plant susceptibility through enhanced SA volatilization to MeSA (Attaran et al. 2009).

2.4.3   Plant Volatiles: C_6-Volatiles i.e. 'Green Leaf Volatiles'

C_6–volatiles or 'Green Leaf Volatiles' (GLVs) are 6-carbon aldehydes, alcohols and esters, released immediately after wounding of plant green tissues (Paré and Tumlinson 1999), whereas their emission from undamaged plant (parts) is negligible. Due to the extremely rapid onset of their emission, GLVs were regarded as general wound signals (Hatanaka 1993; Matsui et al. 2000) but it was found that they carry herbivore-specific information since the herbivore-induced conversion if *cis*- to *trans*-GLVs is due to isomerase-activity of herbivore saliva itself (Allmann and Baldwin 2010). Exposure to GLVs can elicit defense mechanisms in Arabidopsis and other plants (Bate and Rothstein 1998). Application of *cis*-3-hexenyl acetate for instance, results in activation of oxylipin-signaling compounds like JA, as well as the release of terpenes in tomato (Farag and Pare 2002).

Receptors for GLVs, or for other herbivore-induced volatiles, have not yet been identified yet but a mutagenesis screen on Arabidopsis delivered an E-2-hexenal-non-response mutant, *her-1*, encoding a γ-amino butyric acid transaminase (GABA-TP) which degrades GABA, and thereby linked *trans*-2-hexenal-induced responses to the GABA metabolism (Mirabella et al. 2008) and, hence possibly with stress and disease resistance (Park et al. 2010; Renault et al. 2010).

Like JA, GLVs are derived from the octadecanoid pathway and their biosynthesis pathways share 13-HP as common precursor. To form GLVs, 13-HP is cleaved by hydroperoxide lyase (HPL) into C_6-aldehydes and a C_{12}-product that leads to the formation of traumatin, implicated in wound signaling (Zimmerman and Coudron 1979). The C_6-aldehydes can undergo additional modifications such as isomerisation, oxidation/reduction and acylation (D'Auria et al. 2007; Allmann and Baldwin 2010). It is well possible that distinct upstream lipases supply the octadecanoid pathway with either house-keeping JA/GLV precursors; induced GLV-precursors and induced JA-precursors independently (Howe et al. 1996; Li et al. 2002a; Degenhardt et al. 2010; Ellinger et al. 2010).

2.4.4 Plant Volatiles: Terpenes

Terpenes are isoprenoid derivatives and are among the most diverse classes of plant produced secondary compounds and over 30,000 isoprenoids of plant origin have been identified (Connolly and Hill 1991). Terpenoids have are involved in a wide variety of plant processes, many of which are primary such as respiration, photosynthesis, growth, reproduction and adaptation (Harborne 1991). However several volatile terpenes, mostly the short-chain terpenes, are associated with plant defenses and are produced and emitted in response to herbivory. Significant induction of terpene emission usually increases over consecutive photophases, although some are especially emitted during the dark phase (De Moraes et al. 2001), while the emission of GLVs follows induction almost immediately but ceases quite rapidly after the first photophase (Loughrin et al. 1994).

Terpenes are composed of C_5-isoprene units, isopentenyl diphosphates (IPP) and the IPP-isomer dimethylallyl diphosphate (DMAPP), that are condensed by the action of different prenyltransferases in a head-to-tail addition to form prenyl diphosphates of different lengths (Lange et al. 2000). C_5-isoprene is a highly volatile terpenoid that is easily released from the leaf surface of mostly deciduous broad-leaved trees, but not of all plants. Monoterpenes consist of two isoprene units (C_{10}), sesquiterpenes consist of three (C_{15}) and diterpenes are constituted of four isoprene units (C_{20}), but also C_{30}, C_{40} and C_{50} compounds can be made in this way.

The biosynthetic pathways of isoprene/terpenoid olefins are divided over separate routes which take place in different cellular compartments i.e. predominantly the cytosol where sterols and sesquiterpenes are produced via the mevalonate (MVA) pathway; the plastids where monoterpenes and diterpenes are produced via the non-mevalonate-pathway but also give rise to chlorophyll, carotenoids, vitamins and tocopherols, and plant hormones like gibberellins and ABA; and the mitochondria were ubiquinones are formed. The plastidial en cytosolic biosynthesis pathways rely on the precursor IPP.

Sesquiterpenes are almost exclusively produced in the cytosol (Tholl 2006). Early in the MVA-pathway IPP is generated through condensation of acetyl-CoA (Dish et al. 1998) and isomerised to DMAPP. Subsequently, farnesyldiphosphate synthase (FPS) uses IPP and DMAPP to produce C15-farnesyl diphosphate (FPP), the universal precursor in sesquiterpene synthesis. Previously it was thought that the biosynthesis of terpenoids proceeds in the trans-configuration. However, the head-to-tail condensation of IPP to DMAPP can also result in cis-conformations (Kellog and Poulter 1997). Sallaud et al. (2009) were the first to clone and characterize a short chain terpene *cis*-prenyltransferase; Z,Z-FPS and found that it is involved in a plastidial pathway for sesquiterpenes.

Monoterpenes and diterpenes are products of the plastidial 2-C-methyl-Derythritol 4-phosphate (DOXP) pathway, also called non-mevalonate pathway or MEP-pathway. For this pathway, the IPP units are derived from pyruvate and glyceralde-hyde-3-phosphate via the intermediate 1-deoxy-D-xylulose-5-phosphate (Arigoni et al. 1997). Subsequently, isoprene synthase, which is targeted to the plastid, eliminates inorganic pyrophosphate from DMAPP to yield volatile C_5-isoprene (Sasaki et al. 2005; Vickers et al. 2010). Subsequently C_{10}-geranyldiphosphate (GPP) and C_{20}-geranylgeranyl diphosphate (GGPP) are produced via their respective synthases, thereby forming the precursors of mono- and diterpenes. However, in addition to GGP also a *cis*-prenyl transferase has been identified that results in the formation of neryl diphosphate (NDP) which also functions as a precursor for monoterpenes (Schilmiller et al. 2009).

The MVA-pathway and DOXP-pathway were initially thought to be strictly separated within their respective compartments but there have been several studies that provided evidence for IPP-exchange between the plastids and the cytosol (Dudareva et al. 2004; Xie et al. 2008).

Despite a relatively simple skeletal structure, terpenes are extremely variable in their exact chemical structure. Terpene synthases (TPS) are remarkable in the fact that they can make use of a limited number of substrates to produce an enormous structural variety in terpenes. TPSs are cyclases that modify the prenyl diphosphate substrates from the different pathways, within the specific cellular compartments, starting with the removal of the pyrophosphate forming unstable carbocation-intermediates that undergo a cascade of poly-cyclisation reactions. TPSs have been divided in seven subfamilies (TPSa-g) (Bohlmann et al. 1998) but also an additional plastidial, *ent*-kaurene synthase-resembling TPSs has been characterized (Sallaud et al. 2009; Schilmiller et al. 2009). Some TPSs generate a specific product (e.g. van Schie et al. 2007), whereas others can synthesize multiple terpenes from a single substrate (e.g. Schnee et al. 2002). Terpenes can also be further modified by undergoing reactions such a dehydroxylation, dehydrogenation, acylation, carboxylation and methylation giving rise to the enormous diversity of terpenoids as found in plants.

2.4.5 Plant Volatiles: Aromatics and Others

The third class of herbivore-induced plant volatiles is that of the aromatics like indole but also benzenoids like SA-derived MeSA which is probably one of the most generally induced plant volatiles emitted from green tissues (D'Alessandro et al. 2006; Ament et al. 2010; Kollner et al. 2010; Tieman et al. 2010). The odor of flowers is typically due to benzoates. Most aromatics are produced via the phenylproponoid pathway which branches off from the shikimate (Tzin and Galili 2010) and arogenate (Maeda et al. 2010) pathways and starts with phenylalanine as the precursor for an array of products such as lignins; the hormone auxin and part of the SA-pool and for many defense and stress related secondary metabolites like stilbenes, anthocyanins, flavonoids, coumarins and benzoates (Dixon et al. 2002; Vogt 2010). Although the aromatics, the tepenoids and C_6-volatiles represent the most general classes of induced plant volatiles, there are many additional compounds which are often more species specific like nitrilles, oximes and isothiocyanates (Ujvary et al. 1993; van den Boom et al. 2004; Mumm and Dicke 2010) which are derived from glucosinolates and cyanogenesis (Takos et al. 2010; Jørgensen et al. 2011).

Induction of plant defenses needs to be organized in space and time to be effective. In tomatoes infested with the two-spotted spider mite (*T. urticae*) direct and indirect defenses are established successively since defenses marker genes and associated protein activity can be observed already from the first day of the infestation, while significant emission of volatiles is only observed days later (Kant et al. 2004). It is not clear yet if this differential timing reflects a defense program and is organized by plant hormones to optimize the efficiency of the collective response (Kahl et al. 2000) or whether this differential timing is solely due to metabolic constraints.

2.5 Transgenic Approaches to Manipulate Arthropod Responses via Plant Volatiles

One of the first successful attempts to change the volatile bouquet of plants and the subsequent response of arthropods through transgenesis was via engineered down-regulation of *HPL* in potato plants which led to an increase in the performance of the aphid *Myzus persicae* performance (Vancanneyt et al. 2001). In contrast, a similar approach in *Nicotiana attenuata* led to a decrease in the development of the tobacco hornworm *Manduca sexta* which seems to use GLVs as feeding stimulants (Halitschke et al. 2004). Overexpression of *HPL* in Arabidopsis made the plants more attractive to the parasitic wasp *Cotesia glomerata*, leading to higher mortality of parasitized herbivore larvae of the cabbage white butterfly (Shiojiri et al. 2006). Collectively, the data obtained with these transgenic plants confirm an important role of GLVs in arthropod responses and apparently this role can be manipulated via over-expression or silencing of a single biosynthetic gene.

While most, if not all, biosynthetic pathways of terpenoid precursors are well mapped and while many terpene synthases have been cloned and sequenced, the road for manipulation of terpenoid biosynthesis in plants through transgenesis appears to be still wide open. One of the first, very elegant and successful attempts for such manipulation was obtained via the down-regulation of a trichome-specific P450 hydroxylase in tobacco. This resulted in lower levels of the predominant exudate component, the semi-volatile diterpene cembratriene-diol, and a simultaneous big increase in its precursor, cembratriene-ol. Together these changes could decrease the level of colonization by the aphids *M. nicotiana* (Wang et al. 2001). A bottle-neck in the engineering of plant terpene production is that it is often not clear which step or steps in the pathway are rate limiting i.e. over-expression only works when there is sufficient flux through the pathway. However, there are examples where overproduction of terpenoids was successfully done and could be used to manipulate arthropod behavior. A plastid targeted linalool/nerolidol synthase was over-expressed in Arabidopsis resulting in higher emission of linalool with decreased attractiveness to the aphid *M. persicae* when these were offered wildtype plants as alternative in choice tests (Aharoni et al. 2003). When over-expressed with a mitochondrial target peptide, nerolidol was formed from FPP which was subsequently converted to 4,8-dimethyl-1,3(E),7-nonatriene (DMNT) making the plants more attractive to the predatory mite *Phytoseiulus persimilis* (Kappers et al. 2005). Moreover, overproduction of the sesquiterpene patchoulol in tobacco plastids in which FPP synthase was co-expressed, led to deterrence of tobacco hornworms (Wu et al. 2006).

Constitutive over-expression of the maize sesquiterpene synthase *TPS10* in Arabidopsis resulted in higher emission of E-ß-farnesene and E-α-bergamotene leading to a higher attractiveness of the parasitoid *C. marginiventris*, but only after they had learned to associate these volatiles with the presence of prey on the original host (Schnee et al. 2006). Interestingly, production of the structurally simplest terpenoid, the carbon-5 isoprene, has also been achieved in Arabidopsis, while wildtype plants cannot produce this volatile. These transgenic Arabidopsis plants appeared to repel the parasitic wasp *Diadegma semiclausum*, of which the antennae can perceive isoprene (Loivamaki et al. 2008). However, constitutive overproduction of E-ß-farnesene, the principal component of the alarm pheromone of *M. persicae*, did not lead to repellence of aphids. Hence, it was hypothesized that pulsed emission, naturally done by aphids upon danger, might be more effective in this case than constitutive steady-pace emission (Kunert et al. 2010).

Only one study has shown the effects of overproduction of terpenoids below ground. The corn rootworm *Diabrotica virgifera virgifera* induces β-caryophyllene when feeding from the roots of European maize varieties and this leads to the attraction of their natural enemy the predatory nematode *Heterorhabditis megidis*. American maize varieties, however, cannot produce β-caryophyllene probably as a coincidental result of artificial selection. Engineered production of β-caryophyllene in such deficient maize varieties restored the attraction of the predatory nematode (Degenhardt et al. 2009). Together, these studies show that engineering of terpenoids is feasible and that responses of arthropods can be influenced.

However, most, if not all, of these over-expressors show pleiotropic effects varying from changes in the volatiles produced normally to dwarfism or chlorosis. Many of the limitations of the fluxes through the pathways are still poorly understood, as well as transport of precursors and end products. More effective approaches with tissue- or organ-specific promoters, e.g. in glandular trichomes, in combination with mitochondrial- or plastidial-targeting and overproduction of precursors are thus required.

Engineering volatile benzenoids or phenylpropanoids has been very successful, but mostly in flowers with little regard to arthropod, i.e. pollinator, behavior (Dudareva and Pichersky 2000). Kessler et al. (2008) showed via silencing the emission of floral benzyl acetone, the dominant flower volatile of *N. attenuata*, in combination with silencing the production of the anti-herbivore toxin nicotine, that nicotine prevents nectar robbing by nectar-feeding insects that are attracted by flower volatiles. In another study down-regulated emission of MeSA by tomato leaves during herbivory by the spider mite *T. urticae* was obtained through RNA interference of SAMT (salicylic acid methyl transferase). Predatory mites (*P. persimilis*), when given the choice between spider mite-infested SAMT-silenced and infested wild-type plants, preferred the latter, indicating that the absence or presence of single volatiles like MeSA in complex blends can alter the response of predatory mites (Ament et al. 2010).

2.6 Variation in Plant Volatile Production

The kinetics and identity of induced plant volatiles varies highly in time and space and is roughly the resultant of the kinetics of the defense responses and of diurnal rhythms but also of growth conditions, tissue age and type as well as the type and combination of stresses (Fig. 2.3).

2.6.1 Variation Due to Diurnal Rhythms

The quantitative but also the qualitative emission pattern exhibits diurnal photo-periodicity (Loughrin et al. 1994; Turlings et al. 1995). In tomato, emission of terpenes is positively dependent on the amount of light (Maes and Debergh 2003) but also on the presence of JA (Ament et al. 2004), the sesquiterpene α-copaene being a notable exception. The emission of C_6-volatiles as well as cyclic and acyclic terpenes induced by *S. exigua* in cotton *Gossypium hirsutum* was found to follow a diurnal rhythm with high emission during the day but after removal of the caterpillar, the emission of the cyclic terpenes disappeared, while emission of the C_6-volatiles and acyclic terpenes remained keeping their diurnal rhythm albeit in smaller amounts (Loughrin et al. 1994). Diurnal-rhythm dependent emission was also observed in *N. tabacum* after feeding by larvae of *Heliothis virescens*,

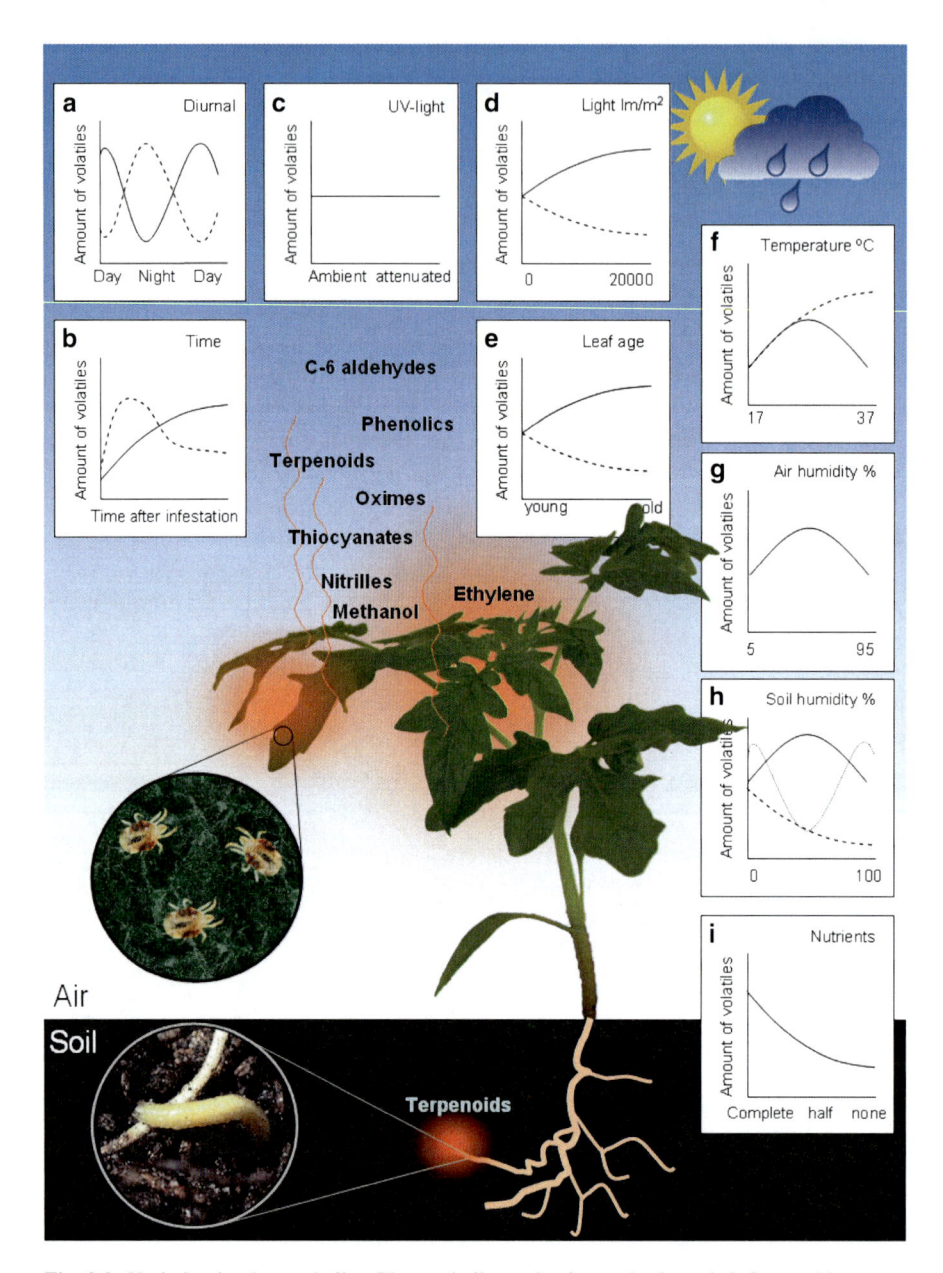

Fig. 2.3 Variation in plant volatiles. Plant volatile production and release is influenced by many different biotic and abiotic factors. Here the generalized relationship between the total amount of volatiles and different environmental parameters, within their natural range, is summarized. (**a**) Most constitutive and induced plant volatiles follow a diurnal pattern: some components peak during the dark period (*dotted line*) but most peak during the light period (Loughrin et al. 1994; De Moraes et al. 2001; Arimura et al. 2004); (**b**) Wounding and herbivore feeding upregulates the release of plant volatiles: while singular treatments usually give rise to a rapid peak followed by a

M. sexta and *Helicoverpa zea* since these induced stronger emission of (E)-2-hexenal and exclusive emission of several other C_6-volatiles during the dark phase and these nocturnal volatiles were used by adult *Helothis virescens* to avoid ovipositioning on plants already containing feeding larvae (De Moraes et al. 2001). Schmelz et al. (2001) showed that applying the same treatment, i.e. mechanical wounding with or without addition of JA or volicitin, at different times of the day had different effects on the emission of volatiles during the subsequent light period. Excised leaves emitted much more sesquiterpenes, and in different ratio's, than the leaves of intact plants and the induction was the strongest when excised leaves were treated in the middle of the dark phase. In contrast, intact plants displayed little or no response to volicitin when these had been treated at the beginning of the light cycle.

2.6.2 Variation Due to Tissue Age and Position

Quantitative emission of induced volatiles can differ greatly depending on tissue type and age. As a rule, on a fresh weight basis young tissue is more active than older tissue. However, it needs to be noted that although young and old leaves can differ greatly in weight, the number of cells between young leaves and old leaves won't differ very much since leaves grow via cell expansion and, hence, the weight difference will be largely due to water. Hence, activity in volatile production per cell and activity in volatile production per gram fresh weight are truly different measures. While often weight corrections are applied for comparing plant volatile production among treatments these will not deliver an accurate measure for comparing the amount of volatiles to which a predator was exposed since predators don't make such 'gram-per-fresh-weight', nor 'gram-per-dry-weight' calculations.

Fig. 2.3 (continued) steady decline in emission over time (*dotted line*) continued herbivore feeding results in continuously enhanced release of volatiles (Kant et al. 2004; Mithofer et al. 2005; Gaquerel et al. 2009); (**c**) Attenuated exposure to UV-light does not significantly alter the release of volatiles (Winter and Rostas 2008); (**d**) There is a positive relationship between the amount of light a plant receives and the amount of volatiles it releases although for some volatiles this relationship is negative (Takabayashi et al. 1994a; Gouinguené and Turlings 2002); (**e**) Younger leaves release more of some but less of other induced volatiles per gram fresh weight compared to older leaves do (Takabayashi et al. 1994b); (**f**) Plants release the highest amounts of volatiles at an optimum temperature, although this optimum differs for different volatiles, and (**g**) optimum air humidity (Gouinguené and Turlings 2002). However the relationship between volatile emission and soil humidity (**h**) is more complex: for some volatiles the relationship is clearly negative (*dashed line*); for some there are two optima (*dotted line*) while for others there is only one (*filled line*). Finally, plants deprived of nutrients release in general less volatiles than plants with complete nutrition (**i**) (Gouinguené and Turlings 2002). Plant volatiles are not only released from above-ground tissues infested by above-ground herbivores, such as the spider mite *T. urticae*, but also below-ground: e.g. the terpenoid β-caryophyllene is induced in roots by root herbivores, such as rootworms, and attracts the herbivore's natural soil-inhabiting enemies (van Tol et al. 2001; Rasmann et al. 2005; Degenhardt et al. 2009). (Photos by Merijn R. Kant except for the rootworm *Diabrotica virgifera virgifera* picture which was taken from the USDA photo archive (UGA1320014))

Hence such corrections are only useful when comparing metabolic activities from a plant physiology point of view but may cause misconceptions in an ecological-behavioural context. In choice experiments the predatory mite *Phytoseiulus persimilis* preferred volatiles from spider mite *T. urticae*-infested young leaves infested over those emitted by old infested leaves. Nevertheless, the qualitative composition of volatiles released by these infested young and old was similar (Takabayashi et al. 1994b). In most plants, volatile production is systemically induced i.e. when one leaf is induced also other leaves will respond to this treatment similarly albeit later in time. For example, *S. exigua* caterpillars eating from older cotton leaves were found to induce the emission of volatiles not only in these leaves but also in the undamaged younger leaves of the same plant (Röse et al. 1996). Such effects are not restricted to green leaf tissues. In sweet-scented tobacco *Nicotiana suaveolens* also the quality and quantity of the floral volatile blend was affected by leaf-feeding *M. sexta* larvae (Effmert et al. 2008). In addition, in a comprehensive analysis of the kinetics of *S. frugiperda*-induced volatiles in soybean it was found that plants which are in the vegetative stage emit ten-fold more volatiles per gram biomass than plant in the reproductive stage. Moreover it was found that young soybean leaves emit much more volatiles than old leaves and that systemic induction in single distal leaves is stronger and faster in the acropetal than in the basipetal direction (Rostás and Eggert 2008). Taken together, local responses affect the metabolism in distal tissues as well, but the magnitude of the response is highly depending on tissue-specific parameters such as age and relative (vascular) position.

2.6.3 Variation Due to Growth Conditions

The emission of plant volatiles depends not only on tissue type, age and time of the day, but is also affected by growth conditions. Systematic studies on the exact relationship between gradients of abiotic factors/stresses and the profile of herbivore-induced volatiles have not often been undertaken. Takabayashi et al. (1994a, b) investigated the impact of light conditions, time of year and water stress on the composition of the spider mite-induced volatile blend of lima beans. Lima beans exposed to low light emitted 5% (E)-β-ocimene in their total blend but in high light this increased to 21%, and in choice tests predatory mites preferred high-light plants over low-light plants and this preference was clearest in summer while absent in winter. Moreover, lima beans grown in very wet soil compared to plants grown under normal moisture levels, but both at 60–70% relative air humidity, produced higher amounts of linalool, (E)-β-ocimene, 4,8-dimethyl-1,3,7-nonatriene (DMNT) and MeSA and were more attractive to predatory mites. Gouinguené and Turlings (2002) tested the impact of different degrees of soil humidity, air humidity, temperature, light and fertilization status on the induced emissions of terpenes, indole and C_6-volatiles of maize plants. Plants standing in dry soil released overall more volatiles than plants in wet soil and emission was maximal at 60% relative air humidity and a temperature between 22°C and 27°C. Emission of volatiles by maize plants appeared photophase-dependent and using richer fertilization regimes the overall quantities of volatiles increased.

Ozone exposure triggered emission of DMNT, 4,8,12-trimethyltrideca-1,3,7,11-tetraene (TMTT) and (Z)-3-hexenyl acetate in lima beans increasing their attractiveness to predatory mites. Moreover, *B. oleracea* plants infested with diamondback moth (*Plutella xylostella*)-and exposed to ozone attracted more predatory mites than uninfested ozone-exposed plants (Vuorinen et al. 2004) while the preference of the parasitoid *Cotesia plutellae* remained unaffected by such treatment although some of the terpenes and C_6-volatiles were oxidized by the ozone (Pinto et al. 2007). This suggested that, while herbivore-induced terpenes might function to quench ozone and reactive oxygen species (ROS), tritrophic interactions are not significantly affected by this (Holopainen 2004). Taken together, while growth conditions and abiotic stress clearly alter induced volatile emission, the volatile-mediated indirect defenses appear quite robust.

2.6.4 Variation Across Genotypes, Species and Different Herbivores

Often plants of different families produce different volatiles when infested with the same herbivore species. A comprehensive across-species analysis of 11 plant species infested with the generalist spider mite *T. urticae* showed that almost all species produce novel compounds upon infestation when compared to their clean controls, including MeSA, terpenes, oximes and nitriles, and that only two species, tobacco and eggplant, alter their emission only quantitatively after induction (Van den Boom et al. 2004). When seven of these species, including eggplant, were used in an olfactory choice assay for *T. urticae*'s natural enemy *P. persimilis* (Fig. 2.4), all elicited a positive response to this predatory mite when clean controls were the alternative. Moreover, in later study in which the relative attractiveness of four *T. urticae*-infested *Gerbera* varieties to *P. persimilis* was compared it appeared that this response was positively correlated with the absolute amount of terpenes and with the level of infestation (Krips et al. 2001). Hence, although volatile production may be variable, the outcome of the tritrophic interaction apparently is not.

Takabayashi et al. (1991) measured the emission of volatiles from two commercial apple varieties after infestation by two different species of spider mites, i.e. *T. urticae* and *Panonychus ulmi*, with similar feeding styles and found that the differences between the two apple varieties infested by the same mite species were bigger than when comparing a single apple variety infested by either one of the two mite species. This suggests that the relatively marginal genetic differences between races of the same plant species already can translate into markedly different blends of induced volatiles. For experimental assessment of volatile production and the

Fig. 2.4 The olfactory choice assay (the "olfactometer"). The olfactometer is an instrument for assessing the response of foraging animals towards different odorous stimuli. The most basic setup comprises of a y-shaped glass tube in which a foraging animal can be exposed to two odors: in the example shown here a setup has been constructed to test the olfactory preferences of predatory mites such as *Iphiseius degenerans* foraging for thrips larvae (**a**) or spider mites. The ends of the arms of the Y-tube (3,7 cm internal diameter) are connected to the respective odor sources and odors are sucked through the tube using a vacuum pump regulated by a flow meter (airflow of 3–15 l/min). Into the tube a y-shaped metal wire (leg 13 cm long; both arms 13 cm long with an angle of 75°) is fixed for the mite to walk on but since the wire is thicker than the mite (**c**) a mirror is placed underneath the glass tube to not loose track of the mite (**b** and **c**) during the test. The ends of the arms of the Y-tube are extended with tubes of the same diameter containing one or two fine-mesh gauze filters to remove turbulence from the laminar air flow. Taken together, with this set-up

subsequent responses of natural enemies mostly crop plants are used. Hence, when variation in induced volatile production is generated during crop breeding by coincidence, i.e. in the absence of targeted selection, one would expect that the degree of variability between cultivars to be different from that between ecotypes of their wild relatives since these have been under natural selection. Turlings et al. (1998) observed considerable differences in the timing of the emission of volatiles by two maize cultivars after inducing them with *Spodoptera littoralis* oral secretions and one of the cultivars produced several terpenoids that the other did not produce at all. Subsequently, Gouinguené et al. (2001) compared the emission of induced volatiles among seven maize cultivars and five of their wild ancestors and included a comparative analysis of eight individuals from a single natural population of a wild teosinte species and observed considerable quantitative differences in ratios of volatiles for all groups except between the eight individuals of the same wild ancestor. A marked difference was the absence or presence of the sesquiterpene β-caryophyllene. In a follow-up study the induced volatiles of 31 maize inbred lines, representing a large portion of the genetic diversity used by breeders, was compared and revealed highly variable odour profiles across genotypes and identified β-caryophyllene as a unique volatile for the European varieties that was not found in any of the American varieties and there appeared to be no relation between the genetic distances of the lines and their odour-profile distances (Degen et al. 2004). Also different natural populations of *N. attenuata* growing in the field appeared variable in their production of volatiles (Halitschke et al. 2000) and it was found that individuals from the same populations vary greatly in the amounts of volatiles they emit and although these plants also accumulated phytohormones in considerably different amounts it was suggested that differences in VOC emission were largely caused by processes downstream of JA signalling (Schuman et al. 2009). Taken together, variation in induced volatile is

---

Fig. 2.4 (continued) a mixture of two odors in the leg of the Y-tube is separated in the two arms and thus, for this set-up to work properly, it is essential that the split of the wire is at the same position as where the two airstreams come together in the Y-tube. Since predatory mites like *I. degenerans* and *Phytoseiulus persimilis* are blind they depend on alternative cues such as herbivore-induced plant odors to find their prey. To test a predatory mite's preference for one odor over the other, e.g. odors of uninfested tomato plants versus the odors of prey-infested infested plants, predatory mites are first deprived from their food for one or several hours. At the start of the experiment a hungry predatory mite is placed at the beginning of the wire using a soft bristle paint brush (**c**) after which the vacuum pump is connected to leg of the Y-tube and the airstream is initiated. The mite is now exposed to a mixture of the two odors. Foraging predatory mites will walk upwind. Hence the mite will walk to the junction of the wire where it has to make a choice for one of the two odor sources. Its choice is scored once the mite has reached the end of one of the two arms. Now the airstream is stopped, the Y-tube opened and the mite is removed (and will not be re-used). Mites that do not reach the end of one of the two arms within 5 min are scored as "no choice" and usually excluded from the analysis. After five mites made a choice, the odors sources are switched, i.e. connected to the opposite arms, to reassure that a possible non-odor related left-right bias is in the setup is accounted for. For a single replicate test on average 20–40 adult predatory mites will be submitted to this choice (Sabelis and Van der Baan 1983; Ament et al. 2004) (Photos by Jan van Arkel (**a**) and Merijn R. Kant (**b**, **c**))

common between and within plant populations, suggesting that insects using such information for finding prey must be able to cope with this variability irrespective of the circumstances. So how capable are foraging natural enemies in discriminating plants infested with different prey types on the basis of plant odors?

2.7   Tritrophic Interactions Mediated by Plant Volatiles

Maize plants infested by a folivorous caterpillar (*S. littoralis*), a stemborer (*Ostrinia nubilalis*), and the maize aphid (*Rhopalosiphum maidis*) release compositionally different volatile blends and this justified the question if specialist parasitoids and predators will be able discriminate between plants with prey and non-prey irrespective of the plant species (Turlings et al. 1998). Indeed it was found that the parasitic wasp *Cardiochiles nigriceps* discriminates between prey-induced volatiles from cotton, tobacco or maize when non-prey-induced volatiles from the same species were the alternative (De Moraes et al. 1998). Moreover, also herbivores are influenced themselves by such volatile information. In choice tests, herbivorous western flower thrips (*Frankliniella occidentalis*) consistently preferred un-induced plants over plants infested with conspecifics or with the chewing herbivore *Heliothis virescens*, or both simultaneously. Hence, herbivores may use the same herbivore-induced volatiles used by natural enemies use to find prey, to avoid competition and possibly predation (Delphia et al. 2007).

Despite all variation there is sufficient evidence that herbivore-induced volatiles mediate indirect defenses also under natural circumstances. Drukker et al. (1995) observed that predatory bugs aggregated near cages containing pear trees infested with their prey the pear psyllid *Cacopsylla* spp. and subsequently, by means of laboratory olfactory choice assays, Scutareanu et al. (1997) showed that these same bugs were attracted to the induced volatiles of *Cacopsylla*-infested pear leaves. But does increased attraction also result in increased predation? In 2001 Kessler and Baldwin showed that mimicking the naturally herbivore-induced emissions from *N. attenuata* in the field via synthetic volatiles increased egg predation by the generalist predator *Geocoris pallens* and, for one of these volatiles, simultaneously decreased lepidopteran oviposition rates. The authors estimated that herbivore-induced volatiles reduced the number of herbivores by more than 90%. In a later field study the same group showed that *N. attenuata* silenced for *LOX3* or *HPL* genes and deficient in α-bergamotene and GLV emission, respectively, were more vulnerable to their common natural herbivores but also attracted new herbivore species, which fed and reproduced successfully while normally ignoring the plant (Kessler et al. 2004) and this was followed by a study of Halitschke et al. (2008) who showed that the same predatory bugs use terpenoids and C_6-volatiles to locate plants with prey and thereby reduce herbivory. Similarly, the release of (*E*)-β-caryophyllene induced by *Diabrotica virgifera virgifera* beetles from the roots of *Z. mays* into the soil appeared to attract the beetle's entomopathogenic nematode *Heterorhabditis megidis* and decreased emergence of adult beetles to less than half (Rasmann et al. 2005). So basically the next question to answer at this stage is if such increased predation or parazitation will really increase a plant's fitness (Kessler and Baldwin 2004) since this is the prerequisite for positive selection on plants in nature to produce specific volatiles for the attraction of natural enemies. In the alternative scenario plant volatiles are produced for other reasons than to attract these natural enemies of herbivores but correlate well with their presence and hence the predators could be simply smart and plastic enough to quickly learn to use variable context-specific information to their own advantage.

Although not obtained from field experiments, there are indications that there are circumstances under which indirect defenses can increase a plant's fitness. Van Loon et al. (2000) showed that the *Arabidopsis* accessions L*er* and Col-0 infested with unparasitized larvae of *P. rapae* produced less seeds than when infested with parasitized larvae suggesting that parazitation can benefit plant fitness. Similar results were obtained with *S. littoralis*-infested maize plants that attract endoparasitoids, i.e. *C. marginiventris* and *Campoletis sonorensis* of the larvae. Parasitism significantly reduced feeding intensity and weight gain of the larvae and plants infested with parasitized larva produced 30% more seeds than plants infested with unparasitized larva (Hoballah and Turlings 2001). Taken together, herbivore-induced volatiles not only benefit prey searching predatory insects but may also increase host-plant fitness via reducing herbivory. However, it maybe still is too early to make generalized statements on this since the overall fitness effect on plants will have a relatively large context-dependent window of variance due to spatiotemporal life-history-complexities of natural communities. Hence designing a solid experimental set-up may be simply constrained too much by time if not by resources. Nevertheless, clearly there are examples where such positive effect on plant fitness could be shown.

2.8 Arthropod Counter-Adaptations to Indirect Defenses

Successful plant defenses will, although this is not a prerequisite, often have a negative impact on herbivorous arthropods and reduce their fitness and survival. In those cases defenses put pressure on herbivores selecting for phenotypes that can cope with plant defenses better and it is well possible that in some cases plants and plant-eaters are involved in evolutionary arms races of consecutive adaptations and counter-adaptations (Berenbaum and Zangerl 1998, 2008). In principle one can think of three kinds of herbivore-adaptations to plant defenses i.e. avoidance of defenses, resistance to defenses or suppression of induced defenses (Utsumi 2011).

2.8.1 Avoidance and Resistance

Herbivores can avoid plant defenses via selecting food, e.g. a plant or plant part, with the lowest level of defenses. For example, the specialist lepidopteran herbivore *Heliothis subflexa* feeds exclusively on the fruits of *Physalis angulata*, most likely because these fruits lack linolenic acid (LA) (De Moraes and Mescher 2004) since LA is the key component for the formation of volicitin (Alborn et al. 1997). In the absence of LA the oral secretions of *H. subflexa* larvae do not form volicitin thereby preventing the induction of defenses; the emission of volatiles and the subsequent attraction of female *Cardiochiles nigriceps* (De Moraes and Mescher 2004). Moreover, *C. nigriceps* larvae require LA for their development and fail to develop in LA-free host larvae (De Moraes and Mescher 2004). Another example of diet selection comes from the cotton bollworm *Helicoverpa armigera* who's feeding pattern on *Arabidopsis thaliana* shows it avoids the areas that have accumulated defensive glucosinolates (Shroff et al. 2008).

Plant defenses, like pesticides, select for resistance to phytotoxins, for example via mutations that lead to target site insensitivity (Feyereisen 1995) or via targeted detoxification mechanisms (Enayati et al. 2005; Siva-Jothy et al. 2005). Detoxification can be obtained via modification and secretion and some specialist insects adapted to sequester plant toxins, like glucosinolates, or their break-down products, and to use them to attract conspecifics or for their own protection (Hopkins et al. 2009). For example, cabbage aphid *Brevicoryne brassicae* raised on a diet containing glucosinolates is toxic to its natural enemies (Kazana et al. 2007). Another example is the polyphagous arctiid moth *Estigmene acrea* that converts the defensive alkaloids from its host plant into the male courtship pheromone hydroxydanaidal (Hartmann et al. 2005) while males of the arctiid moth *Utetheisa ornatrix* even transfer host-derived pyrrolizidine alkaloids to the female during mating making her less vulnerable for predation by spiders (Gonzalez et al. 1999). Interestingly, it was found that the corn earworm *Helicoverpea zea* up-regulates the production of its detoxification enzymes in response to JA and SA, even before the plant has established its down-stream defenses. This early sensing of plant defense signals and subsequent upregulation of cytochrome P450s, which play central roles during detoxification, is thought to protect *H. zea* against the inevitable production of toxins by host plants (Li et al. 2002b).

2.8.2   Suppression of Direct Defenses

Plant-defense suppression of induced defenses is a well established phenomenon occurring for example during compatible plant-pathogen interactions (Nomura et al. 2005; Abramovitch et al. 2006; Metraux et al. 2009) but also herbivores have evolved mechanisms to suppress induced plant defenses (Alba et al. 2011). Pathogens deliver effector molecules into host plants via specialized secretion systems (Abramovitch et al. 2006) and often suppression comes down to effector-mediated interference with receptor kinases (Xiang et al. 2008), transcriptional repression (Kim et al. 2006) but sometimes operate via the plant's own defensive toxins (Bouarab et al. 2002; Ito et al. 2004) or via the plant's own negative regulatory mechanisms such as their protein degradation machinery (Katsir et al. 2008a, b). Moreover, it was suggested that other phytopathogens suppress

SA-defenses via enhancing SA volatilization into volatile MeSA (Attaran et al. 2009). JA and SA are known to antagonize each other's action (Koornneef et al. 2008; Pieterse et al. 2009; Leon-Reyes et al. 2010a, b) and while the biological necessity for this phenomenon is not well understood it is a popular target for microorganisms to manipulate (Zhao et al. 2003; Thatcher et al. 2009; El Oirdi et al. 2011). For example, the bacterium *Pseudomonas syringae* DC3000 suppresses the SA-dependent plant defenses to which it is vulnerable via the phytotoxin coronatine. Coronatine mimics JA-Ile and has high affinity to the SFC^{COI}-complex (Fonseca et al. 2009). Hence, via the coronatine-mediated activation of SFC^{COI}-complex the plant is forced to exhibit a strong JA defense-response thereby antagonizing the SA-response thereby giving rise to bacterial speck disease in tomato (Zhao et al. 2003).

Like pathogens also herbivores were found to suppress induced plant defenses. The regurgitant from the Colorado potato beetle *Leptinotarsa decemlineata* suppresses the wound-induced increase in proteinase-inhibitor transcript levels in wounded tomato and potato leaves albeit via unknown components (Lawrence et al. 2007, 2008). The salivary secretions of the corn earworm *H. zea* contain a protein, glucose oxidase, which appeared to suppress JA-regulated nicotine production in tobacco *Nicotiana tabacum* (Musser et al. 2002) independent from salicylic acid (Musser et al. 2005) and possibly via redox-associated modifications (Thivierge et al. 2010). Moreover, wounded tomato leaves treated with glucose oxidase accumulated levels of JA-dependent trypsin inhibitors that were even lower than the levels of the controls (Musser et al. 2005) and in alfalfa *Medicago truncatula*, glucose oxidase from the saliva of beet armyworm *Spodoptera exigua* larvae was found to suppress transcript levels of two enzymes in the mevalonate and 2C-methyl erythritol 4-phosphate terpenoid pathways (Bede et al. 2006). Thus, glucose oxidase from caterpillar saliva acts on both direct and indirect plant defenses. Diezel et al. (2009) found that glucose oxidase in *S. exigua* saliva was responsible for an increase in SA levels in wild tobacco (*Nicotiana attenuata* L.), in favor of the hypothesis that *S. exigua* caterpillars benefit from decreased JA levels by inducing the SA pathway. In addition, expression and activity of JA-dependent genes was higher in Arabidopsis plants that were fed upon by *S. exigua* caterpillars with impaired salivary secretions compared to plants that were attacked by normal caterpillars but this difference disappeared in Arabidopsis plants that are unable to accumulate SA (Weech et al. 2008).

Also stylet-feeding herbivores were found to suppress host defenses by targeting the antagonistic cross-talk between JA and SA in plants. Feeding on Arabidopsis by phloem-feeding silverleaf whitefly *Bemisia tabaci* induced SA-responsive genes and suppressed JA-responsive genes (Kempema et al. 2007) and the whitefly nymphal developmental rate was higher on plants with a low JA-responsiveness or high SA-responsiveness (Zarate et al. 2007). Moreover, *B. tabaci* suppressed the induction of plant volatiles by two-spotted spider mites *Tetranychus urticae* on lima bean *Phaseolus lunatus* but since in lima beans whitefly feeding resulted in a simultaneous decrease of SA and JA responses a role for the SA-JA antagonism here is doubtful (Zhang et al. 2009).

There is some evidence that also phloem-feeding aphids, like whiteflies, suppress or avoid the JA-regulated defenses to which they are susceptible (Zhu-Salzman et al. 2004; Thompson and Goggin 2006; Walling 2008). For example, attack of sorghum plants by greenbug aphids *Schizaphis graminum* activated SA-dependent genes, although resistance of the plants to aphids was shown to be dependent on JA and not SA. Again, SA/JA cross-talk was suggested to explain this result (Zhu-Salzman et al. 2004). Interestingly, glucose oxidase was also identified in the saliva of aphids (Harmel et al. 2008). The saliva of the green peach aphid *Myzus persicae* contains also other effectors that interfere with plant defenses, for instance by suppressing the flagellin-22 depending oxidative burst. Over-expression of these effectors in plants showed that the activity of individual effectors is not translated into higher aphid fitness *per se*, since the fecundity of aphids feeding from transformed tissue was sometimes reduced rather than increased (Bos et al. 2010) suggesting that their individual action maybe not always is beneficial or only when produced in the appropriate amounts and delivered at the appropriate moment and location.

It was suggested that not only feeding herbivores but also insect eggs may suppress plant defense locally i.e. well before the larvae emerge. Deposition of insect eggs is known to induce direct and indirect defenses in plants (Hilker and Meiners 2006) but Bruessow et al. (2010) discovered that eggs of the butterfly *Pieris brassicae* suppressed JA-dependent defenses in Arabidopsis. This suppression was strong enough to positively affect larval growth of the generalist herbivore *Spodoptera littoralis* and the suppression of JA-related gene induction and enhanced *S. littoralis* performance was not observed in a SA-deficient mutant indicating that SA is required to explain this phenomenon (Bruessow et al. 2010).

2.8.3 Suppression of Induced Plant Volatiles and Indirect Defenses

Given the narrow metabolic association between the biosynthetic and regulatory pathways that give rise to direct and indirect defenses it can be expected that herbivores that suppress direct defenses will in the majority of cases automatically also suppress indirect defenses. However, this regulatory coupling is only a meta-bolic-physiological constraint since direct and indirect defenses can, in principle, operate independently within an ecological context. Therefore, although glucose oxidase from *S. exigua* larvae was found to suppress induced transcript levels of terpenoid biosynthetic enzymes (Bede et al. 2006) this cannot be taken for evidence that the actual indirect defenses, i.e. the attraction of natural enemies, were also suppressed. For example, it was shown that the tobacco spider mite *T. evansi* suppresses induction of tomato volatiles while the indirect defense of the plant, i.e. the attraction of a predatory mite, remained intact (Sarmento et al. 2011). Moreover, it was shown that the two-spotted spider mite *T. urticae* harbors distinct genotypes of which some do, and others don't, induce tomato *Solanum lycopersicum* volatiles sufficient for foraging predatory mites to find plants with prey (Takabayashi et al. 2000). Matsushima et al. (2006) showed that also two different forms of the closely related Kanzawa spider mite *Tetranychus kanzawai* differentially induced the emission of some of the typical spider mite induced volatiles in lima bean as well as SA responses. Distinct *T. urticae* JA-dependent defense-inducing and defense-suppressing genotypes could be extracted from single populations and it was found that the fecundity of inducer mites increased when feeding on the same leaflet as suppressor mites, indicating that suppression of induced defenses can indeed have a beneficial effect on organisms that are sensitive to those defenses (Kant et al. 2008). Also here suppression of induced volatiles was observed but subsequent effects on the behavior of natural enemies were not determined. There are other indications that herbivores can interfere with induced volatile production. Tooker and De Moraes (2007) showed that feeding by larvae of the Hessian fly (*Mayetiola destructor* Say) did not induce volatiles in wheat *Triticum aestivum* plants, whereas a generalist caterpillar did. Moreover, herbivory by two gall-inducing species, the tephritid fly *Eurosta solidaginis* and the gelechiid moth *Gnorimoschema gallaesolidaginis*, as well as the meadow spittlebug, *Philaenus spumarius*, did not induce a significant release of volatiles in goldenrod (*Solidago altissima* L.) plants. In addition, infestation by *E. solidaginis* decreased volatile emission of plants that were subsequently attacked by the generalist caterpillar *Heliothis virescens* (Tooker and De Moraes 2007).

Is suppression of induced volatile production evidence for the suppression of indirect defenses? *T. evansi* harbors genotypes that suppress both JA and SA defenses in tomatoes as for example the activity of induced proteinase inhibitors in *T. evansi*-infested tomatoes was lower in these plants than in un-infested control plants. Moreover, *T. evansi* was found to suppress the induction of JA-dependent volatiles but, surprisingly, this did not reduce the attraction of the predatory mites *Phytoseiulus macropilis* and *P. longipes* to infested plants suggesting that other attractive odors were still produced (Sarmento et al. 2011). Interestingly, whitefly infestation reduced amounts of volatile emission triggered by the beet armyworm in cotton (Rodriguez-Saona et al. 2003) and on lima beans whiteflies were shown to negatively interfere with the attraction of predatory mite *Phytoseiulus persimilis* induced by two-spotted spider mites (Zhang et al. 2009).

Defense suppression may intuitively sound like the right thing to do for an herbivore that induces plant defenses but it has some awkward consequences. A long term consequence is that herbivores eating from suppressed plants will likely accumulate mutations in resistance genes since there is no positive selection on these anymore, reminiscent of mites that loose pesticide resistances when not exposed to these (Nicastro et al. 2010) while a short term consequence is that suppressed plant material may be hard to monopolize since competing herbivores may prosper on suppressed plant material as well (Kant et al. 2008; Sarmento et al. 2011). In contrast, resistant herbivores that induce plant defenses do not have such problems. Indeed, *T. evansi* produces a very dense web to protect and monopolize its feeding site against invasion by competitors, such as *T. urticae*

(Sarmento et al. 2011). Hence it would make more sense if plant-eaters would develop the ability to suppress only those defenses they cannot become resistant to like pathogens have adapted to suppress HR (Bouarab et al. 2002) and gall makers have adapted to manipulate resource flows (Tooker et al. 2008). Since spider mites like *T. evansi* and *T. urticae* are species that develop resistances against toxins relatively easily (Hoy et al. 1998; Van Leeuwen et al. 2010) one may wonder if genotypes that suppress induced defenses have occurred by accident and if the traits leading to resistance or susceptibly to defenses and the traits leading to induction or suppression of defenses are polymorphisms of the same loci or whether they are distinct and can co-occur within individuals. However, the defense suppressing genotypes of *T. urticae* (Kant et al. 2008) and *T. evansi* (Sarmento et al. 2011) appeared to be relatively susceptible to artificially induced defenses of the same type they suppress thereby suggesting discrete traits.

2.9 Plant-Odor Recognition in Arthropods

To understand the evolvability of indirect defenses and to assess the extent to which the phenomenon can be manipulated for IPM purposes it is essential to understand how arthropods deal with odorous and non-odorous information from their environment to make foraging decisions. Whereas it is well known that foraging arthropods heavily rely on their chemical senses to find food, our understanding of the olfactory sensitivity of most species is limited. The only arthropod for which the sensitivity of nearly all of its olfactory receptor cells has been assessed is *Drosophila melanogaster* (de Bruyne et al. 1999, 2001; Galizia et al. 2010; van der Goes van Naters and Carlson 2007; Yao et al. 2005). By approximation insects possess a ten times smaller number of coding units, i.e. olfactory receptor cells that are responsive to a particular group of odor molecules, than vertebrates. *D. melanogaster* encodes for example 61 olfactory receptors (Robertson et al. 2003; Guo and Kim 2007) and possesses 44 types of olfactory receptor cells. The bee *Apis mellifera* appears to represent the higher end of the spectrum with 162 olfactory receptor-coding genes (Robertson and Wanner 2006). Information on predacious insects is more scare and the only estimate for predatory mites, a central model systems in the research field of indirect defenses, is based on the number of olfactory glomeruli and suggests that they contain 14–21 coding units (van Wijk et al. 2006). Behavioral and electrophysiological studies revealed that arthropods are sensitive to a wide variety of common plant volatiles (Bruce et al. 2005) while a limited number of receptor cells has evolved a high sensitivity for specific odorants and among these are the often extremely sensitive pheromone receptor cells (Berg and Mustaparta 1995; Cosse et al. 1998; Baker et al. 2006) and specialized CO_2 receptor systems present in many insects (Grant et al. 1995; Stange and Stowe 1999; Stange 1992). Additionally many arthropods also possess relatively high sensitivities for a number of food related volatiles. *D. melanogaster* is, for example, relatively sensitive to a number of esters associated with rotting fruit (de Bruyne et al. 2001; Hallem and Carlson 2006) and moths are relatively sensitive to a number of terpenoids (Bruce et al. 2005). In some species where the males are highly sensitive to female pheromones, the females possess olfactory receptor cells that are sensitive to plant volatiles and which are absent in males (Heinbockel and Kaissling 1996; King et al. 2000).

If we imagine a foraging arthropod tracking a plant odor in an environment where many non-host plants produce similar volatiles as those that constitute the tracked odor plume, it is tempting to speculate that the animal probably relies on its ability to detect key volatiles which are characteristic for its host or shared by a group of suitable hosts. Although there is some evidence to support this speculation, it appears the exception rather than the rule (Bruce et al. 2005). The clearest example is probably the response of the silkworm, *Bombyx mori*, to cis-jasmone, a volatile produced by intact mulberry leaves. It appeared that other volatiles did not contribute to the chemotaxis elicited by cis-jasmone while cis-jasmone is constitutively produced by green leafs of intact mulberry plants. Hence the authors proposed that this compound functions as a key volatile that silkworms use to identify mulberry (Tanaka et al. 2009). Also the attractiveness of rotting fruit to *D. melanogaster* may be driven by single-component. Rotting fruit releases vinegar and *D. melanogaster* is innately attracted to vinegar. This innate attraction appears to be mediated by the activation of a single olfactory glomerulus, while the loss of attraction to higher concentrations of this odor results from the recruitment of an additional glomerulus (Semmelhack and Wang 2009). Other species have developed highly sensitive olfactory receptors for rare host specific components. Isothiocyanates are for example volatile catabolites derived from glucosinolates characteristic for *Brassicacea* and specialist herbivores such as cabbage aphid *Brevicoryne brassicae (*Nottingham et al. 1991*)* and the cabbage seed weevil, *Ceutorhynchus assimilis* (Blight et al. 1995) are extremely sensitive to them and there are many other, similar, examples (Bjostad and Hibbard 1992; Blight et al. 1995; Guerin et al. 1983; Judd and Borden 1989; Krasnoff and Dussourd 1989; Knight and Light 2001). Whereas a high sensitivity for key compounds often reflects host preference it may also facilitate an animal's ability to reject unsuitable host taxa. This is exemplified by the aphids *Phorodon humuli* and *Aphis fabae* that reject plants that produce isothiocyanates (Nottingham et al. 1991).

Notwithstanding the abovementioned examples of specialized olfactory sensitivities for odors produced by specific plant taxa, the great majority of insect olfactory receptors are sensitive to common plant volatiles and most insects perceive most common plant volatiles (Bruce et al. 2005). Because most volatiles elicit a response in one or more olfactory receptor cells, a blend of odors elicits an activity pattern across these coding units (de Bruyne and Baker 2008). This combinatorial input allows the brain to recognize and discriminate between wide ranges of odors. Hence most arthropods identify plant-odor mixtures of ubiquitous compounds as unique objects i.e. the smell of 'strawberries' is not simply projected as the sum of its components but forms its own discrete object. Sometimes a mixture of a subset of the best perceived compounds, offered in the ratio and concentration at which they occur in the plant, suffices to elicit chemotaxis but not necessarily the same as the intact blend. The aphid *Aphis fabae* is, for example, repelled by nine host-plant compounds while their mixture is perceived as an attractant (Webster et al. 2008) and a mixture of nine well perceived components derived from a complex floral odor was found attractive to hawk moth *Manduca sexta* while the individual components elicited no attraction despite perception (Riffell et al. 2009). Other species may require a more complete mixture before these can elicit chemotaxis. The parasitoid *Cotesia vestalis* is attracted to a mixture of four herbivore induced plant volatiles presented against a background of non-infested cabbage odors whereas none of the components of this mixture acts as an attractant (Shiojiri et al. 2010). The predatory mite *Phytoseiulus persimilis* is attracted to MeSA, a typical spider mite induced lima bean volatile, but it is not attracted to four other volatiles induced by spider mites in the same plant species. Interestingly, a mixture of these five HIPV was not significantly attractive to the predatory mites, but when this mixture was presented against an unattractive background odor of clean lima beans it was highly attractive and could not be discriminated by predatory mites from the odor of spider-mite-infested lima bean. Moreover when this attractive mixture was reduced by one of the spider-mite induced volatiles, DMNT, which elicited no response in its pure form, the mixture lost its attractiveness despite the presence of the "attractant" MeSA (van Wijk et al. 2011). These examples illustrate that components that elicit no response may contribute to the response elicited when they occur in mixtures with other volatiles and that arthropods respond to mixtures as a whole and not to the attractiveness or repellence of their separate components.

Because most arthropods identify plant odors based on mixtures of commonly produced plant volatiles, their relative abundance in odor mixtures might be an important factor to discriminate between different mixtures (Bruce et al. 2005). The Colorado potato beetle *Leptinotarsa decemlineata* is, for example, highly sensitive to the ratio of several GLV, i.e. (E)-3-hexen-1-ol, (E)-2-hexen-1-ol, (Z)-2-hexen-1-ol and (E)-2-hexenal, that form an attractive mixture only if their relative abundance is comparable to the natural plant odor (Visser and Ave 1978) and the predatory bug *Geocoris* ssp. responds to distinct *M. sexta*-derived changes in the isomer ratio of GLVs that emanate from *N. attenuata* (Allmann and Baldwin 2010). Whereas the ratio of components certainly plays an important role in the recognition of plant odor mixtures, these ratios are probably more flexible than the rigid ratio-dependent recognition that plays a role in pheromonal communication in some insects (Wanner et al. 2010) since the composition of plant odor emissions varies with the variation in biotic and abiotic environment. Although some of this variation could provide crucial information and should thus be perceived, a lot of this variation will simply represent noise and should be ignored. If odor representations indeed would dramatically change with small shifts in the relative abundance of their components, plants could easily be selected to become unperceivable to herbivores via slight alterations in the relative abundance of components. Furthermore arthropods endowed with such constrained abilities could not profit from their experience through generalization, since the odor of the same plant species under slightly different conditions would not be perceived as similar, while this ability appears rather common leading to the suggestion that an olfactory system functions as a classification system (Niessing and Friedrich 2010). Experiments on vertebrates suggest that a range of ratio's of two components elicit a correlated activity pattern across output neurons of primary olfactory centers in the brain. As the ratio slowly changes from the situation where the first is more abundant to the situation where the second is abundant there is a sudden unpredictable shift whereupon a new correlated output pattern arises (Niessing and Friedrich 2010). Initially the mixture may, for example, perceptually closely resemble the dominant component while after the shift it may suddenly no longer resemble either component. In insects similar results have been obtained in experiments where the concentration of one component, benzonitrile, in a synthetic odor mixture was varied while the attraction of fruit moth *Cydia molesta* was assessed (Najar-Rodriguez et al. 2010). The concentration could be hundredfold increased without affecting moth attraction but at higher concentrations attraction broke down. The authors concluded that volatile blends in nature might vary quantitatively within a certain range without affecting odor-guided host location.

2.10 The Role of Arthropod Learning and Its Consequences for Indirect Defenses in the Field

In the ecologically relatively simple setting of the Utah dessert it was shown for the first time that mimicking plant volatile production via synthetic HIPV, or via manipulating these using transgenes, indeed the predation intensity of natural predators on the eggs of herbivores can be manipulated (Kessler and Baldwin 2001, 2004; Kessler et al. 2004; Halitschke et al. 2008; Allmann and Baldwin 2010) thereby confirming what many laboratory experiments already had predicted. Hence there have been many attempts to, subsequently, test if and how pure synthetic HIPV can be used to manipulate the movements of foraging natural enemies to improve pest management (Degenhardt et al. 2003). For example, it was suggested that synthetic plant volatiles can be used to attract herbivores for controlling invasive plants (Cosse et al. 2006) or to use some of these volatiles in attract-and-kill strategies against herbivores (Ranger et al. 2010) but most research focused on the possibilities of using HIPV in biological control of herbivores via their natural enemies. There is quite some evidence that synthetic volatiles can increase trap-capturing of several species of predatory insects (James 2003; Simpson et al. 2011) and decrease that of several herbivore species but also reverse effects were observed (Khan et al. 2008; Ali et al. 2011). For example, MeSA has quite consistent attractive properties for several predatory arthropods indoors (Zhu and Park 2005) and outdoors (James and Price 2004; James 2006; Lee 2010; Orre et al. 2010) but not always (Snoeren et al. 2010) and the outcome depends on the dosis (van Wijk et al. 2008) and on the predator's previous experiences (De Boer and Dicke 2004a, b) while it repels herbivores like aphids (Hardie et al. 1994; Glinwood and Pettersson 2000) but again depending on the doses and the absence/presence of additional volatile components (Webster et al. 2008). Although clearly insects respond to synthetic odors, these responses appear to be quite context-dependent and therefore there is sufficient reason to wonder to which extent synthetic analogs of HIPV can be used to force foraging predators to respond to synthetic calls in a sustainable manner. Importantly, natural enemies should not only move towards synthetic plant odors but also increase their predation on herbivores there and given the fact that the natural process works by the grace of this consistent reward, i.e. herbivores as food, one can wonder what will happen when such a reward can not be guaranteed. In principle the answer is easy: first you will get selection against predators that are stupid enough to respond to the signal repeatedly, since these will sooner or later die from starvation. Second, you may evoke a rapid learning response in the predators that are smart enough to not make the same mistake repeatedly. Hence, in both cases the number of predators that respond to the signal will decrease over time. So how much evidence is there for such learning responses in arthropods?

As argued earlier, the selective pressures that shaped the arthropod olfactory system only by exception resulted in systems that only detect ecological relevant key volatiles. Evolution rather appears to have favored a sensory system that is able to detect an enormous variety of odors utilizing the combinatorial input of a limited number of receptors. It would clearly be impossible to hardwire appropriate innate responses to all odor mixtures that might be encountered. It thus comes as no surprise that most arthropods are able to associate odors with reward and punishment. Olfactory learning in arthropods has been studied extensively. Model organisms such as flies (Berry et al. 2008; Kawecki 2010; Pitman et al. 2009), bees (Abramson et al.2010; Menzel 2001), and moths (Ito et al. 2008) are all accomplished olfactory learners.

Unfortunately the learning ability of predatory arthropods that feed on phytophagous arthropods have received far less attention (De Boer and Dicke 2006). Parasitoids readily associate a wide variety of odors with the presence of prey (Dukas 2008). Compared to parasitoids many predatory arthropods have a much broader diet and feed on a variety of herbivores or on herbivores that feed on a variety of plants. The predatory mite *Phytoseiulus persimilis* feeds, for example, on the highly polyphagous spider mite *Tetranychus urticae*. It copes with variability in spider-mite-induced plant odors by learning from experience. Olfactory preference is acquired during development and through associative learning in the adult phase (de Boer and Dicke 2004a; De Boer et al. 2005; Drukker et al. 2000a; Krips et al. 1999; van Wijk et al. 2008).

What is the role of individual HIPV in indirect defenses given the combinatorial perception of odors and the predator's ability to learn? It has often been assumed that one or a number of HIPV may function as predator attractants, particularly if they are attractive in their pure form. Such attractants are known for a wide variety of predator- herbivore-plant systems. For an excellent database see for example El-Sayed (2010). As mentioned before, the notion that arthropods perceive key volatiles in odor mixtures as "attractants" or "repellents" appears to be applicable to a very limited number of highly specialized herbivores only. Most animals perceive odor mixtures as a single whole (Laurent et al. 2001) and such combinatorial encoding of odors ensures that mixtures that differ in only a few commonly induced compounds may already be perceived as two different odors. Thus there is no need for a coevolving signal-receptor system, since most predators possess receptors that are sensitive to most commonly produced HIPV and if they manage to feed on the prey, they will associate the plant odor of that moment with the prey while in the absence of prey they may build a negative association.

Even though most arthropods are able to associated odors with the presence of prey there are instances where they may not be able to utilize this ability. We should discriminate two cases of naivety. First, predators that hatch in the absence of plants infested with their prey which are truly naïve. Second, predators searching for prey after local extermination of a prey patch which are not truly naïve but which might be naïve for the odor of the nearest suitable plant host complex. In both instances predators may rely on an innate preference for HIPV producing plants. In a review of the literature Allison and Hare (2009) found that 55% of the truly naïve predators preferred the odor of herbivore infested plants over the odor of uninfested conspecifics indicating that many predators are innately attracted to plants that produce HIPV. It is however not clear to what extend this innate attraction extends to plants infested with unsuitable prey. Because many plants produce at least a number of the same HIPV and because many predators are sensitive to such volatiles predators may generalize among herbivore induced plant odors. Thereby limiting their search to those plants that are perceptionally similar to those for which they have experience, i.e. generalization may allow experienced predators to make an educated guess when confronted with odors for which they lack experience. Subsequently predators may learn to discriminate between these similar attractive odors depending on reward value associated with each odor (Vet et al. 1998).

Can predators more easily associate odor mixtures that contain HIPV with the absence or presence of prey than odor mixtures without HIPV? The little evidence we have suggests that parasitoids learn just as easily about odor mixtures with and without HIPV. Parasitoids have, for example, been trained to respond to the explosive TNT (Tomberlin et al. 2005), or to perfume (Must de Cartier, Paris) (Dejong and Kaiser 1991). Even more revealing, the Anthocorid predator *Anthocoris nemoralis* is able to associate the HIPV MeSA the presence of prey and while it is not attracted to it when the odor has been associated with prey absence (Drukker et al. 2000b). The predatory mite *P. persimilis* is able to associate the odor of an herbivore infested, HIPV producing plant, with the absence of prey and when trained to associate the odor of an herbivore free plant with the presence of prey it preferred the later over the first (Drukker et al. 2000a; van Wijk et al. 2008). These examples clearly illustrate that HIPV do not confer any meaning, nor do they bias a predators' learning ability towards HIPV producing plants, it simply serves a as a means to discriminate odor mixtures. We can thus predict that any volatile is potentially an effective HIPV as long as predators are able to perceive it and as long it is reliably paired with prey.

2.11   The Evolutionary Dynamics of Plant Signal (dis)Honesty

To assess the evolutionary dynamics of herbivore-induced plant signaling would be a formidable, if not impracticable, task. Not only would it require detailed insight in the genetic architecture of genes involved in generating and perceiving plant signals and their mode of inheritance but it would also require insight in the mating structure of the signaler and receiver populations and in the fitness consequences of signaling and responding. As will be argued below, these fitness consequences are bound to be dependent on frequencies of genotypes and phenotypes in interacting populations at three trophic levels and changes in these frequencies occur at a pace set by the plants, i.e. the signalers, because their generation times are longer than those of the arthropods that live on them. Does this imply that it is better to ignore the eco-evolutionary dynamics of plant signals altogether? We argue against such a view for two reasons. First, predictive modeling provides a rigorous method to assess eco-evolutionary consequences of *a priori* assumptions and these assumptions are falsifiable. Second, predictions on evolutionary dynamics of plant signals in a tritrophic context help to assess the conditions under which HIPV maintain or loose their meaning as signals that betray herbivores to their predators, thereby stabilizing or destabilizing opportunistic plant-predator alliances.

As a first example of predictive modeling – much inspired by theory on the evolution of mimicry and 'green beards' (van Baalen and Jansen 2001, 2003; Jansen and van Baalen 2006) – consider the following plant signaling strategies: (1) no release of signals, (2) constitutive release of signals, (3) herbivore-induced release of signals in large amounts independent of herbivore density or (4) in amounts increasing with herbivore density. Plants may mutate to produce signals in different forms (e.g. A, B, C etc.), and signaling plants may receive assistance from predators in their combat against herbivores, whereas plant fitness is a decreasing function of the amount of herbivory. Assume, for the sake of simplicity, that the risk of herbivory is constant and that predators respond to a signal depending on the mean honesty in reflecting presence or density of herbivores. It can then be shown that a population of non-signaling plants can be invaded by plants sending alarm signals only when induced by herbivory and that a population of plants sending a constitutive (i.e. herbivory-independent) signal (e.g. A = false signal) can be invaded by plants sending a different signal only when induced by herbivory (e.g. B = honest signal). By the time the frequency of the latter plants have sufficiently increased, plants sending the same signal constitutively (thus a false B signal or 'cheater') will easily invade the population but with their subsequent increase in frequency will erode the signal honesty, i.e. the degree to which the signal reliably indicates the presence or density of herbivores. At some point, the honesty of this signal will have eroded so much that plants sending another, more honest signal (e.g. C), will invade. This will give rise to waves in the frequency of honest plants sending signals indicating herbivory, followed by waves of dishonest plants sending the same signal, yet without incurring damage from herbivory, then followed by a new wave of honestly (herbivore-induced) signaling plants and so on. Thus, theory predicts that frequency-dependent selection will drive cycles of honest and dishonest signals and thereby determine the rise and fall of signal honesty. On average, plant-predator alliances will be maintained, but the signals mediating these alliances will change all the time. Much the same argument can be given to show that the meaning of signals may change in space, depending on the extent to which interactions and signaling between individual organisms at different trophic levels are local.

The main message of this example on predictive modeling is that plants may manipulate communication with the herbivore's enemies because the interests of sender and receiver only partially overlap: plants gain by acquiring the enemies of herbivores and these enemies gain by finding herbivores. Thus, in an environment with plants sending alarm only when induced by herbivory, a mutant plant may gain by sending the same alarm signal even when there are no or only few herbivores on that plant, thereby cheating itself into receiving early protection against herbivory. The key question is therefore, whether such mutant plants occur in plant populations. Recently, such a 'cry wolf' strategy has been identified in a Japanese variety of cabbage. Whereas most plants, including other cabbage varieties, produce more herbivore-induced volatiles when there are more herbivores on the plant, this cabbage variety produces a maximal amount of these volatiles irrespective of the number of herbivorous larvae of the diamondback moth (Shiojiri et al. 2010). Since the parasitoids of these herbivores cannot assess the number of hosts on a plant from a distance, they have to rely on the alarm signals of the plant. If most plants send such alarms in amounts proportional to the herbivore damage incurred, this plant genotype gains by acquiring enemies of the herbivores because they need time to inspect the surface of the plant and then to ultimately find out that there are only few herbivores to exploit. Thus, while 'cry wolf'-like genotypes occur in agricultural crop plants, the existence of such squeamish plant genotypes in natural plant populations still has to be shown. If so, these strategies are predicted to trigger frequency-dependent selection causing changes in the chemical composition of plant signals mediating alliances between plants and the enemies of its herbivores.

An important assumption underlying the predictive model discussed above is that predators determine to which signal they respond depending on the mean honesty of the signals in the population at large. This presupposes fast learning of the fitness consequences of responses to the signals available in the environment as a whole. Even though individual arthropods are known to learn at time scales less than an hour in small-scale laboratory settings, this does not necessarily apply to the individuals in the their home range in the field, let alone at the spatial scale covered by the whole population. Unpublished data from field experiments in c. 200 m^2 *Eucalyptus* stands in Viçosa, Brazil (Arne Janssen, personal communication) showed that it takes about a week for predatory arthropods to locate prey eggs more frequently when offered together with an otherwise mildly repellent odor than when offered without such an odor. Evidently, individual predators in small laboratory arenas learn at a much faster pace than populations of individuals in the field. This is exactly why there is scope for 'cry wolf' plants to lure predators for the time they are still naïve with respect to the fitness consequences of their foraging decisions.

Another striking feature of 'cry wolf' signaling by the Japanese variety of cabbage is that maximum signal release is induced by herbivory, yet independent of herbivore density (Shiojiri et al. 2010). Thus, rather than constitutive production, signal release depends on herbivore presence, rather than their density. This is to be expected because the release of distinct chemical signals makes a plant apparent because the information is free to be used by all organisms in the community (Sabelis and de Jong 1988; Sabelis et al. 2001, 2007). For example, other herbivores may use plant alarm signals to spot their host plant. For example, diamondback moths prefer to oviposit on plants infested by caterpillars of cabbage white butterflies and profit from the fact that their natural enemy, a parasitic wasp, does not innately recognize the odors from cabbage plants attacked by both herbivores whereas they do recognize odours from plants with the diamond caterpillars alone (Shiojiri et al. 2002; Takabayashi et al. 2006). It may be that there is no way out for an herbivore-attacked plant than to take some future risk on the negative side effects of releasing alarm signals, given that the alternative is to be eaten by the currently attacking herbivores. In essence, these negative side effects of apparency due to alarm signals can be seen the as signaling costs, or 'handicaps', that represent the very reason why plant alarms may start out as honest signals (Grafen 1990).

What most, if not all, herbivory-induced plant signals have in common, is that their chemical composition is rather complex. Moreover, as argued before, the perception of odor blends seems not to be a simple sum of responses to individual components, but rather to be based on properties of the odor blend as a whole. Thus, small changes in the odor blend may allow the signal to be perceived as new (Van Wijk et al. 2008, 2010). Compared to the case where predator responses are a sum of responses to individual components, plants can more easily generate mutants that do not only release a new signal, but whose signal is also perceived as new. This, in turn, is likely to lead to increased complexity of plant signals and theory on the evolution of cooperation has shown that the more complex the signal, the more likely it is that cooperative alliances evolve and persist (Traulsen and Nowak 2007). We therefore do not only predict that chemical alarm 'languages' of plants change over generations, but also that they become more complex due to frequency dependent selection.

Herbivorous arthropods are not helpless bystanders when plants send out alarm signals and thereby lure predators. They may avoid plants that are induced to send alarm or they may affect the plant so as to reduce the release of alarm (Takabayashi et al. 2000; Bede et al. 2006; Matsushima et al. 2006; Kant et al. 2008; Sarmento et al. 2011). Thus, natural selection may act on herbivores to alter or even avoid alarm release by the plant. Indeed, 'saboteur' lines of two-spotted spider mite *T. urticae* have been found that are morphologically indistinguishable, feed and reproduce on tomato as well as lines resistant against direct plant defense, yet somehow manage to suppress the production of herbivore-induced volatiles from tomato (Kant et al. 2008). Moreover, the related spider mite species *T. evansi*, known to be a specialist of tomato and capable of reducing proteinase inhibitors below housekeeping levels, did not trigger the production of herbivore-induced terpenoids that are part of the alarm signals of tomato plants (Sarmento et al. 2011).

Possibly, there is an arms race between plant and herbivore: the higher the frequency of plant-alarm-suppressing herbivores, the stronger the selection on plants to prevent herbivores from feeding stealthily and vice versa (Kant et al. 2008).

2.12 Synthesis: The Future for Plant Volatiles in IPM

What can we do with all this information: is there a future role for HIPV application to crops, or for their production via transgenic plants, for improving IPM (Turlings and Ton 2006; Unsicker et al. 2009; Shrivastava et al. 2010)? If we assert that there could be such a role we implicitly assume that (1) the chemical communication systems that evolved between plants and predators are ineffective and can be improved or (2) that crop breading led to plants that lack efficient indirect defenses and, hence, this can be repaired. The latter may be the case for American maize varieties which do not produce β-caryophyllene while their European counterparts produce it in response to herbivory such that natural enemies are attracted to infested plants (Degenhardt et al. 2009). Most studies on HIPV-mediated chemical communication have likewise been conducted using crops, their pests and predators used in biocontrol (Arimura et al. 2005; Mumm and Dicke 2010) and, although a number of these systems lack a co-evolutionary history, this does not seem to hamper their chemical communication. That is in principle good news although one should be aware of the fact that the application of synthetic HIPV to crops in an IPM setting may easily do more harm than good: if no food can be found on plants that are made to smell like plants infested with prey, predators are likely to learn this and seize to explore the scented plants and plants infested with herbivores. Hence, rather than improving IPM, it could undermine it.

Despite some opportunities we argue that herbivore-induced plant volatiles are not the holy grail of IPM. Genetic engineering of plants to make them produce constitutive high amounts of false volatile signals to natural enemies may work in the short run, but is doomed sooner or later: predators may become less reliant bodyguards if signals from transgenic plants are not related to herbivore density but even when signal honesty can be guaranteed herbivores will be selected to become saboteurs of the plant-predator alliance sooner or later. Moreover, we predict frequency-dependent selection to play an important role in maintaining temporal alliances between plants and the enemies of their enemies (Sabelis et al. 2001, 2011) and we stress that the meaning of plant alarm signals may change over time (i.e. at the pace of plant generations) and in space (depending on the scale at which senders and receivers are effective and disperse in the field) (Sabelis and de Jong 1988; van Baalen and Jansen 2003; Jansen and van Baalen 2006; Kobayashi et al. 2006) in natural ecosystems. Given these dynamics, the success of sustainable pest management via HIPV will be short term and variable and where it works it will suffer from the same problem that application of pesticides suffers from: arthropod adaptation.

However, a consensus approach revolving around the improvement of honest herbivore-induced signals, via transgenes and/or herbivore-inducible promoters, in order to produce "very-clear-honest signals" may have a future. By doing so, one could leave the natural process largely intact and make, e.g. via breeding or transgenic manipulation, plants that produce "super-clear" signals simultaneous with their normal HIPV e.g. by 'adding' additional volatiles to induced blends or by increasing their absolute amounts. Possibly such signals are more effective, particularly in systems where the crop and the biocontrol predator lack an evolutionary history, and can be designed to stand out in the background to facilitate faster learning responses. An additional approach using plant volatiles in IPM could be to make use of the repellent properties of some compounds. For example, the aphids *Phorodon humuli* and *Aphis fabae* are highly deterred by isothiocyanates (Nottingham et al. 1991). This approach does however also suffer from two possible weaknesses. The first is that starved arthropods might overcome their aversion. The second is that this approach selects for mutant herbivores without this innate deterrence. As argued earlier: there are only a few mutations that can make an insect resistant to pesticides but they do evolve and there are many mutations that will render the intricate pathway that enables innate deterrence by a specific volatile dysfunctional.

Hence a prerequisite for using herbivore-induced plant volatiles successfully in IPM in a relatively sustainable way is that carnivores that respond to the 'synthetic' calls should be rewarded either with real herbivores or with alternative food. Of course plant growers could by-pass this problem via supplying their crops with a continuous flow of artificially selected stupid carnivores but such an approach can not be called 'sustainable' and for it to be successful one would not really need plant volatiles. Hence, there may be room to improve naturally induced indirect defenses but it is not realistic to assume that we will be able to control and orchestrate indirect defenses beyond what's good for the natural enemy. Honesty in the message of synthetic or engineered signals is the key since there is a fundamental force that we cannot compete with: natural selection.

Acknowledgments JMA is funded via Horizon Breakthrough Projects (93519024) and by the Marie Curie Fp7 program (221212); BCJS and PMB by NWO Earth and Life Sciences (ALW) and TTI Green Genetics (1CC026RP); MWS is funded by The Royal Netherlands Academy of Arts and Sciences (KNAW).

References

Abbott E, Hall D, Hamberger B, Bohlmann J (2010) Laser microdissection of conifer stem tissues: isolation and analysis of high quality RNA, terpene synthase enzyme activity and terpenoid metabolites from resin ducts and cambial zone tissue of white spruce (Picea glauca). BMC Plant Biol 10:106
Abramovitch RB, Anderson JC, Martin GB (2006) Bacterial elicitation and evasion of plant innate immunity. Nat Rev Mol Cell Biol 7(8):601–611

Abramson CI, Giray T, Mixson TA, Nolf SL, Wells H, Kence A (2010) Proboscis conditioning experiments with honeybees, *Apis mellifera caucasica*, with butyric acid and DEET mixture as conditioned and unconditioned stimuli. J Insect Sci 10:1–17

Aharoni A, Giri AP, Deuerlein S, Griepink F, de Kogel WJ, Verstappen FW, Verhoeven HA, Jongsma MA, Schwab W, Bouwmeester HJ (2003) Terpenoid metabolism in wild-type and transgenic Arabidopsis plants. Plant Cell 15:2866–2884

Alba JM, Glas JJ, Schimmel BCJ, Kant MR (2011) Avoidance and suppression of plant defenses by herbivores and pathogens. J Plant Interact 6(2):1–7

Alborn HT, Turlings TCJ, Jones TH, Stenhagen G, Loughrin JH, Tumlinson JH (1997) An elicitor of plant volatiles from beet armyworm oral secretions. Science 276:945–949

Alborn HT, Hansen TV, Jones TH, Bennett DC, Tumlinson JH, Schmelz EA, Teal PEA (2007) Disulfooxy fatty acids from the American bird grasshopper *Schistocerca americana*, elicitors of plant volatiles. Proc Natl Acad Sci U S A 104:12976–12981

Ali JG, Alborn HT, Stelinski LL (2011) Constitutive and induced subterranean plant volatiles attract both entomopathogenic and plant parasitic nematodes. J Ecol 99(1):26–35

Allison JD, Hare JD (2009) Learned and naive natural enemy responses and the interpretation of volatile organic compounds as cues or signals. New Phytol 184(4):768–782

Allmann S, Baldwin IT (2010) Insects betray themselves in nature to predators by rapid isomerization of green leaf volatiles. Science 329(5995):1075–1078

Ament K, Kant MR, Sabelis MW, Haring MA, Schuurink RC (2004) Jasmonic acid is a key regulator of spider mite-induced volatile terpenoid and methyl salicilate emission in tomato. Plant Physiol 135:2025–2037

Ament K, Krasikov V, Allmann S, Rep M, Takken FL, Schuurink RC (2010) Methyl salicylate production in tomato affects biotic interactions. Plant J 62:124–134

Andersson J, Borg-Karlson AK, Wiklund C (2003) Antiaphrodisiacs in pierid butterflies: a theme with variation! J Chem Ecol 29:1489–1499

Anten NPR, Pierik R (2010) Moving resources away from the herbivore: regulation and adaptive significance. New Phytol 188(3):644–645

Arigoni D, Sagner S, Latzel C, Eisenreich W, Bacher A, Zenk MH (1997) Terpenoid biosynthesis from 1-deoxy-D-xylulose in higher plants by intramolecular skeletal rearrangement. Proc Natl Acad Sci U S A 94:10600–10605

Arimura G, Ozawa R, Nishioka T, Boland W, Koch T, Kühnemann F, Takabayashi J (2002) Herbivore-induced volatiles induce the emission of ethylene in neighboring lima bean plants. Plant J 29:87–98

Arimura G, Huber DPW, Bohlmann J (2004) Forest tent caterpillars (*Malacosoma disstria*) induce local and systemic diurnal emissions of terpenoid volatiles in hybrid poplar (*Populus trichocarpa x deltoides*): cDNA cloning, functional characterization, and patterns of gene expression of (−)-germacrene D synthase, PtdTPS1. Plant J 37(4):603–616

Arimura G, Kost C, Boland W (2005) Herbivore-induced, indirect plant defences. Biochimica et Biophysica Acta-Molecular and Cell Biology of Lipids 1734(2):91–111

Attaran E, Zeier TE, Griebel T, Zeier J (2009) Methyl salicylate production and jasmonate signalling are not essential for systemic acquired resistance in *Arabidopsis*. Plant Cell 21:954–971

Babst BA, Ferrieri RA, Thorpe MR, Orians CM (2008) Lymantria dispar herbivory induces rapid changes in carbon transport and partitioning in *Populus nigra*. Entomol Exp Appl 128(1):117–125

Baker TC, Quero C, Ochieng SA, Vickers NJ (2006) Inheritance of olfactory preferences II. Olfactory receptor neuron responses from Heliothis subflexa x Hetliothis virescens hybrid male moths. Brain Behav Evol 68(2):75–89

Baldwin IT, Halitschke R, Paschold A, von Dahl CC, Preston CA (2006) Volatile signaling in plant-plant interactions: "talking trees" in the genomics era. Science 311(5762):812–815

Bate NJ, Rothstein SJ (1998) C-6-volatiles derived from the lipoxygenase pathway induce a subset of defense-related genes. Plant J 16(5):561–569

Bede JC, Musser RO, Felton GW, Korth KL (2006) Caterpillar herbivory and salivary enzymes decrease transcript levels of *Medicago truncatula* genes encoding early enzymes in terpenoid biosynthesis. Plant Mol Biol 60:519–531

Belles JM, Garro R, Pallas V, Fayos J, Rodrigo I, Conejero V (2006) Accumulation of gentisic acid as associated with systemic infections but not with the hypersensitive response in plant-pathogen interactions. Planta 223(3):500–511

Berenbaum MR, Zangerl AR (1998) Chemical phenotype matching between a plant and its insect herbivore. Proc Natl Acad Sci U S A 95:13743–13748

Berenbaum MR, Zangerl AR (2008) Facing the future of plant-insect interaction research: le retour à la "raison d'être". Plant Physiol 146:804–811

Berg BG, Mustaparta H (1995) The significance of major pheromone components and interspecific signals as expressed by receptor neurons in the oriental tobacco budworm moth, *Helicoverpa assulta*. J Comp Physiol A-Sens Neural Behav Physiol 177(6):683–694

Berry J, Krause WC, Davis RL (2008) Olfactory memory traces in *Drosophila* essence of memory. In: Sossin WS, Lacaille J-C, Castelucci VF, Bellville S (eds) Progress in brain research: the essence of memory, vol 169. Elsevier, Amsterdam, pp 293–304

Bjostad LB, Hibbard BE (1992) 6-methoxy-2-benzoxazolinone – a semiochemical for host location by western corn-rootworm larvae. J Chem Ecol 18(7):931–944

Bleeker PM, Diergaarde PJ, Ament K, Guerra J, Weidner M, Schutz S, de Both MTJ, Haring MA, Schuurink RC (2009) The role of specific tomato volatiles in tomato whitefly interaction. Plant Physiol 151(2):925–935

Blight MM, Pickett JA, Wadhams LJ, Woodcock CM (1995) Antennal perception of oilseed rape, Brassica napus (Brassicaceae), volatiles by the cabbage seed weevil *Ceutorhynchus assimilis* (Coleoptera: Curculionidae). J Chem Ecol 21(11):1649–1664

Bohlmann J, Meyer-Gauen G, Croteau R (1998) Plant terpenoid synthases: molecular biology and phylogenetic analysis. Proc Natl Acad Sci U S A 95:4126–4133

Bos JIB, Prince D, Pitino M, Maffei ME, Win J, Hogenhout SA (2010) A functional genomics approach identifies candidate effectors from the aphid species *Myzus persicae* (green peach aphid). PLoS Genet 6(11):e1001216

Bouarab K, Melton R, Peart J, Baulcombe D, Osbourn A (2002) A saponin-detoxifying enzyme mediates suppression of plant defences. Nature 418(6900):889–892

Bown AW, Hall DE, MacGregor KB (2002) Insect footsteps on leaves stimulate the accumulation of 4-aminobutyrate and can be visualized through increased chlorophyll fluorescence and superoxide production. Plant Physiol 129:1430–1434

Browse J (2009) Jasmonate passes muster: a receptor and targets for the defense hormone. Annu Rev Plant Biol 60:183–205

Bruce TJA, Wadhams LJ, Woodcock CM (2005) Insect host location: a volatile situation. Trends Plant Sci 10(6):269–274

Bruessow F, Gouhier-Darimont C, Buchala A, Metraux JP, Reymond P (2010) Insect eggs suppress plant defence against chewing herbivores. Plant J 62:876–885

Carlini CR, Grossi-de-Sa MF (2002) Plant toxic proteins with insecticidal properties. A review on their potentialities as bioinsecticides. Toxicon 90(11):1515–1539

Catinot J, Buchala A, Abou-Mansour E, Métraux J-P (2008) Salicylic acid production in response to biotic and abiotic stress depends on isochorismate in *Nicotiana benthamiana*. FEBS Lett 582:473–478

Chapman RF, Bernays EA (1989) Insect behavior at the leaf surface and learning as aspects of host plant-selection. Experientia 45(3):215–222

Chini A, Fonseca S, Fernández G, Adie B, Chico JM, Lorenzo O, Garciá-Casado G, López-Vidriero I, Lozano FM, Ponce MR, Micol JL, Solano R (2007) The JAZ family of repressors is the missing link in jasmonate signalling. Nature 448:666–671

Connolly JD, Hill RA (1991) Dictionary of terpenoids. Chapman and Hall, London

Cosse AA, Todd JL, Baker TC (1998) Neurons discovered in male *Helicoverpa zea* antennae that correlate with pheromone-mediated attraction and interspecific antagonism. J Comp Physiol A-Sens Neural Behav Physiol 182(5):585–594

Cosse AA, Bartelt RJ, Zilkowski BW, Bean DW, Andress ER (2006) Behaviorally active green leaf volatiles for monitoring the leaf beetle, Diorhabda elongata, a biocontrol agent of saltcedar, Tamarix spp. J Chem Ecol 32(12):2695–2708

Cowan MM (1999) Plant products as antimicrobial agents. Clin Microbiol Rev 12(4):564–582

Creelman RA, Mullet JE (1997) Biosynthesis and action of jasmonates in plants. Annu Rev Plant Physiol Plant Mol Biol 48:355–381

D'Alessandro M, Held M, Triponez Y, Turlings TCJ (2006) The role of indole and other shikimic acid derived maize volatiles in the attraction of two parasitic wasps. J Chem Ecol 32(12):2733–2748

D'Auria JC, Pichersky E, Schaub A, Hansel A, Gershenzon J (2007) Characterization of a BAHD acyltransferase responsible for producing the green leaf volatile (Z)-3-hexen-1-yl acetate in *Arabidopsis thaliana*. Plant J 49:194–207

Dangl JL, Jones JDG (2001) Plant pathogens and integrated defence responses to infection. Nature 411(6839):826–833

De Boer JG, Dicke M (2004a) Experience with methyl salicylate affects behavioural responses of a predatory mite to blends of herbivore-induced plant volatiles. Entomol Exp Appl 110(2):181–189

De Boer JG, Dicke M (2004b) The role of methyl salicylate in prey searching behavior of the predatory mite *Phytoseiulus persimilis*. J Chem Ecol 30(2):255–271

De Boer JG, Dicke M (2006) Olfactory learning by predatory arthropods. Anim Biol 56(2):143–155

De Boer JG, Snoeren TAL, Dicke M (2005) Predatory mites learn to discriminate between plant volatiles induced by prey and nonprey herbivores. Anim Behav 69:869–879

De Bruyne M, Baker TC (2008) Odor detection in insects: volatile codes. J Chem Ecol 34(7):882–897

De Bruyne M, Clyne PJ, Carlson JR (1999) Odor coding in a model olfactory organ: the *Drosophila* maxillary palp. J Neurosci 19(11):4520–4532

De Bruyne M, Foster K, Carlson JR (2001) Odor coding in the *Drosophila* antenna. Neuron 30(2):537–552

De Moraes CM, Mescher MC (2004) Biochemical crypsis in the avoidance of natural enemies by an insect herbivore. Proc Natl Acad Sci U S A 101:8993–8997

De Moraes CM, Lewis WJ, Paré PW, Alborn HT, Tumlinson JH (1998) Herbivore-infested plants selectively attract parasitoids. Nature 393:570–573

De Moraes CM, Mescher MC, Tumlinson JH (2001) Caterpillar-induced nocturnal plant volatiles repel conspecific females. Nature 410(6828):577–580

Degen T, Dillmann C, Marion-Poll F, Turlings TCJ (2004) High genetic variability of herbivore-induced volatile emission within a broad range of maize inbred lines. Plant Physiol 135:1928–1938

Degenhardt J, Gershenzon J, Baldwin IT, Kessler A (2003) Attracting friends to feast on foes: engineering terpene emission to make crop plants more attractive to herbivore enemies. Curr Opin Biotechnol 14(2):169–176

Degenhardt J, Hiltpold I, Kollner TG, Frey M, Gierl A, Gershenzon J, Hibbard BE, Ellersieck MR, Turlings TC (2009) Restoring a maize root signal that attracts insect-killing nematodes to control a major pest. Proc Natl Acad Sci U S A 106:13213–13218

Degenhardt DC, Refi-Hind S, Stratmann JW, Lincoln DE (2010) Systemin and jasmonic acid regulate constitutive and herbivore-induced systemic volatile emissions in tomato, *Solanum lycopersicum*. Phytochemistry 71(17–18):2024–2037

Dejong R, Kaiser L (1991) Odor learning by leptopilina-boulardi, a specialist parasitoid (Hymenoptera, eucoilidae). J Insect Behav 4(6):743–750

Delaney KJ (2008) Injured and uninjured leaf photosynthetic responses after mechanical injury on Nerium oleander leaves, and *Danaus plexippus* herbivory on *Asclepias curassavica* leaves. Plant Ecol 199(2):187–200

Delphia CM, Mescher MC, De Moraes CM (2007) Induction of plant volatiles by herbivores with different feeding habits and the effects of induced defenses on host-plant selection by thrips. J Chem Ecol 33:997–1012

Diezel C, von Dahl CC, Gaquerel E, Baldwin IT (2009) Different lepidopteran elicitors account for cross-talk in herbivory-induced phytohormone signaling. Plant Physiol 150:1576–1586

Dish A, Hennerlin A, Bach TJ, Rohmer M (1998) Mevalonate-derived isopentenyl diphosphate is the biosynthetic precursor of ubiquinone prenyl side chain in tobacco BY-2 cells. Biochem J 331:615–621

Dixon RA, Achnine L, Kota P, Liu CJ, Reddy MSS, Wang LJ (2002) The phenylpropanoid pathway and plant defence – a genomics perspective. Mol Plant Pathol 3(5):371–390

Doss RP, Proebsting WM, Potter SW, Clement SL (1995) Response of Np mutant of pea (*Pisum sativum* L.) to pea weevil (*Bruchus pisorum* L.) oviposition and extracts. J Chem Ecol 21:97–106

Douglas AE (1993) The nutritional quality of phloem sap utilized by natural aphid populations. Ecol Entomol 18(1):31–38

Drukker B, Scutareanu P, Sabelis MW (1995) Do anthocorid predators respond to synomones from psylla-infested pear trees under field conditions. Entomol Exp Appl 77(2):193–203

Drukker B, Bruin J, Jacobs G, Kroon A, Sabelis MW (2000a) How predatory mites learn to cope with variability in volatile plant signals in the environment of their herbivorous prey. Exp Appl Acarol 24(12):881–895

Drukker B, Bruin J, Sabelis MW (2000b) Anthocorid predators learn to associate herbivore-induced plant volatiles with presence or absence of prey. Physiol Entomol 25(3):260–265

Dudareva N, Pichersky E (2000) Biochemical and molecular genetic aspects of floral scents. Plant Physiol 122(3):627–633

Dudareva N, Pichersky E, Gershenzon J (2004) Biochemistry of plant volatiles. Plant Physiol 135:1893–1902

Duffey SS, Stout MJ (1996) Antinutritive and toxic components of plant defense against insects. Arch Insect Biochem Physiol 32(1):3–37

Dukas R (2008) Evolutionary biology of insect learning. Annu Rev Entomol 53:145–160

Dussourd DE, Denno RF (1991) Deactivation of plant defense – correspondence between insect behavior and secretory canal architecture. Ecology 72(4):1383–1396

Effmert U, Dinse C, Piechulla B (2008) Influence of green leaf herbivory by *Manduca sexta* on floral volatile emission by *Nicotiana suaveolens*. Plant Physiol 146:1996–2007

Eigenbrode SD, Espelie KE (1995) Effects of plant epicuticular lipids on insect herbivore. Annu Rev Entomol 40:171–194

El Oirdi M, El Rahman TA, Rigano L, El Hadrami A, Rodriguez MC, Daayf F, Vojnov A, Bouarab K (2011) *Botrytis cinerea* manipulates the antagonistic effects between immune pathways to promote disease development in tomato. Plant Cell 23(6):2405–2421

Ellinger D, Stingl N, Kubigsteltig II, Bals T, Juenger M, Pollmann S, Berger S, Schuenemann D, Mueller MJ (2010) DONGLE and DEFECTIVE IN ANTHER DEHISCENCE1 lipases are not essential for wound- and pathogen-induced jasmonate biosynthesis: redundant lipases contribute to jasmonate formation. Plant Physiol 153(1):114–127

El-Sayed AM (2010) The Pherobase: database of insect pheromones and semiochemicals. http://www.pherobase.com

Enayati AA, Ranson H, Hemingway J (2005) Insect glutathione transferases and insecticide resistance. Insect Mol Biol 14(1):3–8

Fahn A (1988) Secretory-tissues in vascular plants. New Phytol 108(3):229–257

Farag MA, Pare PW (2002) C6-Green leaf volatiles trigger local and systemic VOC emissions in tomato. Phytochemistry 61(2002):545–554

Farmer EE, Ryan CA (1990) Interplant communication: airborne methyl jasmonate induces synthesis of proteinase-inhibitors in plant-leaves. Proc Natl Acad Sci U S A 87(19):7713–7716

Fatouros NE, Broekgaarden C, Bukovinszkine'Kiss G, van Loon JJA, Mumm R, Huigens ME, Dicke M, Hilker M (2008) Male-derived butterfly anti-aphrodisiac mediates induced indirect plant defense. Proc Natl Acad Sci U S A 105:10033–10038

Felton GW, Tumlinson JH (2008) Plant-insect dialogs: complex interactions at the plant-insect interface. Curr Opin Plant Biol 11:457–463

Feyereisen R (1995) Molecular biology of insecticide resistance. Toxicol Lett 82:83–90

Fonseca S, Chini A, Hamberg M, Adie B, Porzel A, Kramell R, Miersch O, Wasternack C, Solano R (2009) (+)-7-iso-Jasmonoyl-L-isoleucine is the endogenous bioactive jasmonate. Nat Chem Ecol 5:344–350

Friedman AR, Baker BJ (2007) The evolution of resistance genes in multi-protein plant resistance systems. Curr Opin Genet Dev 17:493–499

Frost CJ, Appel M, Carlson JE, De Moraes CM, Mescher MC, Schultz JC (2007) Within-plant signalling via volatiles overcomes vascular constraints on systemic signalling and primes responses against herbivores. Ecol Lett 10(6):490–498

Galizia CG, Munch D, Strauch M, Nissler A, Ma SW (2010) Integrating heterogeneous odor response data into a common response model: a door to the complete olfactome. Chem Senses 35(7):551–563

Gaquerel E, Weinhold A, Baldwin IT (2009) Molecular interactions between the specialist herbivore *Manduca sexta* (Lepidoptera, Sphigidae) and its natural host *Nicotiana attenuata*. VIII. An unbiased GCxGC-ToFMS analysis of the plant's elicited volatile emissions. Plant Physiol 149(3):1408–1423

Gfeller A, Dubugnon L, Liechti R, Farmer EE (2010) Jasmonate biochemical pathway. Sci Signal 3(109):cm3

Giordanengo P, Brunissen L, Rusterucci C, Vincent C, van Bel A, Dinant S, Girousse C, Faucher M, Bonnemain JL (2010) Compatible plant-aphid interactions: how aphids manipulate plant responses. C R Biol 333(6–7):516–523

Glinwood RT, Pettersson J (2000) Change in response of Rhopalosiphum padi spring migrants to the repellent winter host component methyl salicylate. Entomol Exp Appl 94(3):325–330

Gols R, Roosjen M, Dijkman H, Dicke M (2003) Induction of direct and indirect plant responses by jasmonic acid, low spider mite densities, or a combination of jasmonic acid treatment and spider mite infestation. J Chem Ecol 29:2651–2666

Gomez S, Ferrieri RA, Schueller M (2010) Methyl jasmonate elicits rapid changes in carbon and nitrogen dynamics in tomato. New Phytol 188(3):835–844

Gonzalez A, Rossini C, Eisner M, Eisner T (1999) Sexually transmitted chemical defense in a moth (*Utetheisa ornatrix*). Proc Natl Acad Sci U S A 96(10):5570–5574

Gouinguené SP, Turlings TCJ (2002) The effects of abiotic factors on induced volatile emissions in corn plants. Plant Physiol 129(3):1296–1307

Gouinguené S, Degen T, Turlings TCJ (2001) Variability in herbivore-induced odour emissions among maize cultivars and their wild ancestors (teosinte). Chemoecology 11:9–16

Grafen A (1990) Biological signals as handicaps. J Theor Biol 144:517–546

Grant AJ, Wigton BE, Aghajanian JG, Oconnell RJ (1995) Electrophysiological responses of receptor neurons in mosquito maxillary palp sensilla to carbon-dioxide. J Comp Physiol A-Sens Neural Behav Physiol 177(4):389–396

Gross N, Wasternack C, Kock M (2004) Wound-induced RNaseLE expression is jasmonate and systemin independent and occurs only locally in tomato (*Lycopersicon esculentum* cv. Lukullus). Sour: Phytochem 65(10):1343–1350

Guerin PM, Stadler E, Buser HR (1983) Identification of host plant attractants for the carrot fly, psila-rosae. J Chem Ecol 9(7):843–861

Guo S, Kim J (2007) Molecular evolution of *Drosophila* odorant receptor genes. Mol Biol Evol 24(5):1198–1207

Halitschke R, Baldwin IT (2003) Antisense LOX expression increases herbivore performance by decreasing defense responses and inhibiting growth-related transcriptional reorganization in *Nicotiana attenuata*. Plant J 36:794–807

Halitschke R, Kessler A, Kahl J, Lorenz A, Baldwin IT (2000) Ecophysiological comparison of direct and indirect defenses in *Nicotiana attenuata*. Oecologia 124:408–417

Halitschke R, Schittko U, Pohnert G, Boland W, Baldwin IT (2001) Molecular interactions between the specialist herbivore *Manduca sexta* (Lepidoptera, Sphingidae) and its natural host *Nicotiana attenuata*. III. Fatty acid-amino acid conjugates in herbivore oral secretions are necessary and sufficient for herbivore specific plant responses. Plant Physiol 125:711–717

Halitschke R, Ziegler J, Keinanen M, Baldwin IT (2004) Silencing of hydroperoxide lyase and allene oxide synthase reveals substrate and defense signaling crosstalk in *Nicotiana attenuata*. Plant J 40:35–46

Halitschke R, Stenberg JA, Kessler D, Kessler A, Baldwin IT (2008) Shared signals – 'alarm calls' from plants increase apparency to herbivores and their enemies in nature. Ecol Lett 11(1):24–34

Hallem EA, Carlson JR (2006) Coding of odors by a receptor repertoire. Cell 125(1):143–160

Hamilton JG, Zangerl AR, DeLucia EH, Berenbaum MR (2001) The carbon-nutrient balance hypothesis: its rise and fall. Ecol Lett 4(1):86–95

Hanley ME, Lamont BB, Fairbanks MM, Rafferty CM (2007) Plant structural traits and their role in anti-herbivore defence. Perspect Plant Ecol Evol Syst 8(4):157–178

Hao PY, Liu CX, Wang YY, Chen RZ, Tang M, Du B, Zhu LL, He G (2008) Herbivore-induced callose deposition on the sieve plates of rice: An important mechanism for host resistance. Plant Physiol 146(4):1810–1820

Harborne JB (1991) Ecological chemistry and biochemistry of plant terpenoids. Clarendon, Oxford, pp 399–426

Hardie J, Isaacs R, Pickett JA, Wadhams LJ, Woodcock CM (1994) Methyl salicylate and (−)-(1R,5S)-myrtenal are plant-derived repellents for black bean aphid, *Aphis fabae* scop (Homoptera, Aphididae). J Chem Ecol 20(11):2847–2855

Harmel N, Létocart E, Cherqui A, Giordanengo P, Mazzucchelli G, Guillonneau F, De Pauw E, Haubruge E, Francis F (2008) Identification of aphid salivary proteins: a proteomic investigation of *Myzus persicae*. Insect Mol Biol 17(2):165–174

Hartmann T, Theuring C, Beuerle T, Klewer N, Schulz S, Singer MS, Bernays EA (2005) Specific recognition, detoxification and metabolism of pyrrolizidine alkaloids by the polyphagous arctiid *Estigmene acrea*. Insect Biochem Mol Biol 35:391–411

Hassan S, Behm CA, Mathesius U (2010) Effectors of plant parasitic nematodes that re-program root cell development. Funct Plant Biol 37(10):933–942

Hatanaka A (1993) The biogeneration of green odor by green leaves. Phytochemistry 34:1201–1218

Heinbockel T, Kaissling KE (1996) Variability of olfactory receptor neuron responses of female silkmoths (Bombyx mori L) to benzoic acid and (+/−)-linalool. J Insect Physiol 42(6):565–578

Hematy K, Cherk C, Somerville S (2009) Host-pathogen warfare at the plant cell wall. Curr Opin Plant Biol 12(4):406–413

Herms DA, Mattson WJ (1992) The dilemma of plants: to grow or defend. Q Rev Biol 67:283–335

Higginson AD, Delf J, Ruxton GD, Speed MP (2011) Growth and reproductive costs of larval defence in the aposematic lepidopteran *Pieris brassicae*. J Anim Ecol 80:384–392

Hilker M, Meiners T (2006) Early herbivore alert: insect eggs induce plant defense. J Chem Ecol 32:1379–1397

Hilker M, Kobs C, Varama M, Schrank K (2002) Insect egg deposition induces *Pinus sylvestris* to attract egg parasitoids. J Exp Biol 205:455–461

Hilker M, Stein C, Schro'Der R, Varama M, Mumm R (2005) Insect egg deposition induced defence response in *Pinus sylvestris*. Characterization of the elicitor. J Exp Biol 208:1849–1854

Hoballah MEF, Turlings TCJ (2001) Experimental evidence that plants under caterpillar attack may benefit from attracting parasitoids. Evol Ecol Res 3:553–565

Hoballah ME, Kollner TG, Degenhardt J, Turlings TCJ (2004) Costs of induced volatile production in maize. Oikos 105(1):168–180

Holopainen JK (2004) Multiple functions of inducible plant volatiles. Trends Plant Sci 9:529–533

Hopke J, Donath J, Blechert S, Boland W (1994) Herbivore-induced volatiles: the emission of acyclic homoterpenes from leaves of *Phaseolus lunatus* and *Zea mays* can be triggered by a β-glucosidase and jasmonic acid. FEBS Lett 352:146–150

Hopkins RJ, van Dam N, van Loon JJA (2009) Role of glucosinolates in insect-plant relationships and multitrophic interactions. Annu Rev Entomol 54:57–83

Howe GA, Jander G (2008) Plant immunity to insect herbivores. Annu Rev Plant Biol 59:41–66

Howe GA, Lightner J, Browse J, Ryan CA (1996) An octadecanoid pathway mutant (JL5) of tomato is compromised in signaling for defense against insect attack. Plant Cell 8(11):2067–2077

Hoy CW, Head GP, Hall FR (1998) Spatial heterogeneity and insect adaptation to toxins. Annu Rev Entomol 43:571–594

Hyun Y, Choi S, Hwang HJ, Yu J, Nam SJ, Ko J, Park JY, Seo YS, Kim EY, Ryu SB, Kim WT, Lee YH, Kang H, Lee I (2008) Cooperation and functional diversification of two closely related galactolipase genes for jasmonate biosynthesis. Dev Cell 14(2):183–192

Ito S, Eto T, Tanaka S, Yamauchi N, Takahara H, Ikeda T (2004) Tomatidine and lycotetraose, hydrolysis products of alpha-tomatine by *Fusarium oxysporum* tomatinase, suppress induced defense responses in tomato cells. FEBS Lett 571(1–3):31–34

Ito I, Ong RCY, Raman B, Stopfer M (2008) Sparse odor representation and olfactory learning. Nat Neurosci 11(10):1177–1184

James DG (2003) Synthetic herbivore-induced plant volatiles as field attractants for beneficial insects. Environ Entomol 32:977–982

James DG (2006) Methyl salicylate is a field attractant for the goldeneyed lacewing, *Chrysopa oculata*. Biocontrol Sci Technol 16(1):107–110

James DG, Price TS (2004) Field-testing of methyl salicylate for recruitment and retention of beneficial insects in grapes and hops. J Chem Ecol 30:1613–1628

Jansen VAA, van Baalen M (2006) Altruism through beard chromodynamics. Nature 440:663–666

Jongsma MA, Bakker PL, Peters J, Bosch D, Stiekema WJ (1995) Adaptation of spodoptera-exigua larvae to plant proteinase-inhibitors by induction of gut proteinase activity insensitive to inhibition. Proc Natl Acad Sci U S A 92(17):8041–8045

Jørgensen K, Vinther-Morant A, Morant M, Jensen NB, Olsen CE, Kannangara R, Motawia MS, Lindberg-Møller B, Bak S (2011) Biosynthesis of the cyanogenic glucosides linamarin and lotaustralin in cassava: isolation, biochemical characterization, and expression pattern of CYP71E7, the oxime-metabolizing cytochrome P450 enzyme. Plant Physiol 155(1):282–292

Judd GJR, Borden JH (1989) Distant olfactory response of the onion fly, delia-antiqua, to host-plant odor in the field. Physiol Entomol 14(4):429–441

Kahl J, Siemens DH, Aerts RJ, Gäbler R, Kühnemann F, Preston CA, Baldwin IT (2000) Herbivore-induced ethylene suppresses a direct defense but not a putative indirect defense against an adapted herbivore. Planta 210:336–342

Kaloshian I, Walling LL (2005) Hemipterans as plant pathogens. Annu Rev Phytopathol 43:491–521

Kang JH, Liu GH, Shi F, Jones AD, Beaudry RM, Howe GA (2010) The tomato odorless-2 mutant is defective in trichome-based production of diverse specialized metabolites and broad-spectrum resistance to insect herbivores. Plant Physiol 154(1):262–272

Kant MR, Ament K, Sabelis MW, Haring MA, Schuurink RC (2004) Differential timing of spider mite-induced direct and indirect defenses in tomato plants. Plant Physiol 135(1):483–495

Kant MR, Sabelis MW, Haring MA, Schuurink RC (2008) Intraspecific variation in a generalist herbivore accounts for differential induction and impact of host plant defences. Proc R Soc B: Biol Sci 275:443–452

Kant MR, Bleeker PM, Van Wijk M, Schuurink RC, Haring MA (2009) Plant volatiles in defence. In: Plant innate immunity. Advances in Botanical Research, Vol. 51, Burlington: Academic Press, pp 613–666

Kappers IF, Aharoni A, van Herpen TW, Luckerhoff LL, Dicke M, Bouwmeester HJ (2005) Genetic engineering of terpenoid metabolism attracts bodyguards to *Arabidopsis*. Science 309:2070–2072

Katsir L, Chung HS, Koo AJK, Howe GA (2008a) Jasmonate signaling: a conserved mechanism of hormone sensing. Curr Opin Plant Biol 11:428–435

Katsir L, Schilmiller AL, Staswick PE, He SY, Howe GA (2008b) COI1 is a critical component of a receptor for jasmonate and the bacterial virulence factor coronatine. Proc Natl Acad Sci U S A 105:7100–7105

Kawecki TJ (2010) Evolutionary ecology of learning: insights from fruit flies. Popul Ecol 52 (1):15–25

Kazana E, Pope TW, Tibbles L, Bridges M, Pickett JA, Bones AM, Powell G, Rossiter JT (2007) The cabbage aphid: a walking mustard oil bomb. Proc R Soc B: Biol Sci 274:2271–2277

Kellog BA, Poulter CD (1997) Chain elongation in the isoprenoid biosynthetic pathway. Curr Opin Chem Biol 1:570–578

Kempema LA, Cui X, Holzer FM, Walling LL (2007) Arabidopsis transcriptome changes in response to phloem-feeding silverleaf whitefly nymphs. Similarities and distinctions in responses to aphids. Plant Physiol 143:849–865

Kessler A, Baldwin IT (2001) Defensive function of herbivore-induced plant volatile emissions in nature. Science 291(5511):2141–2144

Kessler A, Baldwin IT (2004) Herbivore-induced plant vaccination. Part I. The orchestration of plant defenses in nature and their fitness consequences in the wild tobacco Nicotiana attenuata. Plant J 38(4):639–649

Kessler A, Halitschke R, Baldwin IT (2004) Silencing the jasmonate cascade: induced plant defenses and insect populations. Science 305(5684):665–668

Kessler D, Gase K, Baldwin IT (2008) Field experiments with transformed plants reveal the sense of floral scents. Science 321(5893):1200–1202

Khan ZR, James DG, Midega CAO, Pickett JA (2008) Chemical ecology and conservation biological control. Biol Control 45(2):210–224

Kim KC, Fan BF, Chen ZX (2006) Pathogen-induced Arabidopsis WRKY7 is a transcriptional repressor and enhances plant susceptibility to Pseudomonas syringae. Plant Physiol 142:1180–1192

King JR, Christensen TA, Hildebrand JG (2000) Response characteristics of an identified, sexually dimorphic olfactory glomerulus. J Neurosci 20(6):2391–2399

Kleiner KW, Raffa KF, Dickson RE (1999) Partitioning of C-14-labeled photosynthate to allelochemicals and primary metabolites in source and sink leaves of aspen: evidence for secondary metabolite turnover. Oecologia 119(3):408–418

Knight AL, Light DM (2001) Attractants from Bartlett pear for codling moth, Cydia pomonella (L.), larvae. Naturwissenschaften 88(8):339–342

Kobayashi Y, Yamamura N, Sabelis MW (2006) Evolution of talking plants in a tritrophic context: conditions for uninfested plants to attract predators prior to herbivore attack. J Theor Biol 243:361–374

Kollner TG, Lenk C, Zhao N, Seidl-Adams I, Gershenzon J, Chen F, Degenhardt J (2010) Herbivore-induced SABATH methyltransferases of maize that methylate anthranilic acid using S-Adenosyl-L-Methionine. Plant Physiol 153(4):1795–1807

Koornneef A, Pieterse CMJ (2008) Cross talk in defense signaling. Plant Physiol 146(3):839–844

Krasnoff SB, Dussourd DE (1989) Dihydropyrrolizine attractants for arctiid moths that visit plants containing pyrrolizidine alkaloids. J Chem Ecol 15(1):47–60

Krips OE, Willems PEL, Gols R, Posthumus MA, Dicke M (1999) The response of Phytoseiulus persimilis to spider mite-induced volatiles from Gerbera: influence of starvation and experience. J Chem Ecol 25(12):2623–2641

Krips OE, Willems PEL, Gols R, Posthumus MA, Gort G, Dicke M (2001) Comparison of cultivars of ornamental crop Gerbera jamesonii on production of spider mite-induced volatiles, and their attractiveness to the predator Phytoseiulus persimilis. J Chem Ecol 27:1355–1372

Kunert G, Reinhold C, Gershenzon J (2010) Constitutive emission of the aphid alarm pheromone, (E)-beta-farnesene, from plants does not serve as a direct defense against aphids. BMC Ecol 10:23

Labandeira CC (1997) Insect mouthparts: ascertaining the paleobiology of insect feeding strategies. Annu Rev Ecol Syst 28:153–193

Lam E (2004) Controlled cell death, plant survival and development. Nat Rev Mol Cell Biol 5(4):305–315

Lange BM, Wildung MR, Stauber EJ, Sanchez C, Pouchnik D, Croteau R (2000) Probing essential oil biosynthesis and secretion by functional evaluation of expressed sequence tags from mint glandular trichomes. Proc Natl Acad Sci U S A 97:2934–2939

Langenheim JH (1994) Higher-plant terpenoids – a phytocentric overview of their ecological roles. J Chem Ecol 20(6):1223–1280

Laurent G, Stopfer M, Friedrich RW, Rabinovich MI, Volkovskii A, Abarbanel HDI (2001) Odor encoding as an active, dynamical process: experiments, computation, and theory. Annu Rev Neurosci 24:263–297

Lawrence SD, Novak NG, Blackburn MB (2007) Inhibition of proteinase inhibitor transcripts by *Leptinotarsa decemlineata* regurgitant in *Solanum lycopersicum*. J Chem Ecol 33:1041–1048

Lawrence SD, Novak NG, Ju CJT, Cooke JEK (2008) Potato, Solanum tuberosum, defense against Colorado potato beetle, *Leptinotarsa decemlineata* (Say): microarray gene expression profiling of potato by Colorado potato beetle regurgitant treatment of wounded leaves. J Chem Ecol 34:1013–1025

Lee JC (2010) Effect of methyl salicylate-based lures on beneficial and pest arthropods in strawberry. Environ Entomol 39(2):653–660

Leon-Reyes A, Du YJ, Koornneef A, Proietti S, Korbes AP, Memelink J, Pieterse CMJ, Ritsema T (2010a) Ethylene signaling renders the jasmonate response of *Arabidopsis* insensitive to future suppression by salicylic acid. Mol Plant Microbe Interact 23(2):187–197

Leon-Reyes A, Van der Does D, De Lange ES, Delker C, Wasternack C, Van Wees SCM, Ritsema T, Pieterse CMJ (2010b) Salicylate-mediated suppression of jasmonate-responsive gene expression in *Arabidopsis* is targeted downstream of the jasmonate biosynthesis pathway. Planta 232(6):1423–1432

Lerdau M, Gershenzon J (1997) Allocation theory and chemical defense. In: Bazzaz FA, Grace J (eds) Plant resource allocation. Academic, San Diego, pp 265–291

Li CY, Williams MM, Loh YT, Lee GI, Howe GA (2002a) Resistance of cultivated tomato to cell content-feeding herbivores is regulated by the octadecanoid-signaling pathway. Plant Physiol 130(1):494–503

Li X, Schuler MA, Berenbaum MR (2002b) Jasmonate and salicylate induce expression of herbivore cytochrome P450 genes. Nature 419:712–715

Li L, Zhao Y, McCaig BC, Wingerd BA, Wang J, Whalon ME, Pichersky E, Howe GA (2004) The tomato homolog of CORONATINE-INSENSITIVE1 is required for the maternal control of seed maturation, jasmonate-signaled defense responses, and glandular trichome development. Plant Cell 16:126–143

Loivamaki M, Mumm R, Dicke M, Schnitzler JP (2008) Isoprene interferes with the attraction of bodyguards by herbaceous plants. Proc Natl Acad Sci U S A 105:17430–17435

Loughrin JH, Manukian A, Heath RR, Turlings TCJ, Tumlinson JH (1994) Diurnal cycle of emission of induced volatile terpenoids herbivore-injured cotton plants. Proc Natl Acad Sci U S A 91(25):11836–11840

Maeda H, Shasany AK, Schnepp J, Orlova I, Taguchi G, Cooper BR, Rhodes D, Pichersky E, Dudareva N (2010) RNAi suppression of arogenate dehydratase1 reveals that phenylalanine is synthesized predominantly via the arogenate pathway in *Petunia* petals. Plant Cell 22(3):832–849

Maes K, Debergh PC (2003) Volatiles emitted from in vitro grown tomato shoots during abiotic and biotic stress. Plant Cell Tissue Organ Cult 75:73–78

Maffei ME (2010) Sites of synthesis, biochemistry and functional role of plant volatiles. S Afr J Bot 76(4):612–631

Markovich NA, Kononova GL (2003) Lytic enzymes of Trichoderma and their role in plant defense from fungal diseases: a review. Appl Biochem Microbiol 39(4):341–351

Matsui K, Kurishita S, Hisamitsu A, Kajiwara T (2000) A lipid-hydrolysing activity involved in hexenal formation. Biochem Soc Trans 28:857–860

Matsushima R, Ozawa R, Uefune M, Gotoh T, Takabayashi J (2006) Intraspecific variation in the Kanzawa spider mite differentially affects induced defensive response in lima bean plants. J Chem Ecol 32:2501–2512

Mattiacci L, Dicke M, Posthumus MA (1995) β-Glucosidase: an elicitor of herbivore-induced plant odor that attracts host-searching parasitic wasps. Proc Natl Acad Sci U S A 92:2036–2040

McCloud ES, Baldwin IT (1997) Herbivory and caterpillar regurgitants amplify the wound-induced increases in jasmonic acid but not nicotine in *Nicotiana sylvestris*. Planta 203(4):430–435

Menzel R (2001) Searching for the memory trace in a mini-brain, the honeybee. Learn Mem 8(2):53–62

Metraux JP, Jackson RW, Schnettler E, Goldbach RW (2009) Plant pathogens as suppressors of host defense. In: Plant innate immunity. Advances in Botanical Research, Vol. 51, Burlington: Academic Press, pp 39–89

Miles PW (1972) The saliva of Hemiptera. Adv Insect Physiol 9:183–240

Miles PW (1999) Aphid saliva. Biol Rev 74(1):41–85

Mirabella R, Rauwerda H, Struys EA, Jakobs C, Triantaphylides C, Haring MA, Schuurink RC (2008) The *Arabidopsis* her1 mutant implicates GABA in E-2-hexenal responsiveness. Plant J 53:197–213

Mithöfer A, Wanner G, Boland W (2005) Effects of feeding *Spodoptera littoralis* on lima bean leaves. II. Continuous mechanical wounding resembling insect feeding is sufficient to elicit herbivory-related volatile emission. Plant Physiol 137:1160–1168

Mumm R, Dicke M (2010) Variation in natural plant products and the attraction of bodyguards involved in indirect plant defense. Can J Zool-Revue Canadienne De Zoologie 88(7):628–667

Musser RO, Hum-Musser SM, Eichenseer H, Peiffer M, Ervin G, Murphy JB, Felton GW (2002) Herbivory: caterpillar saliva beats plant defences – A new weapon emerges in the evolutionary arms race between plants and herbivores. Nature 416:599–600

Musser RO, Cipollini DF, Hum-Musser SM, Williams SA, Brown JK, Felton GW (2005) Evidence that the caterpillar salivary enzyme glucose oxidase provides herbivore offense in solanaceous plants. Arch Insect Biochem Physiol 58:128–137

Nabity PD, Zavala JA, DeLucia EH (2009) Indirect suppression of photosynthesis on individual leaves by arthropod herbivory. Ann Bot 103(4):655–663

Najar-Rodriguez AJ, Galizia CG, Stierle J, Dorn S (2010) Behavioral and neurophysiological responses of an insect to changing ratios of constituents in host plant-derived volatile mixtures. J Exp Biol 213(19):3388–3397

Newingham BA, Callaway RM, BassiriRad H (2007) Allocating nitrogen away from a herbivore: a novel compensatory response to root herbivory. Oecologia 153(4):913–920

Nicastro RL, Sato ME, Da Silva MZ (2010) Milbemectin resistance in *Tetranychus urticae* (Acari: Tetranychidae): selection, stability and cross-resistance to abamectin. Exp Appl Acarol 50(3):231–241

Niessing J, Friedrich RW (2010) Olfactory pattern classification by discrete neuronal network states. Nature 465(7294):47–U53

Nombela G, Beitia F, Muñiz M (2000) Variation in tomato host response to *Bemisia tabaci* (Hemiptera: Aleyrodidae) in relation to acyl sugar content and presence of the nematode and potato aphid resistance gene Mi. Bull Entomol Res 90:161–167

Nomura K, Melotto M, He SY (2005) Suppression of host defense in compatible plant-*Pseudomonas syringae* interactions. Curr Opin Plant Biol 8:361–368

Nottingham SF, Hardie J, Dawson GW, Hick AJ, Pickett JA, Wadhams LJ et al (1991) Behavioral and electrophysiological responses of aphids to host and nonhost plant volatiles. J Chem Ecol 17(6):1231–1242

Oliver JE, Doss RP, Williamson RT, Carney JR, De Vilbiss ED (2000) Bruchins-mitogenic 3-(hydroxypropanoil) esters of long chain diols from weevils of the bruchidae. Tetrahedron 39:7633–7641

Orre GUS, Wratten SD, Jonsson M, Hale RJ (2010) Effects of an herbivore-induced plant volatile on arthropods from three trophic levels in brassicas. Biol Control 53(1):62–67

Otsuka M, Kenmoku H, Ogawa M, Okada K, Mitsuhashi W, Sassa T, Kamiya Y, Toyomasu T, Yamaguchi S (2004) Emission of ent-kaurene, a diterpenoid hydrocarbon precursor for gibberellins, into the headspace from plants. Plant Cell Physiol 45(9):1129–1138

Paré PM, Tumlinson JH (1999) Plant volatiles as defense against insect herbivores. Plant Physiol 121:325–331

Paré PW, Alborn HT, Tumlinson JH (1998) Concerted biosynthesis of an insect elicitor of plant volatiles. Proc Natl Acad Sci U S A 95:13971–13975

Park SW, Kaimoyo E, Kumar D, Mosher S, Klessig DF (2007) Methyl salicylate is a critical mobile signal for plant systemic acquired resistance. Science 318(5847):113–116

Park DH, Mirabella R, Bronstein PA, Preston GM, Haring MA, Lim CK, Collmer A, Schuurink RC (2010) Mutations in gamma-aminobutyric acid (GABA) transaminase genes in plants or *Pseudomonas syringae* reduce bacterial virulence. Plant J 64(2):318–330

Paschold A, Halitschke R, Baldwin IT (2007) Co(i)-ordinating defenses: NaCOI1 mediates herbivore-induced resistance in Nicotiana attenuata and reveals the role of herbivore movement in avoiding defenses. Plant J 51(1):79–91

Peiffer M, Felton GW (2009) Do caterpillars secrete "oral secretions"? J Chem Ecol 35:326–335

Peiffer M, Tooker JF, Luthe DS, Felton GW (2009) Plants on early alert: glandular trichomes as sensors for insect herbivores. New Phytol 184:644–656

Penuelas J, Llusia J (2004) Plant VOC emissions: making use of the unavoidable. Trends Ecol Evol 19(8):402–404

Penuelas J, Llusia J, Asensio D, Munne-Bosch S (2005) Linking isoprene with plant thermotolerance, antioxidants and monoterpene emissions. Plant Cell Environ 28(3):278–286

Perez JL, French JV, Summy KR, Baines AD, Little CR (2009) Fungal phyllosphere communities are altered by indirect interactions among trophic levels. Microb Ecol 57(4):766–774

Pieterse CMJ, Leon-Reyes A, Van der Ent S, Van Wees SCM (2009) Networking by small-molecule hormones in plant immunity. Nat Chem Biol 5:308–316

Pinto DM, Nerg AM, Holopainen JK (2007) The role of ozone-reactive compounds, terpenes, and green leaf volatiles (glvs), in the orientation of *Cotesia plutellae*. J Chem Ecol 33:2218–2228

Pitman JL, Dasgupta S, Krashes MJ, Leung B, Perrat PN, Waddell S (2009) There are many ways to train a fly. Fly 3(1):3–9

Raguso RA (2004) Why are some floral nectars scented? Ecology 85(6):1486–1494

Raguso RA, Levin RA, Foose SE, Holmberg MW, McDade LA (2003) Fragrance chemistry, nocturnal rhythms and pollination "syndromes" in Nicotiana. Phytochemistry 63(3):265–284

Ranger CM, Reding ME, Persad AB, Herms DA (2010) Ability of stress-related volatiles to attract and induce attacks by Xylosandrus germanus and other ambrosia beetles. Agricultural and Forest Entomology 12(2):177–185

Rasmann S, Köllner TG, Degenhardt J, Hiltpold I, Toepfer S, Kuhlmann U, Gershenzon J, Turlings TCJ (2005) Recruitment of entomopathogenic nematodes by insect-damaged maize roots. Nature 434:732–737

Renault H, Roussel V, El Amrani A, Arzel M, Renault D, Bouchereau A, Deleu C (2010) The Arabidopsis pop2-1 mutant reveals the involvement of GABA transaminase in salt stress tolerance. BMC Plant Biol 10:20

Reynolds AM, Bohan DA, Bell JR (2007) Ballooning dispersal in arthropod taxa: conditions at take-off. Biol Lett 3(3):237–240

Riffell JA, Lei H, Christensen TA, Hildebrand JG (2009) Characterization and coding of behaviorally significant odor mixtures. Curr Biol 19(4):335–340

Robertson HM, Wanner KW (2006) The chemoreceptor superfamily in the honey bee, Apis mellifera: expansion of the odorant, but not gustatory, receptor family. Genome Res 16(11):1395–1403

Robertson HM, Warr CG, Carlson JR (2003) Molecular evolution of the insect chemoreceptor gene superfamily in Drosophila melanogaster. Proc Natl Acad Sci U S A 100:14537–14542

Rodriguez-Saona C, Crafts-Brandner SJ, Canas LA (2003) Volatile emissions triggered by multiple herbivore damage: Beet armyworm and whitefly feeding on cotton plants. J Chem Ecol 29(11):2539–2550

Rogers BT, Peterson MD, Kaufman TC (2002) The development and evolution of insect mouthparts as revealed by the expression patterns of gnathocephalic genes. Evol Dev 4(2):96–110

Röse USR, Manukian A, Heath RR, Tumlinson JH (1996) Volatile semiochemicals released from undamaged cotton leaves: a systemic response of living plants to caterpillar damage. Plant Physiol 111:487–495

Rostás M, Eggert K (2008) Ontogenetic and spatio-temporal patterns of induced volatiles in Glycine max in the light of the optimal defence hypothesis. Chemoecology 18:29–38

Runyon JB, Mescher MC, De Moraes CM (2006) Volatile chemical cues guide host location and host selection by parasitic plants. Science 311(5795):1964–1967

Sabelis MW, de Jong MCM (1988) Should all plants recruit bodyguards? Conditions for a polymorphic ESS of synomone production in plants. Oikos 53:247–252

Sabelis MW, van de Baan HE (1983) Location of distant spider mite colonies by phytoseiid predators: demonstration of specific kairomones emitted by Tetranychus urticae and Panonychus ulmi (Acari: Phytoseiidae, Tetranychidae). Entomol Exp Appl 33:303–314

Sabelis MW, Janssen A, Kant MR (2001) The enemy of my enemy is my ally. Science 291:2104–2105

Sabelis MW, Takabayashi J, Janssen A, Kant MR, van Wijk M, Sznajder B, Aratchige NS, Lesna I, Belliure B, Schuurink RC (2007) Ecology meets plant physiology: herbivore-induced plant responses and their indirect effects on arthropod communities. In: Ohgushi T, Craig TP, Price PW (eds) Ecological communities: plant mediation in indirect interaction webs. Cambridge University Press, Cambridge, pp 188–217

Sabelis MW, Janssen A, Takabayashi J (2011) Can plants establish stable alliances with their enemies' enemies? J Plant Interact 6(2–3):71–75

Sallaud C, Rontein D, Onillon S, Jabès F, Duffé P, Giacalone C, Thoraval S, Escoffier C, Herbette G, Leonhardt N, Causse M, Tissier A (2009) A novel pathway for sesquiterpene biosynthesis from Z, Z-farnesyl pyrophosphate in the wild tomato Solanum habrochaites. Plant Cell 21:301–317

Sarmento RA, Lemos F, Bleeker PM, Schuurink RC, Pallini A, Oliveira MGA, Lima E, Kant M, Sabelis MW, Janssen A (2011) A herbivore that manipulates plant defence. Ecol Lett 14:229–236

Sasaki K, Ohara K, Yazaki K (2005) Gene expression and characterization of isoprene synthase from Populus alba. FEBS Lett 579:2514–2518

Schilmiller AL, Schauvinhold I, Laerson M, Xu R, Charbonneau AL, Schmidt A, Wilkerson C, Last RL, Pichersky E (2009) Monoterpenes in the glandular trichomes of tomato are synthesized from a neryl diphosphate precursor rather than geranyl diphosphate. Proc Natl Acad Sci U S A 106:10865–10870

Schmelz EA, Alborn HT, Tumlinson JH (2001) The influence of intact-plant and excised-leaf bioassay designs on volicitin- and jasmonic acid-induced sesquiterpene volatile release in Zea mays. Planta 214:171–179

Schmelz EA, Alborn HT, Tumlinson JH (2003) Synergistic interactions between volicitin, jasmonic acid and ethylene mediated insect-induced volatile emission in Zea mays. Physiol Plant 117:403–412

Schmelz EA, Carrol MJ, Le Clere S, Phipps SM, Meredith J, Chourey PS, Alborn HT, Teal PEA (2006) Fragments of ATP synthase mediate plant perception of insect attack. Proc Natl Acad Sci U S A 103:8894–8899

Schmelz EA, Le Clere S, Carrol MJ, Alborn HT, Teal PEA (2007) Cowpea (Vigna unguiculata) chloroplastic ATP synthase is the source of multiple plant defense elicitors during insect herbivory. Plant Physiol 144:793–805

Schmelz EA, Engelberth J, Alborn HT, Tumlinson JH, Teal PEA (2009) Phytohormone-based activity mapping of insect herbivore-produced elicitors. Proc Natl Acad Sci U S A 106(2):653–657

Schnee C, Köllner TG, Gershenzon J, Degenhardt J (2002) The maize gene terpene synthase 1 encodes a sesquiterpene synthase catalyzing the formation of (E)-beta-farnesene, (E)-nerolidol, and (E, E)-farnesol after herbivore damage. Plant Physiol 130:2049–2060

Schnee C, Kollner TG, Held M, Turlings TC, Gershenzon J, Degenhardt J (2006) The products of a single maize sesquiterpene synthase form a volatile defense signal that attracts natural enemies of maize herbivores. Proc Natl Acad Sci U S A 103:1129–1134

Schuman MC, Heinzel N, Gaquerel E, Svatos A, Baldwin IT (2009) Polymorphism in jasmonate signaling partially accounts for the variety of volatiles produced by Nicotiana attenuata plants in a native population. New Phytol 183(4):1134–1148

Schuurink RC, Haring MA, Clark DG (2006) Regulation of volatile benzenoid biosynthesis in petunia flowers. Trends Plant Sci 11:20–25

Schwachtje J, Minchin PEH, Jahnke S, van Dongen JT, Schittko U, Baldwin IT (2006) SNF1-related kinases allow plants to tolerate herbivory by allocating carbon to roots. Proc Natl Acad Sci U S A 103(34):12935–12940

Scutareanu P, Drukker B, Bruin J, Posthumus MA, Sabelis MW (1997) Volatiles from Psylla-infested pear trees and their possible involvement in attraction of anthocorid predators. J Chem Ecol 23:2241–2260

Semmelhack JL, Wang JW (2009) Select *Drosophila* glomeruli mediate innate olfactory attraction and aversion. Nature 459(7244):218–U100

Shepherd RW, Bass WT, Houtz RL, Wagner GJ (2005) Phylloplanins of tobacco are defensive proteins deployed on aerial surfaces by short glandular trichomes. Plant Cell 17(6):1851–1861

Shikano I, Ericsson JD, Cory JS, Myers JH (2010) Indirect plant-mediated effects on insect immunity and disease resistance in a tritrophic system. Basic and Applied Ecology 11(1):15–22

Shiojiri K, Takabayashi J, Yano S, Takafuji A (2002) Oviposition preferences of herbivores are affected by tritrophic interaction webs. Ecol Lett 5:186–192

Shiojiri K, Kishimoto K, Ozawa R, Kugimiya S, Urashimo S, Arimura G, Horiuchi J, Nishioka T, Matsui K, Takabayashi J (2006) Changing green leaf volatile biosynthesis in plants: an approach for improving plant resistance against both herbivores and pathogens. Proc Natl Acad Sci U S A 103:16672–16676

Shiojiri K, Ozawa R, Kugimiya S, Uefune M, van Wijk M, Sabelis MW (2010) Herbivore-specific, density-dependent induction of plant volatiles: honest or "cry wolf" signals? PLoS One 5(8):e12161

Shrivastava G, Rogers M, Wszelaki A, Panthee DR, Chen F (2010) Plant volatiles-based insect pest management in organic farming. Crit Rev Plant Sci 29(2):123–133

Shroff R, Vergara F, Muck A, Svatos A, Gershenzon J (2008) Nonuniform distribution of glucosinolates in Arabidopsis thaliana leaves has important consequences for plant defense. Proc Natl Acad Sci U S A 105:6196–6201

Shulaev V, Silverman P, Raskin I (1997) Airborne signalling by methyl salicylate in plant pathogen resistance. Nature 385(6618):718–721

Simmons AT, Gurr GM (2005) Trichomes of *Lycopersicon* species and their hybrids: effects on pest and natural enemies. Agric For Entomol 8:1–11

Simpson M, Gurr GM, Simmons AT, Wratten SD, James DG, Leeson G, Nicol HI (2011) Insect attraction to synthetic herbivore-induced plant volatile-treated field crops. Agric For Entomol 13(1):45–57

Siva-Jothy MT, Moret Y, Rolff J (2005) Insect immunity: an evolutionary ecology perspective. In: Advances in insect physiology, vol 32. Elsevier Science & Technology, Oxford, pp 1–48

Smith JL, De Moraes CM, Mescher MC (2009) Jasmonate- and salicylate-mediated plant defense responses to insect herbivores, pathogens and parasitic plants. Pest Manag Sci 65(5):497–503

Snoeren TAL, Mumm R, Poelman EH, Yang Y, Pichersky E, Dicke M (2010) The herbivore-induced plant volatile methyl salicylate negatively affects attraction of the parasitoid *Diadegma semiclausum*. J Chem Ecol 36(5):479–489

Solomon M, Belenghi B, Delledonne M, Menachem E, Levine A (1999) The involvement of cysteine proteases and protease inhibitor genes in the regulation of programmed cell death in plants. Plant Cell 11(3):431–443

Spoel SH, Johnson JS, Dong X (2007) Regulation of tradeoffs between plant defenses against pathogens with different lifestyles. Proc Natl Acad Sci U S A 104(47):18842–18847

Stange G (1992) High-resolution measurement of atmospheric carbon-dioxide concentration changes by the labial palp organ of the moth *Heliothis armigera* (Lepidoptera, Noctuidae). J Comp Physiol A-Sens Neural Behav Physiol 171(3):317–324

Stange G, Stowe S (1999) Carbon-dioxide sensing structures in terrestrial arthropods. Microsc Res Tech 47(6):416–427

Steppuhn A, Baldwin IT (2007) Resistance management in a native plant: nicotine prevents herbivores from compensating for plant protease inhibitors. Ecol Lett 10(6):499–511

Stowe KA, Marquis RJ, Hochwender CG, Simms EL (2000) The evolutionary ecology of tolerance to consumer damage. Annu Rev Ecol Syst 31:565–595

Strauss SY, Rudgers JA, Lau JA, Irwin RE (2002) Direct and ecological costs of resistance to herbivory. Trends Ecol Evol 17(6):278–285

Takabayashi J, Dicke M, Posthumus MA (1991) Induction of indirect defense against spider-mites in uninfested lima-bean leaves. Phytochemistry 30:1459–1462

Takabayashi J, Dicke M, Posthumus MA (1994a) Volatile herbivore-induced terpenoids in plant mite interactions – variation caused by biotic and abiotic factors. J Chem Ecol 20(6):1329–1354

Takabayashi J, Dicke M, Takahashi S, Posthumus MA, Vanbeek TA (1994b) Leaf age affects composition of herbivore-induced synomones and attraction of predatory mites. J Chem Ecol 20(2):373–386

Takabayashi J, Shimoda T, Dicke M, Ashihara W, Takafuji A (2000) Induced response of tomato plants to injury by green and red strains of Tetranychus urticae. Exp Appl Acarol 24:377–383

Takabayashi J, Sabelis MW, Janssen A, Shiojiri K, van Wijk M (2006) Can plants betray the presence of multiple herbivore species to predators and parasitoids? The role of learning in phytochemical information networks. Ecol Res 21:3–8

Takos A, Lai D, Mikkelsen L, Abou Hachem M, Shelton D, Motawia OCE, Wang TL, Martin C, Rook F (2010) Genetic screening identifies cyanogenesis-deficient mutants of Lotus japonicus and reveals enzymatic specificity in hydroxynitrile glucoside metabolism. Plant Cell 22(5):1605–1619

Tanaka K, Uda Y, Ono Y, Nakagawa T, Suwa M, Yamaoka R et al (2009) Highly selective tuning of a silkworm olfactory receptor to a key mulberry leaf volatile. Curr Biol 19(11):881–890

Thatcher LF, Manners JM, Kazan K (2009) *Fusarium oxysporum* hijacks COI1-mediated jasmonate signaling to promote disease development in *Arabidopsis*. Plant J 58:927–939

Thivierge K, Prado A, Driscoll BT, Bonneil E, Thibault P, Bede JC (2010) Caterpillar- and salivary-specific modification of plant proteins. J Proteome Res 9(11):5887–5895

Tholl D (2006) Terpene synthases and the regulation, diversity and biological roles of terpene metabolism. Curr Opin Plant Biol 9:1–8

Thompson GA, Goggin FL (2006) Transcriptomics and functional genomics of plant defence induction by phloem-feeding insects. J Exp Bot 57:755–766

Tieman D, Zeigler M, Schmelz E, Taylor MG, Rushing S, Jones JB, Klee HJ (2010) Functional analysis of a tomato salicylic acid methyl transferase and its role in synthesis of the flavor volatile methyl salicylate. Plant J 62(1):113–123

Tjallingii WF, Garzo E, Fereres A (2010) New structure in cell puncture activities by aphid stylets: a dual-mode EPG study. Entomol Exp Appl 135(2):193–207

Tomberlin JK, Tertuliano M, Rains G, Lewis WJ (2005) Conditioned Microplitis croceipes cresson (Hymenoptera: Braconidae) detect and respond to 2,4-DNT: development of a biological sensor. J Forensic Sci 50(5):1187–1190

Tooker JF, De Moraes CM (2007) Feeding by Hessian fly [*Mayetiola destructor* (Say)] larvae does not induce plant indirect defences. Ecol Entomol 32:153–161

Tooker JF, Rohr JR, Abrahamson WG, De Moraes CM (2008) Gall insects can avoid and alter indirect plant defenses. New Phytol 178:657–671

Traulsen A, Nowak MA (2007) Chromodynamics of cooperation in finite populations. PLoS One 2(3):e270. doi:10.1371/journal.pone.0000270

Truitt CL, Wei HX, Pare PW (2004) A plasma membrane protein from Zea mays binds with the herbivore elicitor volicitin. Plant Cell 16:523–532

Tumlinson JH, Lait CG (2005) Biosynthesis of fatty acid amide elicitors of plant volatiles by insect herbivores. Arch Insect Biochem Physiol 58:54–68

Turlings TCJ, Ton J (2006) Exploiting scents of distress: the prospect of manipulating herbivore-induced plant odours to enhance the control of agricultural pests. Curr Opin Plant Biol 9(4):421–427

Turlings TCJ, Tumlinson JH, Lewis WJ (1990) Exploitation of herbivore-induced plant odors by host-seeking parasitic wasps. Science 250:1251–1253

Turlings TCJ, Loughrin JH, McCall PJ, Röse USR, Lewis WJ, Tumlinson JH (1995) How caterpillar-damaged plants protect themselves by attracting parasitic wasps. Proc Natl Acad Sci U S A 92:4169–4174

Turlings TCJ, Bernasconi M, Bertossa R, Bigler F, Caloz G, Dorn S (1998) The induction of volatile emissions in maize by three herbivore species with different feeding habits: possible consequences for their natural enemies. Biol Control 11:122–129

Tzin V, Galili G (2010) New insights into the shikimate and aromatic amino acids biosynthesis pathways in plants. Mol Plant 3(6):956–972

Ujvary I, Dickens JC, Kamm JA, McDonough LM (1993) Natural product analogs – stable mimics of aldehyde pheromones. Arch Insect Biochem Physiol 22(3–4):393–411

Unsicker SB, Kunert G, Gershenzon J (2009) Protective perfumes: the role of vegetative volatiles in plant defense against herbivores. Curr Opin Plant Biol 12(4):479–485

Utsumi S (2011) Eco-evolutionary dynamics in herbivorous insect communities mediated by induced plant responses. Popul Ecol 53(1):23–34

Van Baalen M, Jansen VAA (2001) Dangerous liaisons: the ecology of private interest and common good. Oikos 95:211–224

Van Baalen M, Jansen VAA (2003) Common language or Tower of Babel? On the evolutionary dynamics of signals and their meanings. Proc R Soc Lond B 270:69–76

Van den Boom CE, Van Beek TA, Posthumus MA, De Groot A, Dicke M (2004) Qualitative and quantitative variation among volatile profiles induced by *Tetranychus urticae* feeding on plants from various families. J Chem Ecol 30:69–89

Van der Goes van Naters W, Carlson JR (2007) Receptors and neurons for fly odors in *Drosophila*. Curr Biol 17(7):606–612

Van der Vossen EAG, Gros J, Sikkema A, Muskens M, Wouters D, Wolters P, Pereira A, Allefs S (2005) The Rpi-blb2 gene from *Solanum bulbocastanum* is a Mi-1 gene homolog conferring broad-spectrum late blight resistance in potato. Plant J 44(2):208–222

Van Leeuwen T, Vontas J, Tsagkarakou A, Dermauw W, Tirry L (2010) Acaricide resistance mechanisms in the two-spotted spider mite *Tetranychus urticae* and other important Acari: a review. Insect Biochem Mol Biol 40(8):563–572

Van Loon JJA, De Boer JG, Dicke M (2000) Parasitoid-plant mutualism: parasitoid attack of herbivore increases plant reproduction. Entomol Exp Appl 97:219–227

Van Schie CCN, Haring MA, Schuurink RC (2007) Tomato linalool synthase is induced in trichomes by jasmonic acid. Plant Mol Biol 64:251–263

Van Tol RWHM, van der Sommen ATC, Boff MIC, van Bezooijen J, Sabelis MW, Smits PH (2001) Plants protect their roots by alerting the enemies of grubs. Ecol Lett 4(4):292–294

Van Wijk M, Wadman WJ, Sabelis MW (2006) Morphology of the olfactory system in the predatory mite *Phytoseiulus persimilis*. Exp Appl Acarol 40(3–4):217–229

Van Wijk M, De Bruijn PJA, Sabelis MW (2008) Predatory mite attraction to herbivore-induced plant odors is not a consequence of attraction to individual herbivore-induced plant volatiles. J Chem Ecol 34(6):791–803

Van Wijk M, de Bruijn PJ, Sabelis MW (2010) The predatory mite Phytoseiulus persimilis does not perceive odor mixtures as strictly elemental objects. J Chem Ecol 36:1211–1225

Van Wijk M, De Bruijn PJA, Sabelis MW (2011) Complex odor from plants under attack: herbivore's enemies react to the whole, not its parts. PLoS One 6(7):e21742

Vancanneyt G, Sanz C, Farmaki T, Paneque M, Ortego F, Castanera P, Sanchez-Serrano JJ (2001) Hydroperoxide lyase depletion in transgenic potato plants leads to an increase in aphid performance. Proc Natl Acad Sci U S A 98:8139–8144

Vet LEM, De Jong AG, Franchi E, Papaj DR (1998) The effect of complete versus incomplete information on odour discrimination in a parasitic wasp. Anim Behav 55:1271–1279

Vickers CE, Possell M, Hewitt CN, Mullineaux PM (2010) Genetic structure and regulation of isoprene synthase in Popular (Populus spp.). Plant Mol Biol 73:547–558

Villada ES, Garzo-González E, López-Sesé AI, Fereres-Castiel A, Gómez-Guillamón ML (2009) Hypersensitive response to *Aphis gossypii* Glover in melon genotypes carrying the Vat gene. J Exp Bot 60(11):3269–3277

Visser JH, Ave DA (1978) General green leaf volatiles in the olfactory orientation of the Colorado beetle, *Leptinotarsa decemlineata*. Entomol Exp Appl 24(3):738–749

Vlot AC, Dempsey DA, Klessig DF (2009) Salicylic acid, a multifaceted hormone to combat disease. Annu Rev Phytopathol 47:177–206

Vogt T (2010) Phenylpropanoid biosynthesis. Mol Plant 3(1):2–20

Vos P, Simons G, Jesse T, Wijbrandi J, Heinen L, Hogers R, Frijters A, Groenendijk J, Diergaarde P, Reijans M, Fierens-Onstenk J, de Both M, Peleman J, Liharska T, Hontelez J, Zabeau M (1998) The tomato Mi-1 gene confers resistance to both root-knot nematodes and potato aphids. Nat Biotechnol 16:1365–1369

Vuorinen T, Nerg A-M, Ibrahim MA, Reddy GVP, Holopainen JK (2004) Emission of Plutella xylostella-induced compounds from cabbages grown at elevated CO2 and orientation behavior of the natural enemies. Plant Physiol 135:1984–1992

Wagner GJ (1991) Secreting glandular trichomes: more than just hairs. Plant Physiol 96:675–679

Wagner GJ, Wang E, Shepherd RW (2004) New approaches for studying and exploiting an old protuberance, the plant trichome. Ann Bot 93(1):3–11

Walling LL (2008) Avoiding effective defenses: strategies employed by phloem-feeding insects. Plant Physiol 146:859–866

Wang EM, Wang R, DeParasis J, Loughrin JH, Gan SS, Wagner GJ (2001) Suppression of a P450 hydroxylase gene in plant trichome glands enhances natural-product-based aphid resistance. Nat Biotechnol 19:371–374

Wang X, Zhou GX, Xiang CY, Du MH, Chen JA, Liu SS, Lou YG (2008) b-glucosidase treatment and infestation by the rice brown planthopper *Nilaparvata lugens* elicit similar signalling pathways in rice plants. Chin Sci Bull 53:53–57

Wanner KW, Nichols AS, Allen JE, Bunger PL, Garczynski SF, Linn CE (2010) Sex pheromone receptor specificity in the European corn borer moth, *Ostrinia nubilalis*. PLoS One 5(1):e8685

Wasternack C (2007) Jasmonates: an update on biosynthesis, signal transduction and action in plant stress response, growth and development. Ann Bot 100(4):681–697

Weber H (2002) Fatty acid-derived signals in plants. Trends Plant Sci 7(5):217–224

Webster B, Bruce T, Dufour S, Birkemeyer C, Birkett M, Hardie J, Pickett J (2008) Identification of volatile compounds used in host location by the black bean aphid, *Aphis fabae*. J Chem Ecol 34(9):1153–1161

Weech MH, Chapleau M, Pan L, Ide C, Bede JC (2008) Caterpillar saliva interferes with induced *Arabidopsis thaliana* defence responses via the systemic acquired resistance pathway. J Exp Bot 59:2437–2448

Wildermuth MC, Dewdney J, Wu G, Ausubel FM (2001) Isochorismate synthase is required to synthesize salicylic acid for plant defence. Nature 414:562–571

Will T, Tjallingii WF, Thonnessen A, van Bel AJE (2007) Molecular sabotage of plant defense by aphid saliva. Proc Natl Acad Sci U S A 104(25):10536–10541

Will T, Kornemann SR, Furch ACU, Tjallingii WF, van Bel AJE (2009) Aphid watery saliva counteracts sieve-tube occlusion: a universal phenomenon? J Exp Biol 212(20):3305–3312

Willmer PG, Gordon SC, Wishart J, Hughes JP, Matthews IM, Woodford JAT (1998) Flower choices by raspberry beetles: cues for feeding and oviposition. Anim Behav 56(4):819–827

Winter TR, Rostas M (2008) Ambient ultraviolet radiation induces protective responses in soybean but does not attenuate indirect defense. Environ Pollut 155(2):290–297

Wu S, Schalk M, Clark A, Miles RB, Coates R, Chappell J (2006) Redirection of cytosolic or plastidic isoprenoid precursors elevates terpene production in plants. Nat Biotechnol 24:1441–1447

Xiang TT, Zong N, Zou Y, Wu Y, Zhang J, Xing WM, Li Y, Tang XY, Zhu LH, Chai JJ, Zhou JM (2008) Pseudomonas syringae effector AvrPto blocks innate immunity by targeting receptor kinases. Curr Biol 18:74–80

Xie Z, Kapteyn J, Gang DR (2008) A systems biology investigation of the MEP/terpenoid and shikimate/phenylpropanoid pathways points to multiple levels of metabolic control in sweet basil glandular trichomes. Plant J 54:349–361

Yao CA, Ignell R, Carlson JR (2005) Chemosensory coding by neurons in the coeloconic sensilla of the Drosophila antenna. J Neurosci 25(37):8359–8367

Yoshinaga N, Aboshi T, Ishikawa C, Fukui M, Shimoda M, Nishida R, Lait C, Tumlinson J, Mori N (2007) Fatty acid amides, previously identified in caterpillars, found in the cricket Teleogryllus taiwanemma and fruit fly Drosophila melanogaster larvae. J Chem Ecol 33:1376–1381

Yoshinaga N, Aboshia T, Abea H, Nishidaa R, Alborn HT, Tumlinsonb JH, Moria N (2008) Active role of fatty acid amino acid conjugates in nitrogen metabolism in *Spodoptera litura* larvae. Proc Natl Acad Sci U S A 105:18058–18063

Zarate SI, Kempema LA, Walling LL (2007) Silverleaf whitefly induces salicylic acid defenses and suppresses effectual jasmonic acid defenses. Plant Physiol 143:866–875

Zhang PJ, Zheng SJ, van Loon JJA, Boland W, David A, Mumm R, Dicke M (2009) Whiteflies interfere with indirect plant defense against spider mites in Lima bean. Proc Natl Acad Sci U S A 106:21202–21207

Zhao Y, Thilmony R, Bender CL, Schaller A, Yang He S, Howe GA (2003) Virulence systems of Pseumonas syringae pv. tomato promote bacterial speck disease in tomato by targeting the jasmonate signaling pathway. Plant J 36:485–499

Zhao N, Guan J, Ferrer JL, Engle N, Chern M, Ronald P, Tschaplinski TJ, Chen F (2010) Biosynthesis and emission of insect-induced methyl salicylate and methyl benzoate from rice. Plant Physiol Biochem 48(4):279–287

Zhu JW, Park KC (2005) Methyl salicylate, a soybean aphid-induced plant volatile attractive to the predator Coccinella septempunctata. J Chem Ecol 31(8):1733–1746

Zhu-Salzman K, Salzman RA, Ahn J, Koiwa H (2004) Transcriptional regulation of sorghum defense against a phloem-feeding aphid. Plant Physiol 134:420–431

Zimmerman DC, Coudron CA (1979) Identification of traumatin, a wound hormone, as 12-oxo-trans-10-dodecenoic acid. Plant Physiol 63:536–541

Chapter 3
Physiological Adaptations of the Insect Gut to Herbivory

Félix Ortego

3.1 Insects Feeding on Plants

Plant biomass is the most abundant resource in terrestrial communities, and terrestrial green plants and the herbivorous (phytophagous) insects that feed on them account for more than half of all living species (Scudder 2009). Still, herbivory appears to have represented a challenge that most insect orders have not been able to adapt, since phytophagous insects are only represented in nine (Coleoptera, Lepidoptera, Diptera, Hymenoptera, Hemiptera, Orthoptera, Phasmida, Thysanoptera, and Collembola) of the 29 orders of insects. Nonetheless, once an insect group has overcome the initial difficulties and can exploit the resources provided by the plants, the herbivorous habit seems to have promoted diversification (Futuyma and Agrawal 2009).

The evolution of phytophagous insects has been closely related to that of plants (Grimaldi and Engel 2005). The first insects appeared by the early Devonian (around 410 Mya) and these primitive wingless insects, similar to current collembolans, most probably lived on terrestrial or semi-aquatic media feeding on decomposing material. During the Carboniferous (355–290 Mya) and the Permian (290–250 Mya) periods winged insects evolved and diversified, including the first phytophagous insects that probably fed on fluid tissues, including vascular sap, from tree ferns (Labandeira and Phillips 1996). Most of them did not survive the great Permian extinction, and were replaced by more advanced descendants that gave place to the orders Thysanoptera and Hemiptera. The first fossil records of plants damaged by chewing insects also come from the Carboniferous, probably by the ancestors of current orthopterans (Chaloner et al. 1991). During the Triassic (250–205 Mya) and the Jurassic (205–145 Mya) most of the modern insect orders

F. Ortego (✉)
Dpto. de Biología Medioambiental, Centro de Investigaciones Biológicas, CSIC,
Ramiro de Maeztu 9, 28040 Madrid, Spain
e-mail: ortego@cib.csic.es

G. Smagghe and I. Diaz (eds.), *Arthropod-Plant Interactions: Novel Insights and Approaches for IPM*, Progress in Biological Control 14,
DOI 10.1007/978-94-007-3873-7_3, © Springer Science+Business Media B.V. 2012

became established, some of which (Coleoptera, Diptera, Hymenoptera and Lepidoptera) colonized emerging gymnosperm plants. The fossil record of plant-insect associations during this period documents the presence of some specialized types of insect herbivory, such as leafmining, leaf galls and wood boring (Chaloner et al. 1991). However, the big radiation of phytophagous insects occurred during the Cretaceous (145–65 Mya), coincident with the angiosperms becoming dominant in most terrestrial environments (Farrell 1998; Mayhew 2007).

Plant-feeding insects have developed a remarkable diversity of morphological, physiological and behavioural adaptations depending on their feeding habits and lifestyles (chewing or sucking insects, scale insects, gall-inducing insects, etc.) (Berbays 1991; Gullan and Kosztarab 1997; Raman et al. 2005). In particular, phytophagous insects have required specific physiological adaptations of the gut to digest plant material (Felton 1996; Terra and Ferreira 2005), and to counteract or adapt to plant chemical defenses, including secondary metabolites (allelochemicals) and insecticidal proteins (Duffey and Stout 1996; Baldwin et al. 2001; Pieterse and Dicke 2007).

This review focussed on the physiological adaptations of the gut in insects feeding on living plant material. Thus, pollen feeders were not considered as well as those feeding on dead or decaying plant material. Special attention has been given to the digestion of plant tissues by chewing and phloem-feeding insects and to the physiological adaptations of these insects to deal with plant secondary metabolites and insecticidal proteins within the gut lumen, excluding the detoxification enzymes that operate once the allelochemicals go through the gut wall.

3.2 Digestive Physiology of Phytophagous Insects

The quantity and quality of dietary proteins and carbohydrates is considered an important limiting factor for insect performance (Broadway and Duffey 1988; Felton 1996; Lee 2007). Insects require these nutrients in suitable ratios to meet their optimal nutrient requirements (Raubenheimer and Simpson 1999). This is especially relevant for phytophagous insects because of the low protein/high carbohydrate contents of most plant tissues (Dadd 1985). Moreover, the amino acid content of foliar proteins varies among plant species, contributing also to nutritional differences (Yeoh et al. 1992). Most phytophagous insects deal with the low protein levels of plant tissues by increasing food consumption (Slansky and Wheeler 1991). Insect herbivores may also regulate the intake of protein/carbohydrate ingested (Waldbauer and Friedman 1991; Chambers et al. 1995; Behmer 2009), and exhibit specific post-ingestive physiological adaptations (Felton 1996; Bede et al. 2007).

Phytophagous insects rely on digestive proteases, carbohydrases and lipases to hydrolyze proteins, carbohydrates and lipids to absorbable end-products (Terra and Ferreira 2005). The digestion of proteins is carried out by different types of proteases (serine, cysteine and aspartic proteases) that hydrolyze internal peptide bonds in proteins. The resulting oligopeptides are then digested by exopeptidases (aminopeptidases and carboxypeptidases), which remove amino acids from the amino and carboxyl ends of peptide chains. Hydrocarbons are major constituents of cell walls and starch granules within plant cells. Cell walls are disrupted by pectinases, hemicellulases and cellulases produced by the insect or by symbiotic organisms. The digestion of starch is mediated by α-amylases to produce soluble oligosaccharides, which are further hydrolyzed to dimers (such as sucrose, cellobiose and maltose) and monosaccharides by a complex of carbohydrases. Storage and membrane lipids are hydrolyzed by lipases and phospholipases to fatty acids and other lipophilic substances.

It has been hypothesized that the general organization of the digestive process correlates with their phylogenetic position rather than with dietary habits, as a result of the adaptation of ancestral insects to a particular type of feeding (Terra and Ferreira 1994, 2005). Most insects of the Holometabola (Coleoptera, Hymenoptera, Diptera and Lepidoptera) and Hemimetabola (Orthoptera and Phasmida) groups possess a secreted extracellular matrix formed by proteins, chitin, and proteoglycans that surrounds the food bolus in the lumen of the midgut called peritrophic membrane (PM) (Lehane 1997). Suggested functions for the PM include: protection of the midgut epithelium from abrasive food particles; compartmentalization of the digestive process to increase the efficiency of the digestion of protein and carbohydrate polymers; and a barrier to microorganisms (Lehane 1997; Terra 2001; Hegedus et al. 2009). Based on investigations with a number of insects (Terra and Ferreira 1994; Terra et al. 1996a), an ecto-endoperitrophic flow model has been proposed that allows the digestive enzymes to cross the PM into the endoperitrophic space to hydrolyze dietary proteins and hydrocarbons; the resulting oligomers then undergo further hydrolysis by oligomer hydrolases free in the ectoperitrophic space or bound to microvillar membranes; and the ultimate products are usually monomers that are uptaken by sugar and amino acid transporters located in the brush border membrane. Carbohydrate digestion in Orthoptera occurs mainly in the crop, under the action of midgut enzymes, whereas protein digestion and final carbohydrate digestion take place at the anterior midgut (Teo and Woodring 1994). On the other hand, the crop is generally absent or very slightly developed in most phytophagous species of Coleoptera, Hymenoptera and Lepidoptera, as well as larvae of Diptera, and enzymatic activity is mainly restricted to the midgut (Terra et al. 1985; Ortego et al. 1996; Novillo et al. 1997). The digestion of proteins and carbohydrates by adults of the most evolved flies (Cyclorrhapha) is initiated in the crop and continued in the midgut (Espinoza-Fuentes and Terra 1987; San Andres et al. 2007).

The Hemiptera and Thysanoptera are characterized by the absence of the PM, probably associated with the phloem feeding habits of a putative common ancestor (Silva et al. 2004). The content of polymeric molecules in the phloem sap is very low and, therefore, luminal digestion and the PM may have been lost upon adapting to this type of food. However, their midgut cells are not in direct contact with the ingested fluids, due to the existence of an extra-cellular lipoprotein membrane called perimicrovillar membrane which prevents the binding of undigested material onto the surface of microvilli and compartmentalizes the enzymes involved in terminal sugar and protein digestion (Terra et al. 1996a). The major problems facing sap-sucking insects are the low concentrations of essential amino acids and the large amounts of sucrose in the sap, with an osmolarity up to three times that of the insect hemolymph (Sandström and Moran 1999; Ashford et al. 2000). There are strong evidences that endogenous symbiotic bacteria of the genus *Buchnera*, located in the cytoplasm of specialized insect cells (bacteriocytes) in the aphid haemocoel and transferred vertically to eggs in the female reproductive organs (Dillon and Dillon 2004; Douglas 2006), are able to use the non-essential amino acids absorbed by the aphid to synthesize essential amino acids (Shigenobu et al. 2000; Gündüz and Douglas 2009). Amino acids may then be absorbed according to a hypothesized mechanism that depends on the presence of amino acid K^+ symports on the surface of the perimicrovillar membranes and of amino acid carriers and potassium pumps on the microvillar membranes (Silva and Terra 1994; Silva et al. 1995). The osmotic pressure of the ingested phloem sap is reduced in the midgut of the aphid *Acyrthosiphon pisum* by the combined action of α-glucosidases that catalyze the release of fructose from sucrose and the incorporation of the glucose moiety into long-chain oligosaccharides voided via honeydew, and by the rapid absorption of fructose (Ashford et al. 2000; Cristofoletti et al. 2003; Douglas 2006). Other physiological roles of the perimicrovillar membrane include: making amino acid absorption from dilute diets easier by binding them in a reversible way (Terra 1990); and immobilizing hydrolytic enzymes (Cristofoletti et al. 2003). The ability to digest polymers was regained in seed-feeding hemipterans by producing enzymes in the secreted saliva and using enzymes derived from lysosomes in the gut (Colebatch et al. 2001).

The types of digestive proteases in the insect gut are also phylogenetically determined (Terra et al. 1996b). Thus, most lepidopteran, orthopteran and hymenopteran species and some coleopterans possess alkaline midgut fluids and a digestive system based largely on serine proteases (trypsin, chymotrypsin and elastase) and exopeptidases (Ortego et al. 1996; Teo 1997; Wolfson and Murdock 1990; Johnson and Rabosky 2000). The majority of coleopterans and hemipterans have slightly acidic midguts and cysteine and aspartyl proteases and exopeptidases provide the major midgut proteolytic activity (Murdock et al. 1987; Cristofoletti et al. 2003; Carrillo et al. 2011). Most dipterans studied so far possess aspartyl and serine proteases and exopeptidases (San Andres et al. 2007). Nevertheless, phytophagous insects have been reported to adapt its digestive system to specific food sources by: (1) synthesizing specific digestive enzymes (Chougule et al. 2005); and (2) regulating the level of synthesis and/or secretion of those enzymes (Broadway and Duffey 1986).

Screening of cDNA libraries from the midguts of phytophagous insects has revealed multigene families potentially involved in proteolytic digestion (Gatehouse et al. 1997; Oliveira-Neto et al. 2004; Díaz-Mendoza et al. 2005). Expression levels and protease activity profiles have been reported to vary through larval development (Novillo et al. 1999; Chougule et al. 2005), but also depending on the type of plant consumed. Thus, the proteolytic profile of larvae of the cotton bollworm, *Helicoverpa armigera*, a polyphagous species that is able to feed on many plant species, was different after feeding on chickpea, pigeonpea, cotton or okra (Patankar et al. 2001). Moreover, expression analysis confirmed that gut proteases of *H. armigera* larvae fed on different plants exhibited differential expression (Chougule et al. 2005). Similar diet-related changes on digestive proteases have been reported for other polyphagous species (Hinks and Erlandson 1995), but also for oligophagous insects feeding only on a few plant species, such as the Colorado potato beetle, *Leptinotarsa decemlineata* (Overney et al. 1997), and the red palm weevil, *Rhynchophorus ferrugineus* (Alarcon et al. 2002). Changes in digestive enzymes may compensate for variable protein quality and/or quantity in host plants. A correlation between the quantity of dietary protein and the level of proteolytic activity in the gut has been reported for different lepidopteran species (Broadway and Duffey 1986; Shinbo et al. 1996; Woods and Kingsolver 1999). However, it has been demonstrated that phytophagous insects may also increase the levels of protease activity in response to low protein diet to make more efficient use of the limited protein that is available (Neal 1996). Another important factor to be considered is the nutritive value of dietary proteins, as highlighted by Felton (1996) that showed that larvae of the corn earworm, *Heliothis zea,* feeding on low quality protein (alkylated by chlorogenoquinone) exhibited a notable increase in trypsin activity that resulted in enhanced protein digestion. Multigene families of digestive enzymes have also evolved for the hydrolysis of carbohydrates and lipids (Horne et al. 2009; Morris et al. 2009; Pauchet et al. 2010), and phytophagous insects are able to modulate their expression depending on the host plant (Silva et al. 1999, 2001a; Kotkar et al. 2009). The regulation of these digestive processes, including enzyme synthesis and secretion, is controlled by secretagogue and/or endocrine mechanisms, (Terra and Ferreira 2005; Bede et al. 2007).

The pH and redox potential of the insect gut lumen determines the optimal conditions for enzyme activity and the quality and quantity of nutrients that can be digested (Johnson and Felton 1996; Terra and Ferreira 2005). Gut pH tends to correlate with phylogenetic lineages, while redox conditions are more variable and partly dependent on diet (Johnson and Felton 1996; Clark 1999). The pH of the midgut among herbivorous insects is usually in the slightly acidic to neutral range, with the exception of the alkaline midgut of lepidopteran larvae, the acid crop of fruit-feeding Tephritidae, and the acid posterior region of the midgut of heteropteran Hemiptera (Terra and Ferreira 1994). Lepidopteran serine proteases and α-amylases are evolutionarily adapted for effective functioning in the alkaline digestive system (Terra and Ferreira 2005; Pytelková et al. 2009). The alkalinization of the lumen in lepidopterans is apparently produced by the secretion of K^+ ions from the hemolymph (Dow 1992), and there are evidences that leaf proteins and cell wall polysaccharides are more soluble under alkaline pH and therefore of higher nutritional quality (Felton and Duffey 1991). The acid region in the midgut of dipteran insects is assumed to have been retained from their putative ancestral bacteria-feeding habit, and the acid posterior midgut of Hemiptera may be related to the presence of cysteine and aspartic proteases (Terra and Ferreira 2005). Changes in feeding state and diet pose substantial acid–base challenges to phytophagous insects, because their high rates of food consumption and the differences in acid–base composition among plants (Harrison 2001). Diet and time since feeding have been shown to affect the midgut pH in some species (Schultz and Lechowicz 1986), but in most cases no evidences of association between diet and pH were found (Appel and Maines 1995; Appel and Joern 1997). These results indicate an active regulation (acid–base transport) to stabilize the lumen pH after feeding, though the mechanism is unclear (Harrison 2001).

3.3 Physiological Adaptations to Plant Chemical Defenses

Salivary enzymes have been proposed as the first-line of defenses to deal with noxious plant chemicals in lepidopteran and hemipteran insects. In caterpillars, glucose oxidase produces H_2O_2 that may inhibit ingested plant oxidative enzymes (e.g.,peroxidase, polyphenol oxidase and lipoxygenase) that reduce the nutritive quality of the diet (Eichenseer et al. 1999), and interfere with induced plant defensive pathways (Musser et al. 2002; Weech et al. 2008). In addition, glucose oxidase converts glucose to gluconate, which cannot be utilized by the insect, helping caterpillars to cope with excess dietary carbohydrates when feeding on plants (Babic et al. 2008). Aphid saliva contains non-enzymic reducing compounds and oxidases (polyphenol oxidases and peroxidases) that deactivate defensive phytochemicals by oxidative polymerization (Miles 1999), and calcium-binding proteins that have the ability to prevent the occlusion of the sieve tubes (Will et al. 2007).

Protection from ingested plant allelochemicals by binding, ultrafiltration, and/or ionic exclusion are additional physiological roles for the PM in herbivorous insects (Barbehenn 2001). Bernays and Chamberlain (1980) reported that the tolerance of the polyphagous grasshopper *Schistocerca gregaria* to potentially toxic dietary tannins was partially due to the adsorptive properties of the PM. Moreover, the PM might act as a barrier to impede the passage of large complexes formed between hydrophilic allelochemicals (such as tannic acid) with proteins, lipids, or polyvalent metal ions in the digestive tract of lepidopteran insects (Barbehenn and Martin 1992; Barbehenn 2001); and between lipophilic allelochemicals (such as digitoxin) and lipids in the gut fluid of grasshoppers (Barbehenn 1999). In addition, the permeability of the PM to tannins might also be reduced by charge exclusion (Barbehenn and Martin 1994). Finally, carbohydrates and certain amino acids associated with the PM may confer antioxidant protection against reactive oxygen species generated in the gut of phytophagous lepidopterans, protecting the midgut epithelium from damage by dietary prooxidants (Summers and Felton 1996; Barbehenn and Stannard 2004).

It has been suggested that lepidopteran insects maintain a high gut pH to inactivate potentially harmful plant allelochemicals (Berenbaum 1980). This may be the case of tannins that disrupt insect digestion by cross-linking and precipitating enzymes and dietary proteins, but these complexes are dissociated at alkaline pH

(Felton and Duffey 1991). Besides, some lepidopteran species contain in their guts compounds that act as surfactants (Aboshi et al. 2010), helping also to counteract the potential of tannins to precipitate proteins (Martin and Martin 1984). However, the high pH found in the gut lumen of lepidopteran larvae promotes the oxidative activation of tannins and other phenolic compounds, and the formation of detrimental reactive oxygen species (Appel 1993). Ingested ascorbate and up-regulation of antioxidant enzymes is utilized by caterpillars as a protection against oxidative radicals generated by ingested pro-oxidant allelochemicals (Barbehenn et al. 2001; Krishnan and Kodrík 2006). The high pH and redox potential of the lepidopteran gut might also alter the effectiveness of some insecticidal proteins, through the oxidative cleavage of disulphide bonds that may be critical for their activity (Duffey and Stout 1996).

Insects feeding on plants are adapted to circumvent the effects of insecticidal proteins of plant origin that specifically target their digestive enzymes by: (1) hyperproduction of proteases; (2) up-regulation of "inhibitor-insensitive" enzymes; and (3) proteolysis of insecticidal proteins (Jongsma and Bolter 1997). Overproduction of digestive enzymes compensates for inhibitory effects of protease and α-amylase inhibitors in lepidopteran (Markwick et al. 1998; Pytelkova et al. 2009), coleopteran (Girard et al. 1998; Silva et al. 2001b) and hemipteran species (Carrillo et al. 2011). Adaptation of insects to plant hydrolytic inhibitors may also occur by the induction of enzymes insensitive to inhibition (Jongsma et al. 1995; Broadway 1997; Cloutier et al. 2000). Newly expressed enzymes may even be of a different type than those that are being targeted. When larvae of the red flour beetle, *Tribolium castaneum*, ingested cysteine protease inhibitors, there was a shift from cysteine proteases to serine proteases in the protease profile of the midgut (Oppert et al. 2005). Likewise, the inhibition of trypsin activity in larvae of the beet armyworm, *Spodoptera exigua*, after ingestion of a barley trypsin inhibitor, was compensated with a significant induction of exopeptidases (Lara et al. 2000). Additionally, hydrolytic enzyme inhibitors from host and non-host plants are susceptible to proteolysis by insect gut proteases (Michaud et al. 1995; Silva et al. 2001c; Yang et al. 2009). The success of phytophagous insects to overcome plant protease and α-amylase inhibitor defences depends on the differential regulation of multiple genes encoding digestive enzymes (Bown et al. 1997; Zhu-Salzman et al. 2003; Chougule et al. 2005; Chi et al. 2009). Digestive compensatory responses have been shown to depend both on the quality and the quantity of the hydrolytic enzyme inhibitors present in the host plants (De Leo et al. 1998; Rivard et al. 2004). Moreover, the time-course for remodeling the digestive enzymatic complement appears to follow a sequential strategy. Thus, larvae of *H. armigera* adapt to the presence of protease inhibitors in their diet by an initial general upregulation of protease-encoding genes, followed by a down-regulation of genes that encode proteases sensitive to the inhibitor, whereas genes encoding putative inhibitor-insensitive proteases continue to be up-regulated (Bown et al. 2004). Mazumdar-Leighton and Broadway (2001) indicated that in the larvae of the corn earworm, *Heliothis zea*, and the black cutworm, *Agrotis ipsilon*, constitutive trypsins are translational products of a pre-existing pool of mRNA, whereas induced trypsins are transcriptionally regulated following ingestion of protease inhibitors. The ability to regulate gene expression of digestive enzymes allows insects to overcome plant defenses, minimizing at the same time the metabolic cost associated with their production (Ortego et al. 2001; Chi et al. 2009).

The analysis of the transcriptome of phytophagous insects is revealing new aspects of the physiology of the insect gut that may be involved in the adaptation to plant defenses, such as: amino acid replacement in the vicinity of the active site of digestive enzymes as an adaptation to the presence of dietary ketones (Lopes et al. 2009); cysteine protease-inhibiting activity in midgut fluid that protects the PM from degradation by host plant cysteine proteases (Li et al. 2009); and expression at very high levels of catalytically inactive serine protease homologues that may act as baits within the gut lumen sequestering protease inhibitors (Prabhakar et al. 2007; Simpson et al. 2007). However, studies analyzing both sides of the plant-insect interaction at the whole genome level are necessary for a better understanding of the plant-insect interaction, including the physiological adaptations of the insect gut to herbivory.

References

Aboshi T, Yoshinaga N, Nishida R, Mori N (2010) Phospholipid biosynthesis in the gut of *Spodoptera litura* larvae and effects of tannic acid ingestion. Insect Biochem Mol Biol 40:325–330

Alarcon FJ, Martinez TF, Barranco P, Cabello T, Diaz M, Moyano FJ (2002) Digestive proteases during development of larvae of red palm weevil, *Rhynchophorus ferrugineus* (Olivier, 1790) (Coleoptera: Curculionidae). Insect Biochem Mol Biol 32:265–274

Appel HM (1993) Phenolics in ecological interactions: the importance of oxidation. J Chem Ecol 19:1521–1552

Appel HM, Joern A (1997) Gut physicochemistry of grassland grasshoppers. J Insect Physiol 44:693–700

Appel HM, Maines LW (1995) The influence of host plant on gut conditions of gypsy moth (*Lymantria dispar*) caterpillars. J Insect Physiol 41:241–246

Ashford DA, Smith WA, Douglas AE (2000) Living on a high sugar diet: the fate of sucrose ingested by a phloem-feeding insect, the pea aphid *Acyrthosiphon pisum*. J Insect Physiol 46:335–341

Babic B, Poisson A, Darwish S, Lacasse J, Merkx-Jacques M, Despland E, Bede JC (2008) Influence of dietary nutritional composition on caterpillar salivary enzyme activity. J Insect Physiol 54:286–296

Baldwin IT, Halitschke R, Kessler A, Schittko U (2001) Merging molecular and ecological approaches in plant-insect interactions. Curr Opin Plant Biol 4:351–358

Barbehenn RV (1999) Non-absorption of ingested lipophilic and amphiphilic allelochemicals by generalist grasshoppers: the role of extractive ultrafiltration by the peritrophic envelope. Arch Insect Biochem Physiol 42:130–137

Barbehenn RV (2001) Roles of peritrophic membranes in protecting herbivorous insects from ingested plant allelochemicals. Arch Insect Biochem Physiol 47:86–99

Barbehenn RV, Martin MM (1992) The protective role of the peritrophic membrane in the tannin-tolerant larvae of *Orgyia leucostigma* (Lepidoptera). J Insect Physiol 38:973–978

Barbehenn RV, Martin MM (1994) Tannin sensitivity in larvae of *Malacosoma disstria* (Lepidoptera): roles of the peritrophic envelope and midgut oxidation. J Chem Ecol 20:1985–2001

Barbehenn RV, Stannard J (2004) Antioxidant defense of the midgut epithelium by the peritrophic envelope in caterpillars. J Insect Physiol 50:783–790

Barbehenn RV, Bumgarner SL, Roosen EF, Martin MM (2001) Antioxidant defenses in caterpillars: role of the ascorbate-recycling system in the midgut lumen. J Insect Physiol 47:349–357

Bede JC, McNeil JN, Tobe SS (2007) The role of neuropeptides in caterpillar nutritional ecology. Peptides 28:185–196

Behmer ST (2009) Insect herbivore nutrient regulation. Annu Rev Entomol 54:165–187

Berbays EA (1991) Evolution of insect morphology in relation to plants. Philos Trans R Soc Lond B 333:257–264

Berenbaum M (1980) Adaptive signifi-cance of midgut pH in larval Lepidoptera. Am Nat 115:138–146

Bernays EA, Chamberlain DJ (1980) A study of tolerance of ingested tannin in *Schistocerca gregaria*. J Insect Physiol 26:415–420

Bown DP, Wilkinson HS, Gatehouse JA (1997) Differentially regulated inhibitor-sensitive and insensitive protease genes from the phytophagous insect pest, *Helicoverpa armigara*, are members of complex multigene families. Insect Biochem Mol Biol 27:625–638

Bown DP, Wilkinson HS, Gatehouse JA (2004) Regulation of expression of genes encoding digestive proteases in the gut of a polyphagous lepidopteran larva in response to dietary protease inhibitors. Physiol Entomol 29:278–290

Broadway RM (1997) Dietary regulation of serine proteinases that are resistant to serine proteinase inhibitors. J Insect Physiol 43:855–874

Broadway RM, Duffey SS (1986) The effect of dietary protein on the growth and digestive physiology of larval *Heliothis zea* and *Spodoptera exigua*. J Insect Physiol 32:673–680

Broadway RM, Duffey SS (1988) The effect of plant protein quality on insect digestive physiology and the toxicity of plant proteinase inhibitors. J Insect Physiol 34:1111–1117

Carrillo L, Martinez M, Alvarez-Alfageme F, Castañera P, Smagghe G, Diaz I, Ortego F (2011) A barley cysteine-proteinase inhibitor reduces the performance of two aphid species in artificial diets and transgenic *Arabidopsis* plants. Transgenic Res 20:305–319

Chaloner WG, Scott AC, Stephenson J (1991) Fossil evidence for plant-arthropod interactions in the Palaeozoic and Mesozoic. Philos Trans R Soc Lond B 333:177–186

Chambers PG, Simpson SJ, Raubenheimer D (1995) Behavioural mechanisms of nutrient balancing in *Locusta migratoria* nymphs. Anim Behav 50:1513–1523

Chi YH, Salzman RA, Balfe S, Ahn JE, Sun W, Moon J, Yun DJ, Lee SY, Higgins TJV, Pittendrigh B, Murdock LL, Zhu-Salzman K (2009) Cowpea bruchid midgut transcriptome response to a soybean cystatin – costs and benefits of counter-defence. Insect Mol Biol 18:97–110

Chougule NP, Giri AP, Sainani MN, Gupta VS (2005) Gene expression patterns of *Helicoverpa armigera* gut proteases. Insect Biochem Mol Biol 35:355–367

Clark TM (1999) Evolution and adaptive significance of larval midgut alkalinisation in the insect superorder Mecopterida. J Chem Ecol 21:1945–1960

Cloutier C, Jean C, Fournier M, Yelle S, Michaud D (2000) Adult Colorado potato beetles, *Leptinotarsa decemlineata* compensate for nutritional stress on oryzacystatin I-transgenic potato plants by hypertrophic behavior and over-production of insensitive proteases. Arch Insect Biochem Physiol 44:69–81

Colebatch G, East P, Cooper P (2001) Preliminary characterisation of digestive proteases of the green mirid, *Creontiades dilutus* (Hemiptera: Miridae). Insect Biochem Mol Biol 31:415–423

Cristofoletti PT, Ribeiro AF, Deraison C, Rahbé Y, Terra WR (2003) Midgut adaptation and digestive enzyme distribution in a phloem feeding insect, the pea aphid *Acyrthosiphon pisum*. J Insect Physiol 49:11–24

Dadd RH (1985) Nutrition: organisms. In: Kerkut GA, Gilbert LI (eds) Comparative insect physiology, biochemistry and pharmacology, vol 4. Pergamon Press, New York, pp 313–390

De Leo F, Bonadé-Bottino MA, Ceci LR, Gallerani R, Jouanin L (1998) Opposite effects on *Spodoptera littoralis* larvae of high expression level of a trypsin proteinase inhibitor in transgenic plants. Plant Physiol 118:997–1004

Díaz-Mendoza M, Ortego F, de Lacoba MG, Magaña C, de la Poza M, Farinós GP, Castañera P, Hernandez-Crespo P (2005) Diversity of trypsins in the Mediterranean corn borer *Sesamia nonagrioides* (Lepidoptera: Noctuidae), revealed by nucleic acid sequences and enzyme purification. Insect Biochem Mol Biol 35:1005–1020

Dillon RJ, Dillon VM (2004) The gut bacteria of insects: nonpathogenic interactions. Annu Rev Entomol 49:71–92

Douglas AE (2006) Phloem-sap feeding by animals: problems and solutions. J Exp Bot 57:747–754

Dow JAT (1992) pH gradients in lepidopteran midgut. J Exp Biol 172:355–375

Duffey SS, Stout MJ (1996) Antinutritive and toxic components of plant defense against insects. Arch Insect Biochem Physiol 32:3–37

Eichenseer H, Mathews MC, Bi JL, Murphy JB, Felton GW (1999) Salivary glucose oxidase: multifunctional roles for *Helicoverpa zea*? Arch Insect Biochem Physiol 42:99–109

Espinoza-Fuentes FP, Terra WR (1987) Physiological adaptations for digesting bacteria. Water fluxes and distribution of digestive enzymes in Musca domestica larval midgut. Insect Biochem 17:809–817

Farrell BD (1998) "Inordinate fondness" explained: why are there so many beetles? Science 281:555–559

Felton GW (1996) Nutritive quality of plant protein: sources of variation and insect herbivore responses. Arch Insect Biochem Physiol 32:107–130

Felton GW, Duffey SS (1991) Reassessment of the role of gut alkalinity and detergency in insect herbivory. J Chem Ecol 17:1821–1836

Futuyma DJ, Agrawal AA (2009) Macroevolution and the biological diversity of plants and herbivores. Proc Natl Acad Sci USA 106:18054–18061

Gatehouse LN, Shannon AL, Burgess EPJ, Christeller JT (1997) Characterization of major midgut proteinase cDNAs from *Helicoverpa armigera* larvae and changes in gene expression in response to four proteinase inhibitors in the diet. Insect Biochem Mol Biol 27:929–944

Girard C, Le Métayer M, Zaccomer B, Bartlet E, Williams I, Bonadé-Bottino M, Pham-Delegue MH, Jouanin L (1998) Growth stimulation of beetle larvae reared on a transgenic oilseed rape expressing a cysteine proteinase inhibitor. J Insect Physiol 44:263–270

Grimaldi D, Engel MS (2005) Evolution of the insects. Cambridge University Press, Cambridge, p 755

Gullan PJ, Kosztarab M (1997) Adaptations in scale insects. Annu Rev Entomol 42:23–50

Gündüz EA, Douglas AE (2009) Symbiotic bacteria enable insect to use a nutritionally inadequate diet. Proc R Soc B 276:987–991

Harrison JF (2001) Insect acid–base physiology. Annu Rev Entomol 46:221–250

Hegedus D, Erlandson M, Gillott C, Toprak U (2009) New insights into peritrophic matrix synthesis, architecture, and function. Annu Rev Entomol 54:285–302

Hinks CF, Erlandson MA (1995) The accumulation of haemolymph proteins and activity of digestive proteinases of grasshoppers (*Melanoplus sanguinipes*) fed wheat, oats or kochia. J Insect Physiol 41:425–433

Horne I, Haritos VS, Oakeshott JG (2009) Comparative and functional genomics of lipases in holometabolous insects. Insect Biochem Mol Biol 39:547–567

Johnson KS, Felton GW (1996) Potential influence of midgut pH and redox potential on protein utilization in insect herbivores. Arch Insect Biochem Physiol 32:85–105

Johnson KS, Rabosky D (2000) Phylogenetic distribution of cysteine proteinases in beetles: evidence for an evolutionary shift to an alkaline digestive strategy in Cerambycidae. Comp Biochem Physiol 126B:609–619

Jongsma MA, Bolter C (1997) The adaptation of insects to plant protease inhibitors. J Insect Physiol 43:885–895

Jongsma MA, Bakker PL, Peters J, Bosch D, Stiekema WJ (1995) Adaptation of *Spodoptera exigua* larvae to plant proteinase inhibitors by induction of gut proteinase activity insensitive to inhibition. Proc Natl Acad Sci USA 92:8041–8045

Kotkar HM, Sarate PJ, Tamhane VA, Gupta VS, Giri AP (2009) Responses of midgut amylases of *Helicoverpa armigera* to feeding on various host plants. J Insect Physiol 55:663–670

Krishnan N, Kodrík D (2006) Antioxidant enzymes in *Spodoptera littoralis* (Boisduval): are they enhanced to protect gut tissues during oxidative stress? J Insect Physiol 52:11–20

Labandeira CC, Phillips TL (1996) Insect fluid-feeding on Upper Pennsylvanian tree ferns (Palaeodictyoptera, Marattiales) and the early history of the piercing-and-sucking functional feeding group. Ann Entomol Soc Am 89:157–183

Lara P, Ortego F, Gonzalez-Hidalgo E, Castañera P, Carbonero P, Diaz I (2000) Adaptation of *Spodoptera exigua* (Lepidoptera: Noctuidae) to barley trypsin inhibitor BTI-CMe expressed in transgenic tobacco. Transgenic Res 9:169–178

Lee PL (2007) The interactive effects of protein quality and macronutrient imbalance on nutrient balancing in an insect herbivore. J Exp Biol 210:3236–3244

Lehane MJ (1997) Peritrophic matrix structure and function. Annu Rev Entomol 42:525–550

Li C, Song X, Li G, Wang P (2009) Midgut cysteine protease-inhibiting activity in *Trichoplusia ni* protects the peritrophic membrane from degradation by plant cysteine proteases. Insect Biochem Mol Biol 39:726–734

Lopes AR, Sato PM, Terra WR (2009) Insect chymotrypsins: chloromethyl ketone inactivation and substrate specificity relative to possible coevolutional adaptation of insects and plants. Arch Insect Biochem Physiol 70:188–203

Markwick NP, Laing WA, Christeller JT, McHenry JZ, Newton MR (1998) Overproduction of digestive enzymes compensates for inhibitory effects of protease and α-amylase inhibitors fed to three species of leafrollers (Lepidoptera: Tortricidae). J Econ Entomol 91:1265–1276

Martin MM, Martin JS (1984) Surfactants: their role in preventing the precipitation of proteins by tannins in insect guts. Oecologia 61:342–345

Mayhew PJ (2007) Why are there so many insect species? Perspectives from fossils and phylogenies. Biol Rev 82:425–454

Mazumdar-Leighton S, Broadway RM (2001) Transcriptional induction of diverse midgut trypsins in larval *Agrotis ipsilon* and *Helicoverpa zea* feeding on the soybean trypsin inhibitor. Insect Biochem Mol Biol 31:645–657

Michaud D, Cantin L, Vrain TC (1995) Carboxy-terminal truncation of oryzacysatin II by oryzacytatin-insensitive insect digestive proteinases. Arch Biochem Biophys 322:469–474

Miles PW (1999) Aphid saliva. Biol Rev 74:41–85

Morris K, Lorenzen MD, Hiromasa Y, Tomich J, Oppert C, Elpidina EN, Vinokurov K, Jurat-Fuentes JL, Fabrick J, Oppert B (2009) *Tribolium castaneum* larval gut transcriptome and proteome: a resource for the study of the coleopteran gut. J Proteome Res 8:3889–3898

Murdock LL, Brookhart G, Dunn PE, Foard DE, Kelley S, Kitch L, Shade RE, Shukle RH, Wolfson JL (1987) Cysteine digestive proteinases in Coleoptera. Comp Biochem Physiol 87B:783–787

Musser RO, Hum-Musser SM, Eichenseer H, Peiffer M, Ervin G, Murphy JB, Felton GW (2002) Caterpillar saliva beats plant defences: a new weapon emerges in the evolutionary arms race between plants and herbivores. Nature 416:599–600

Neal JJ (1996) Brush border membrane and amino acid transport. Arch Insect Biochem Physiol 32:55–64

Novillo C, Castañera P, Ortego F (1997) Characterization and distribution of chymotrypsin-like and other digestive proteases in Colorado potato beetle larvae. Arch Insect Biochem Physiol 36:181–201

Novillo C, Castañera P, Ortego F (1999) Isolation and characterization of two digestive trypsin-like proteinases from larvae of the stalk corn borer, *Sesamia nonagrioides*. Insect Biochem Mol Biol 29:177–184

Oliveira-Neto OB, Batista JAN, Rigden DJ, Fragoso RR, Silva RO, Gomes EA, Franco OL, Dias SC, Cordeiro CMT, Monnerat RG, Grossi de Sa MF (2004) A diverse family of serine protease genes expressed in cotton boll weevil (*Anthonomus grandis*): implications for the design of pest-resistant transgenic cotton plants. Insect Biochem Mol Biol 34:903–918

Oppert B, Morgan TD, Hartzer K, Kramer KJ (2005) Compensatory proteolytic responses to dietary proteinase inhibitors in the red flour beetle, *Tribolium castaneum* (Coleoptera: Tenebrionidae). Comp Biochem Physiol C 140:53–58

Ortego F, Novillo C, Castañera P (1996) Characterization and distribution of digestive proteases of the stalk corn borer, *Sesamia nonagrioides* Lef. (Lepidoptera: Noctuidae). Arch Insect Biochem Physiol 33:163–180

Ortego F, Novillo C, Sánchez-Serrano JJ, Castañera P (2001) Physiological response of Colorado potato beetle and beet armyworm larvae to depletion of wound-inducible proteinase inhibitors in transgenic potato plants. J Insect Physiol 47:1291–1300

Overney S, Fawe A, Yelle S, Michaud D (1997) Diet-related plasticity of the digestive proteolytic system in larvae of the Colorado potato beetle (*Leptinotarsa decemlineata* Say). Arch Insect Biochem Physiol 36:241–250

Patankar AG, Giri AP, Harsulkar AM, Sainani MN, Deshpande VV, Ranjekar PK, Gupta VS (2001) Complexity in specificities and expression of *Helicoverpa armigera* gut proteinases explains polyphagous nature of the insect pest. Insect Biochem Mol Biol 31:453–464

Pauchet Y, Wilkinson P, Vogel H, Nelson DR, Reynolds SE, Heckel DG, ffrench-Constant RH (2010) Pyrosequencing the *Manduca sexta* larval midgut transcriptome: messages for digestion, detoxification and defence. Insect Mol Biol 19:61–75

Pieterse CMJ, Dicke M (2007) Plant interactions with microbes and insects: from molecular mechanisms to ecology. Trends Plant Sci 12:564–569

Prabhakar S, Chen M-S, Elpidina EN, Vinokurov KS, Smith CM, Oppert B (2007) Molecular characterization of digestive proteinases and sequence analysis of midgut cDNA transcripts of the yellow mealworm, *Tenebrio molitor* L. Insect Mol Biol 16:455–468

Pytelková J, Hubert J, Lepšík M, Šobotník J, Šindelka R, Křížková I, Horn M, Mareš M (2009) Digestive α-amylases of the flour moth *Ephestia kuehniella* – adaptation to alkaline environment and plant inhibitors. FEBS J 276:3531–3546

Raman A, Schaefer CW, Withers TM (2005) Galls and gall-inducing arthropods: an overview of their biology, ecology and evolution. In: Raman A, Schaefer CW, Withers TM (eds) Biology, ecology, and evolution of gall-inducing arthropods. Science Publishers, Enfield, pp 1–33

Raubenheimer D, Simpson SJ (1999) Integrating nutrition: a geometrical approach. Entomol Exp Appl 91:67–82

Rivard D, Cloutier C, Michaud D (2004) Colorado potato beetles show differential digestive compensatory responses to host plants expressing distinct sets of defense proteins. Arch Insect Biochem Physiol 55:114–123

San Andres V, Ortego F, Castañera P (2007) Effects of gamma-irradiation on midgut proteolytic activity of the Mediterranean fruit fly, *Ceratitis capitata* (Diptera: Tephritidae). Arch Insect Biochem Physiol 65:11–19

Sandström J, Moran N (1999) How nutritionally imbalanced is phloem sap for aphids? Ent Exp Appl 91:203–210

Schultz JC, Lechowicz MJ (1986) Host plant, larval age and feeding behavior influence midgut pH in the gypsy moth (*Lymantria dispar*). Oecologia 71:133–137

Scudder GGE (2009) The importance of insects. In: Foottit RG, Adler PH (eds) Insect biodiversity: science and society. Wiley-Blackwell, Chichester, pp 7–32

Shigenobu S, Watanabe H, Hattori M, Sasaki Y, Ishikawa H (2000) Genome sequence of the endocellular bacterial symbiont of aphids *Buchnera* sp. APS. Nature 407:81–86

Shinbo H, Konno K, Hirayama C, Watanabe K (1996) Digestive sites of dietary proteins and absorptive sites of amino acids along the midgut of the silkworm, *Bombyx mori*. J Insect Physiol 42:1129–1138

Silva CP, Terra WR (1994) Digestive and absorptive sites along the midgut of the cotton seed sucker bug *Dystercus peruvianus* (Hemiptera: Pyrrhocoridae). Insect Biochem Mol Biol 24:493–505

Silva CP, Ribeiro AF, Gulbenkian S, Terra WR (1995) Organization, origin and function of the outer microvillar (perimicrovillar) membranes of *Dysdercus peruvianus* (Hemiptera) midgut cells. J Insect Physiol 41:1093–1103

Silva CP, Terra WR, Xavier-Filho J, Grossi de Sá MF, Lopes AR, Pontes E (1999) Digestion in larvae of *Callosobruchus maculatus* and *Zabrotes subfasciatus* (Coleoptera: Bruchidae) with emphasis on a-amylases and oligosaccharidases. Insect Biochem Mol Biol 29:355–366

Silva CP, Terra WR, Xavier-Filho J, Grossi de Sa MF, Isejima EM, DaMatta RA, Miguens FC, Bifano TD (2001a) Digestion of legume starch granules by larvae of *Zabrotes subfasciatus* (Coleoptera: Bruchidae) and the induction of α-amylases in response to different diets. Insect Biochem Mol Biol 31:41–50

Silva CP, Terra WR, deSá MFG, Samuels RI, Isejima EM, Bifano TD, Almeida JS (2001b) Induction of digestive α-amylases in larvae of *Zabrotes subfasciatus* (Coleoptera: Bruchidae) in response to ingestion of common bean α-amylase inhibitor 1. J Insect Physiol 47:1283–1290

Silva CP, Terra WR, Lima RM (2001c) Differences in midgut serine proteinases from larvae of the bruchid beetles *Callosobruchus maculatus* and *Zabrotes subfasciatus*. Arch Insect Biochem Physiol 47:18–28

Silva CP, Silva JR, Vasconcelos FF, Petretski MDA, DaMatta RA, Ribeiro AF, Terra WR (2004) Occurrence of midgut perimicrovillar membranes in paraneopteran insect orders with comments on their function and evolutionary significance. Arthropod Struct Dev 33:139–148

Simpson RM, Newcomb RD, Gatehouse HS, Crownhurst RN, Changné D, Gatehouse LN, Markwick NP, Beuning LL, Murray C, Marshall SD, Yauk Y-K, Nain B, Gleave AP, Christeller JT (2007) Expressed sequence tags from the midgut of *Epiphyas postvittana* (Walker) (Lepidoptera: Torticidae). Insect Mol Biol 16:675–690

Slansky FJ, Wheeler GS (1991) Food consumption and utilization responses to dietary dilution with cellulose and water by velvetbean caterpillars, *Anlicarsia gemmnatalis*. Physiol Entomol 16:99–116

Summers CB, Felton GW (1996) Peritrophic envelope as a functional antioxidant. Arch Insect Biochem Physiol 32:131–142

Teo LH (1997) Tryptic and chymotryptic activities in different parts of the gut of the field cricket *Gryllus bimaculatus* (Orthoptera: Gryllidae). Ann Entomol Soc Am 90:69–74

Teo LH, Woodring JP (1994) Comparative total activities of digestive enzymes in different gut regions of the house cricket, *Acheta domesticus* L. (Orthoptera: Gryllidae). Ann Entomol Soc Am 87:886–890

Terra WR (1990) Evolution of digestive systems in insects. Annu Rev Entomol 35:181–200

Terra WR (2001) The origin and functions of the insect peritrophic membrane and peritrophic gel. Arch Insect Biochem Physiol 47:47–61

Terra WR, Ferreira C (1994) Insect digestive enzymes: properties, compartmentalization and function. Comp Biochem Physiol 109B:1–62

Terra WR, Ferreira C (2005) Biochemistry of digestion. In: Gilbert LI, Iatrou K, Gill SS (eds) Comprehensive molecular insect science, vol 4. Elsevier, Oxford, pp 171–224

Terra WR, Ferreira C, Bastos F (1985) Phylogenetic considerations of insect digestion. Disaccharidases and the spatial organization of digestion in the *Tenebrio molitor* larvae. Insect Biochem 15:443–449

Terra WR, Ferreira C, Baker JE (1996a) Compartmentalization of digestion. In: Lehane MJ, Billingsley PF (eds) Biology of the insect midgut. Chapman & Hall, London, pp 206–235

Terra WR, Ferreira C, Jordão BP, Dillon RJ (1996b) Digestive enzymes. In: Lehane MJ, Billingsley PF (eds) Biology of the insect midgut. Chapman & Hall, London, pp 153–194

Waldbauer GP, Friedman S (1991) Self-selection of optimal diets by insects. Annu Rev Entomol 36:43–63

Weech MH, Chapleau M, Pan L, Ide C, Bede JC (2008) Caterpillar saliva interferes with induced *Arabidopsis thaliana* defence responses via the systemic acquired resistance pathway. J Exp Bot 59:2437–2448

Will T, Tjallingii WF, Thönnessen A, van Bel AJE (2007) Molecular sabotage of plant defense by aphid saliva. Proc Natl Acad Sci USA 104:10536–10541

Wolfson JL, Murdock LL (1990) Diversity in digestive proteinase activity among insects. J Chem Ecol 16:1089–1102

Woods HA, Kingsolver JG (1999) Feeding rate and the structure of protein digestion and absorption in lepidopteran midguts. Arch Insect Biochem Physiol 42:74–87

Yang L, Fang Z, Dicke M, van Loon JJ, Jongsma MA (2009) The diamondback moth, *Plutella xylostella*, specifically inactivates Mustard Trypsin Inhibitor 2 (MTI2) to overcome host plant defence. Insect Biochem Mol Biol 39:55–61

Yeoh HH, Wee YC, Watson L (1992) Leaf protein contents and amino acid patterns of dicotyledonous plants. Biochem Syst Ecol 20:657–663

Zhu-Salzman K, Koiwa H, Salzman RA, Shade RE, Ahn JE (2003) Cowpea bruchid *Callosobruchus maculatus* uses a three-component strategy to overcome a plant defensive cysteine protease inhibitor. Insect Mol Biol 12:135–145

Chapter 4
Successes and Failures in Plant-Insect Interactions: Is it Possible to Stay One Step Ahead of the Insects?

Angharad Gatehouse and Natalie Ferry

> *. . . animals annually consume an amount of produce that sets calculation at defiance; and, indeed, if an approximation could be made to the quantity thus destroyed, the world would remain skeptical of the result obtained, considering it too marvelous to be received as truth'.* – John Curtis, 1860 (entomologist).

Arthropods are the most widespread and diverse group of animals, with an estimated four to six million species worldwide (Novotny et al. 2002). Whilst only a small percentage of arthropods are classified as phytophagous pests they cause major devastation of crops, destroying around 14% of the world annual crop production, contributing to 20% of losses of stored grains and causing around US$100 billion of damage each year (Nicholson 2007). Thus herbivorous insects and mites are a major threat to food production for livestock and human consumption. Larval forms of lepidopterans are considered the most destructive insects, with about 40% of all insecticides directed against heliothine species (Brooks and Hines 1999), although, many species within the Orders Acrina, Coleoptera, Diptera, Hemiptera, Orthoptera and Thysanoptera are also considered agricultural pests with significant economic impact. Insect pests may cause direct damage by feeding on crop plants in the field or infesting stored products, so competing with humans for plants as a food resource. Some cause indirect damage, especially the sap feeding insects by transmitting viral diseases or secondary microbial infections of crop plants (Ferry and Gatehouse 2010).

Innumerable examples exist of insect pests that are highly injurious to agricultural production. The most notable for their destructive capacity being the migratory locust (*Locusta migratoria*), Colorado potato beetle (*Leptinotarsa decemlineata*), boll weevil (*Anthonomus grandis*), Japanese beetle (*Popillia japonica*), aphids, which are among the most destructive pests on earth as vectors of plant viruses (many species

A. Gatehouse (✉)
School of Biology, Newcastle University, Ridley Building, Newcastle upon Tyne NE1 7RU, UK
e-mail: a.m.r.gatehouse@newcastle.ac.uk

N. Ferry
School of Environment and Life Sciences, University of Salford, Peel Park, Salford M5 4WT, UK
e-mail: n.ferry@salford.ac.uk

G. Smagghe and I. Diaz (eds.), *Arthropod-Plant Interactions: Novel Insights and Approaches for IPM*, Progress in Biological Control 14, DOI 10.1007/978-94-007-3873-7_4, © Springer Science+Business Media B.V. 2012

Fig. 4.1 Monument to the
cotton boll weevil
(Source: Ferry and Gatehouse
2010)

in ten families of the Aphidoidea), and the western corn rootworm (*Diabrotica virgifera virgifera),* also called the 'billion dollar bug' due to its economic impact in the US alone. Curiously one of these pests, the cotton boll weevil, responsible for near destruction of the cotton industry in North America, is ultimately also responsible for subsequent diversification of agriculture in many regions thus warranting a monument in the town of Enterprise, Alabama, in profound appreciation of its role in bringing to an end the State's dependence on a poverty crop (Fig. 4.1).

The global challenge facing agriculture is to secure high and quality crop yields and to make agricultural production environmentally sustainable. Control of insect pests would go some way towards achieving this goal.

4.1 Insect Control Strategies: Successes, Failures or Perhaps a Bit of Both?

The aim of this chapter is to discuss the major control strategies for insect pests of crops deployed to date and to attempt to evaluate how successful they have been – of course this will depend, to some extent, upon how success is measured.

For example, considerably different conclusions would be drawn if one judged a technology in terms of pure economic returns, versus in terms of the technology having a minimal environmental impact. Thus criteria need to be drawn up by which to measure 'success'. For simplicity, success here will be evaluated as a balance of: insect mortality (and thus crop protection), insect resistance, and non-target effects. It is not the aim of this chapter to provide an exhaustive review of all technologies, as this has been done elsewhere by experts in the respective fields, but to focus on the evaluation of transgenic crops and molecular breeding approaches; however, insecticide use, biological control and plant breeding will be briefly discussed as comparators. References to significant papers in these fields will be given in the text.

4.1.1 Insecticides

Insecticides have been, and still are, a highly effective method to control pests quickly when they threaten to destroy crops. The chemical nature of the insecticides used has evolved over time. In early farming practices inorganic chemicals were used for insect control, however with the advances in synthetic organic chemistry that followed two world wars the synthetic insecticides were born. In the 1940s the neuro-toxic organochlorine, DDT, was the pesticide of choice, but following its indiscriminate use it was reported to bio-accumulate in the food chain were it affected the fertility of higher organisms such as birds. Rachel Carson first highlighted this in her book *Silent Spring* published in 1962; whilst her presumptions have since been proven to be wrong the book was never-the-less an important signature event in the birth of the environmental movement. This pesticide was subsequently replaced by the comparatively safer organophosphate and carbamate-based pesticides [both acetyl cholinesterase (AChE) inhibitors] and many of these were replaced in turn by the even safer pyrethroid-based pesticides (axonic poisons). Synthetic pyrethroids continue to be used today despite the fact that they are broad spectrum.

A major limiting factor regarding a high dependency on insecticides is the occurrence of resistance in insect populations. In fact, resistance to insecticides has now been reported in more than 500 species (Nicholson 2007). Furthermore, resistance has evolved to every major class of chemical. The underlying causes of insecticide resistance are many-fold. Due to wide usage and narrow target range, arthropods have been put under a high degree of selection pressure (Feyereisen 1995). Insecticide resistance may be characterized by:

(a) metabolic detoxification (up-regulation of esterases, glutathione-S-transferases, and monooxygenases),
(b) decreased target site sensitivity (via mutation of the target receptor), and
(c) sequestration or lowered insecticide availability.

In addition, cross-resistance to different classes of chemicals has occurred due to the fact that many insecticides target a limited number of sites in the insect nervous system (Raymond-Delpech et al. 2005). The five target sites in insects comprise: nicotinic acetylcholine receptors (e.g., imidacloprid), voltage-gated sodium channels (e.g., DDT, pyrethroids), γ-aminobutyric acid receptors (e.g., fipronil), glutamate receptors (e.g., avermectins), and AChE (e.g., organophosphates and carbamates). The world insecticide market is dominated by compounds that inhibit the enzyme AChE. Together AChE inhibitors and insecticides acting on the voltage-gated sodium channel, in particular the pyrethroids, account for approximately 70% of the world market (Nauen et al. 2001).

Unfortunately as insecticide target sites are conserved between invertebrates and vertebrates, insecticides often have undesirable non-target effects and unacceptable ecological impacts. Insecticides are implicated in the poisoning of non-target insects, other arthropods, marine life, birds, and humans (Fletcher et al. 2000). The poisoning of non-target organisms has obvious implications for biodiversity. Despite these major drawbacks with the technology there is little doubt that crop yields could not be sustained at current levels without insecticide treatment. Thus a major goal in crop protection must be to replace insecticides with a sustainable alternative while still maintaining high levels of crop protection. With a projected increase in world population to ten billion over the next four decades this must be viewed as an immediate priority for agriculture.

4.1.2 Biopesticides

In parallel to the development of modern insecticides, biopesticides, which are naturally occurring substances that may be used to control insect pests, have also been developed. The most commonly used being microbial pesticides, which may consist of bacteria, viruses or entomopathogenic fungi, and nematodes. The specific microbial toxins produced by the soil-dwelling bacterium *Bacillus thuringiensis* (*Bt*) are increasingly being adopted as biopesticides. In fact, microbial *Bt*-based sprays are used in organic agriculture and as a part of Integrated Pest Management (IPM). This shift is due in part to a demand for increased safety both for humans and for the environment.

4.1.3 Biological Control

Biological control of pests in agriculture is a method of controlling pests that relies on predation, parasitism, or other natural mechanisms such as the release of pathogens (Wäckers et al. 2005). Classic examples include the use of ladybirds to control aphids or the introduction of parasitic wasps to control lepidopteran pests. When used alone, only a minority of biological control programmes succeed in bringing the target pest population under effective control. Biological control is, therefore, usually employed with chemical, cultural, genetic or other methods in an integrated pest management (IPM) strategy (Gurr and Kvedaras 2010).

4.1.4 Integrated Pest Management (IPM)

Integrated pest management (IPM) has also been proposed as a sustainable control system for insects. Several control systems are combined, including the judicious application of chemicals and biopesticides, use of trap crops, biological control, rotation, good husbandry and cultural control to manage all the pests of a particular crop (Gatehouse and Gatehouse 1999). Increasing crop varietal resistance is critical to IPM.

It is debatable as to whether IPM could sustain crop production at sufficiently high levels to feed the current world population without chemical insecticides. In order to feed an increasing world population more food must be produced in the future, and on either the same amount, or less land (Ferry and Gatehouse 2010). It is questionable if such farming methods will be as productive as will be necessary to meet these increased demands (Amman 2010) unless they embrace the use of transgenic (genetically modified) crops.

4.2 Plant Breeding and Varietal Resistance – Exploiting Plant Endogenous Defenses

Plants are challenged constantly by many different potential pathogens. There are hundreds of thousands of viral, bacterial, and fungal species in the world, thousands of which infect plants (Chrispeels and Sadava 2003). Any one pathogen can severely depress the yield of a given crop. Pathogens may reduce yield by causing tissue lesions; by reducing leaf, root, or seed growth; or by clogging up vascular tissue and causing wilt. Even in the absence of obvious symptoms pathogens can still be a major metabolic drain that reduces productivity. Plant breeders have often relied on genetic disease resistance traits to manage pathogens of particular crops. In classical plant breeding this has relied on crosses between elite crops and wild relatives (that are more genetically diverse) to introduce new disease resistance traits into the crops. Extensive backcrossing of the elite line is then required to eliminate the undesirable traits in the wild relative and thus makes traditional breeding a time consuming process, with a time of ca. 15 years required before a new resistant variety is available for release to growers.

Essentially, plants have two major types of induced disease resistance, basal defense and resistance (R) gene mediated defense. All plants have basal defense, this is a general immune response to pathogens and other environmental stresses.

R gene mediated defense is more specific and is only found in certain plant species. The *R* gene mediated defense involves recognition of a specific pathogen effector by a plant ligand receptor. These pathogen effectors can suppress plant basal defense, making any plant without the *R* gene defense susceptible. The ligand-effector recognition can result in a dramatic immune response such as cell death. Both types of plant defenses (*R* and basal) involve signaling via three major plant hormones: salicylic acid, jasmonic acid, and ethylene (ETH). In some instances defense responses are induced distal to the site of infection and this is referred to as systemic acquired resistance (SAR). At least three nonspecific induced defense pathways are describe which are triggered by these specific signaling molecules:

(a) the salicylic acid (SA) dependent pathway is induced by necrosis inducing pathogens and triggers systemic acquired resistance (SAR),
(b) a second pathway is triggered by nonpathogenic rhizobacteria, it is dependent on jasmonic acid (JA) and ethylene (ETH) and constitutes induced systemic resistance (ISR).
(c) JA and ETH regulate a third pathway that is effective against a different set of pathogens and not affected by ISR.

Most of the inducible defense related genes are regulated by these signaling pathways (Delaney et al. 1994; Sticher et al. 1997; Van Loon 1997; Reymond and Farmer 1998; Knoester et al. 1998; Ananieva and Ananiev 1999).

The most extensively studied gene-for-gene interactions mediate resistance against plant pathogens but R-genes are also thought to mediate gene-for-gene interactions with insect pests. In other words, a particular R-gene confers resistance against specific biotypes of a pest that carry a corresponding avirulence (Avr) gene (Flor 1955), including aphids, Hessian flies, midges, and nematodes (Puterka and Peters 1989; Roberts 1995; Zantoko and Shukle 1997; Milligan et al. 1998; Rossi et al. 1998; Stuart et al. 1998; Sardesai et al. 2001; Brotman et al. 2002). In tomato, *Solanum lycopersicum,* the *Mi* gene, a member of a large family of *R* genes, mediate resistance to potato aphids, whiteflies, and root-knot nematodes (Kaloshian and Walling 2005). However, to date, resistant germplasm has not had anywhere near a similar impact in insect control as it has in pathogen control, although it has a significant role to play in IPM programmes.

Alternative strategies for protecting crops from insect pests, that are not dependent on the expression of single resistance genes, seek to exploit the induced endogenous resistance mechanisms exhibited by plants to most insect herbivores. Such induced defences are exemplified by the wounding response, first identified as the local and systemic synthesis of proteinase inhibitors (PIs), which block insect digestion in response to plant damage (Gatehouse 2002). Recent research has shown that induced defences also involve the plant's ability to produce toxic or repellent secondary metabolites as direct defences, and volatile molecules, which play an important role in indirect defence (Kessler and Baldwin 2002). Insect herbivores activate induced defences both locally and systemically via signalling pathways involving systemin, jasmonate, oligogalacturonic acid and hydrogen peroxide rather than SA mediated responses as with pathogens.

Ecologists have long understood that plants exhibit multi-mechanistic resistance towards herbivores, but the molecular mechanisms underpinning these complicated responses have remained elusive (Baldwin et al. 2001). However, recent studies investigating the plant's herbivore-induced transcriptome, using microarrays and differential display technologies, have provided novel insights into plant-insect interactions. The jasmonic acid cascade plays a central role in transcript accumulation in plants exposed to herbivory (Hermsmeier et al. 2001). A single microarray based study revealed that the model plant *Arabidopsis* undergoes changes in levels of over 700 mRNAs during the defence response (Schenk et al. 2000). In contrast, only 100 mRNAs were up-regulated by spider mite (*Tetranychus urticae*) infestation in lima bean (*Phaseolus lunatus*), although a further 200 mRNAs were up-regulated in an indirect response mediated by feeding-induced volatile signal molecules (Arimura et al. 2000). Thus deciphering of the signals regulating herbivore-responsive gene expression will afford many opportunities to manipulate the response. Signalling molecules such as salicylic acid, jasmonic acid and ethylene do not activate defences independently by linear cascades, but rather establish complex interactions that determine specific responses. Knowledge of these interactions (discussed in detail later) can be exploited in the rational design of transgenic plants with increased insect resistance (Rojo et al. 2003; De Vos et al. 2005; Giri et al. 2006).

4.2.1 Indirect Defence (Volatile Production)

The role of plant volatiles in indirect defence has been described as 'top-down' defence (Baldwin et al. 2001). Some volatiles appear to be common to many different plant species, including C_6 aldehydes, alcohols and esters (green leaf volatiles), C_{10} and C_{15} terpenoids, and indole, whereas others are specific to a particular plant species. Many volatiles are preformed and act in herbivore deterrence; furthermore the wounding response also includes the formation of volatile compounds. Top-down' control of herbivore populations is achieved by attracting predators and parasitoids to the feeding herbivore, mediated by these volatile organic compounds (VOCs). For example, genes involved in the biosynthesis of the maize VOC bouquet are up-regulated by insect feeding (Frey et al. 2000; Shen et al. 2000). In addition, herbivore oviposition has been shown to induce VOC emissions, which attract egg parasitoids (Hilker and Meiners 2002). Herbivore-induced VOCs can also elicit production of defence-related transcripts in plants near the individual under attack (Arimura et al. 2000; Dicke et al. 2003). Manipulation of volatile biosynthesis can affect insect resistance. Transgenic potatoes in which production of hydroperoxide lyase (the enzyme involved in green leaf volatile biosynthesis) was reduced were found to support improved aphid performance and fecundity, suggesting toxicity of these volatiles to *Myzus persicae* (Vancanneyt et al. 2001). In a review of the topic, Degenhardt et al. (2003) discuss the potential of modifying terpene emission with the aim of making crops more attractive to herbivore natural enemies.

The manipulation of volatile cue for crop protection has been the subject of intensive research (e.g. Pickett et al. 1997), however in isolation such strategies can not provide levels of insect pest mortality comparable with that from insecticide use.

4.2.2 Detoxification and Insect Modulation of the Wounding Response

It is noteworthy that insect pests are able to feed on plants despite their defenses, both constitutive and inducible. Many insects are able to detoxify potentially toxic secondary metabolites, using cytochrome P-450 monoxygenases and glutathione S-transfereases. These enzymes are induced by exposure to toxic plant secondary compounds, for example xanthotoxin (a furanocoumarin) induces P-450 expression in corn earworm (Li et al. 2002). More recently, Li et al. (2002) have shown that corn earworm uses signalling molecules from its plant host, jasmonate and salicylate, to activate four of its cytochrome P450 genes, thus making the induction of detoxifying enzymes rapid and specific. Recent strategies based on RNAi technology have shown that it is possible to overcome these insect responses.

It is clear that the ability/potential for insect populations to develop resistance to some current methods for crop protection (insecticides, varietal resistance) is a major limiting factor in crop production; it is also clear that many promising and environmentally benign strategies do not give as high a level of control as chemical treatment. This chapter will now focus on the development and use of transgenic crops for insect control, and ask the question whether they are doomed to follow the insecticide paradigm or will the new deluge of genomic information on crops allow us to design crops not simply based on a reliance on single genes and resistance factors, but to develop crops exhibiting multigenic resistance?

4.3 Transgenic (Genetically Modified) Crops

Transgenic (GM) maize and cotton varieties that express insecticidal proteins derived from *Bacillus thuringiensis* (*Bt*) have become an important component in agriculture worldwide. At present 26.3 million hectares of the 148 million ha of land planted to biotech crops, is planted with insect protected transgenic *Bt* cotton and maize, with stacked traits occupying a further 32.3 million hectares (James 2010). Whilst it is difficult to provide current economic values for these insect-resistant transgenic crops on their own, the global value of biotech seed alone was valued at US $11.2 billion in 2010, with commercial biotech maize, soybean grain and cotton valued at approximately US $150 billion per year (James 2010). In 2007 the economic benefits from *Bt* crops were estimated at US$21 billion and stacked traits a further US$28 billion (James 2007). Significantly Phipps and Park (2002) showed that on a global basis GM technology has reduced pesticide use. These authors estimated that pesticide use was reduced by a total of 22.3 million kg of formulated product in 2000 alone.

4.3.1 *Bacillus thuringiensis Toxins*

Bacillus thuringiensis (*Bt*) is a soil-dwelling bacterium of major agronomic and scientific interest. Whilst the subspecies of this bacterium colonize and kill a large variety of host insects, each strain tends to be highly specific. Toxins for insects in the orders Lepidoptera (butterflies and moths), Diptera (flies and mosquitoes), Coleoptera (beetles and weevils), and Hymenoptera (wasps and bees) have been identified (de Maagd et al. 2001), but interestingly none with activity towards Homoptera (sap suckers) have, as yet, been identified, although a few with activity against nematodes have been isolated (Gatehouse et al. 2002). Further there is little evidence of effective *Bt* toxins against many of the major storage insect pests.

Bt toxins (also referred to as δ-endotoxins; Cry proteins) exert their pathological effects by forming lytic pores in the cell membrane of the insect gut. On ingestion, they are solubilized and proteolytically cleaved in the midgut to remove the C-terminal region, thus generating an 'activated' 65–70 kDa toxin. The active toxin molecule binds to a specific high-affinity receptor in the insect midgut epithelial cells. Following binding the pore forming domain, consisting of α-helices, inserts into the membrane; this results in cell death by colloid osmotic lysis, followed by death of the insect (de Maagd et al. 2001). A number of putative receptors in the insect gut have been identified and include aminopeptidase N proteins (Knight et al. 1994; Sangadala et al. 1994; Gill et al. 1995; Luo et al. 1997), cadherin-like proteins (Vadlamudi et al. 1995; Nagamatsu et al. 1998; Gahan et al. 2001) and glycolipids (Denolf 1996).

Transgenic plants expressing *Bt* toxins were first reported in 1987 (Vaeck et al. 1987) and following this initial study numerous crop species have been transformed with genes encoding a range of different Cry proteins targeted towards different pests species. Since bacterial *cry* genes (genes encoding *Bt* toxins) are rich in A/T content compared to plant genes, both the full-length and truncated versions of these *cry* genes have had to undergo considerable modification of codon usage and removal of polyadenylation sites before successful expression in plants (de Maagd et al. 1999). Crops expressing *Bt* toxins were first commercialized in the mid-1990s, with the introduction of Bt potato and cotton. Currently 26.3 million hectares of land is planted with *Bt* cotton and maize (James 2010). A summary of the major Bt expressing crops commercialised to date is provided in Table 4.1. Although not as yet commercialised, approval for Bt rice (China) and Bt eggplant (India) is pending (James 2010).

Table 4.1 Commercial crops expressing insecticidal proteins from *Bacillus thuringiensis*

Crop	Trait	Target	Company
Bollgard® cotton	Cry1Ac	Lepidoptera	Monsanto
Attribute® maize	Cry1Ab	Lepidoptera	Syngenta
Widestrike® cotton	Cry1Ac + Cry2Ab2	Lepidoptera	Dow Agrosciences
Bollgard II® cotton	Cry1Ac + Cry2Ab2	Lepidoptera	Monsanto
Insect resistant cotton	Cry1Ac + modified CpTI	Lepidoptera	China
Yieldgard Rootworm® maize	Cry3Bb1	Coleoptera	Monsanto
Agrisure RW® maize	Modified version of Cry3A	Coleoptera	Syngenta
Herculex RW® maize	Cry34Ab1 + Cry35Ab1 + HT	Coleoptera Weeds	Dow Agrosciences
SmartStax maize	Cry1A.105 + Cry2Ab + Cry3Bb1 + Cry1Fa2 + Cry35Ab1 + Cry35Ab1 + CP4 epsps + Pat	Lepidoptera Coleoptera Weeds	Monsanto Dow Agrosciences (approval granted)
Agrisure Viptera trait-stacked corn	Vip3A	Lepidoptera Molds	Syngenta (from 2011)

4.3.2 Bt Maize (corn)

Lepidopteran pests such as European corn borer (*Ostrinia nubilalis*), fall army-worm (*Spodoptera frugiperda*) and corn earworm (*Helicoverpa zea*) perennially cause leaf and ear damage to corn. The *Bt* concept was particularly attractive for maize, since it made it possible to combat European corn borer larvae hidden inside the stem of the plant for the first time. *Bt* maize has now been grown on a large scale for over a decade, particularly in the US. In 2007, insect-resistant *Bt* maize was grown on 21% of the total maize cultivation area, and *Bt* maize with a combination of insect and herbicide resistance was grown on a further 28% (James 2007). Various *Bt* maize varieties are also authorized in the EU. In 2007 there was notable cultivation of *Bt* maize primarily in Spain, where it was grown on around 75,000 ha (Ortego et al. 2010).

Transgenic corn hybrids expressing the insecticidal protein Cry1Ab from *B thuringiensis* (*Bt*) var. *kurstaki* were originally developed to control European corn borer, and offer the potential for reducing losses by fall armyworm and corn earworm. Several events of transgenic *Bt* corn have been developed with different modes of toxin expression (Ostlie et al. 1997; Ostlie 2001). Amongst the most promising events were *Bt11* expressing the *cry1Ab* gene from *B. thuringiensis* subsp. *kurstaki* (Novartis Seeds) and MON810 expressing a truncated form of the *cry1Ab* gene from *B. thuringiensis* subsp. *kurstaki* HD-1 (Monsanto Co.). In both events the endotoxins are expressed in vegetative and reproductive structures throughout the season (Armstrong et al. 1995; Williams and Davis 1997). Crops containing either of these events are collectively referred to as having 'YieldGard Technology'. Furthermore a modified *cry9C* gene from *B. thuringiensis* subsp. *tolworthi* strain BTS02618A is expressed in maize (tradename StarLink) marketed by Aventis CropScience. However, StarLink corn was only approved in the US for livestock feed use. In recent years, there has been increasing focus on another maize pest, this time a coleopteran (beetle), the western corn rootworm. Western corn rootworm is one of the most devastating corn rootworm species in North America. Its larvae are root pests and can destroy significant percentages of corn if left untreated. In the US, current estimates show that 30 million acres (120,000 km) of corn (out of 80 million grown) are infested with corn rootworms. The United States Department of Agriculture (USDA) estimates that corn rootworms cause US $1 billion in lost revenue each year. To compound matters, this pest is continually extending its geographical range, including throughout Europe. *Bt* maize with resistance to the western corn rootworm has been authorized in the US since 2003 and has been grown on a large scale since. YieldGard Rootworm uses event MON 863 and expresses the Cry3Bb1 protein from *B. thuringiensis* (subsp. *kumamotoensis*), protecting the plant against root feeding from both western and northern corn rootworm larvae. Products containing both YieldGard Corn Borer (MON810) and YieldGard Rootworm (MON 863) are marketed under the trade name YieldGard Plus (http://www.agbios.com). Corn rootworm-resistant maize is also produced by expression of the *cry34Ab1* and *cry35Ab1* genes from *B. thuringiensis* strain PS149B1 (DOW AgroSciences LLC and Pioneer Hi-Bred International, Inc.).

4.3.3 *Bt Cotton*

Cotton fibers used in textiles around the world come from the seed hairs of *Gossypium hirsutum*. Cotton develops in closed, green capsules known as bolls that burst open when ripe, revealing the white, fluffy fibers. However, cotton is more than just a fiber for textiles. It is also an important source of raw materials used in animal feed and for various processed food ingredients, including cotton-seed oil, protein-rich cottonseed meal (mostly used as animal feed) and even leftover fibers can be used as food additives.

Lepidopteran, particularly heliothine, pests can have an enormously damaging effect on a cotton crop and controlling these insects in conventional farming involves treatment with a number of insecticide sprays. In 1996, Bollgard® cotton (Monsanto) was the first *Bt* cotton to be marketed in the US. Bollgard cotton produces the Cry1Ac toxin from *B. thuringiensis* (subsp. *kurstaki*), which has excellent activity on tobacco budworm and pink bollworm. These two insects are extremely important as both are difficult and expensive to control with traditional insecticides and the damage caused by them directly impacts on the harvestable plant organ, the cotton bolls themselves. Bollgard II® was introduced in 2003, representing the next generation of *Bt* cottons. Bollgard II contains *Cry1Ac* plus a second gene from *Bt* bacteria, which encodes the production of Cry 2Ab (also subsp. *kurstaki*). WideStrike (a Trademark of DowAgrosciences) was registered for use in 2004, and like Bollgard II, expresses two *Bt* toxins but this time Cry1Ac and Cry1F were used in combination. Both Bollgard II and WideStrike have better activity on a wider range of lepidopteran pests than the original Bollgard technology. *Bt* cotton has become widespread, covering a total of 15 million hectares in 2007, or 43% of the world's cotton. Most Bt cotton is grown in the US and China, but it can also be found in India (with an adoption rate of 86%), South Africa, Australia, Argentina, Mexico, and Columbia (http://www.agbios.com). In 2009, it was estimated that approximately 75% of the cotton grown in China was Bt cotton (He, pers comm.), with approximately 20% expressesing Cry1Ac in combination with a plant protease inhibitor, cowpea trypsin inhibitor (CpTI) (He et al. 2009).

4.3.4   Bt Potato

The potato (*Solanum tuberosum* L.) is a major world food crop. Potato is exceeded only by wheat, rice, and maize in terms of world production for human consumption (Ross 1986). Many commercial potato varieties are highly susceptible to damage by the Colorado potato beetle. In 1999, 93% of the 1.1 million potato acres grown in the US were treated with a total of 2.6 million pounds of insecticide (http://www.usda.gov/). To date few traditionally bred varieties have been produced with resistance to this major pest. Unfortunately many of the pesticides currently used are broad spectrum, killing not only the target pest but most of its natural enemies as well. The Cry3A δ-endotoxin from *B. thuringiensis* Berliner subsp. *tenebrionis* is toxic to coleopterans, particularly chrysomelids (Krieg et al. 1983; Bauer 1990; MacIntosh et al. 1990) including the Colorado potato beetle, *Leptinotarsa decemlineata* (Ferro and Gelernter 1989). In 1995, *Bt*.Cry3A (NewLeaf™) potato became the first *Bt*-crop to be commercialized, although this has currently been withdrawn from the US market.

4.3.5   Evolution of Resistance to Bt Cry Toxins in Pest Populations

Perhaps one of the most important issues surrounding cultivation of *Bt* crops relates to the evolution of target pest resistance, which will limit the life-span of the technology. In the case of *Bt* toxins this is a major concern for the organic farming community, since the potential for insect populations to evolve resistance to *Bt* will not only limit the effectiveness of *Bt*-expressing crops but also *Bt*-based biopesticides. *Bt* resistance in insect pests has been reported to evolve within four to five generations in the laboratory (Stone et al. 1989). To date, the mechanism of resistance to Cry toxins in insects has been most commonly ascribed to the loss or inactivation of specific toxin-binding sites on midgut cell membranes (Ferré and Van Rie 2002). Other resistance mechanisms that have been proposed include a defect in the toxin activation by midgut proteases (Oppert et al. 1994; Sayyed et al. 2001), or an increased repair and/or replacement rate of Cry-damaged midgut cells by stem cells (Forcada et al. 1999). Studies have also revealed evidence for novel resistance mechanisms based on active defensive responses (Rahman et al. 2004; Ma et al. 2005). When one considers the ability of insects to evolve resistance to chemical pesticides (Ffrench-Constant et al. 2004) the development of field resistance was inevitable and was recently reported to have already occurred (Tabashnik and Carrière 2009). Analysis of monitoring data shows that some field populations of *Helicoverpa zea* have evolved resistance to Cry1Ac, the toxin produced by first-generation *Bt* cotton (Tabashnik et al. 2008). Nonetheless, resistance of *H. zea* to Cry1Ac has not caused widespread crop failures in the field for several reasons (Tabashnik et al. 2008). First, the documented resistance is spatially limited. Second, from the outset, insecticide sprays have been used to improve control of *H. zea* on *Bt* cotton because Cry1Ac alone is not sufficiently effective to manage this pest. Finally, transgenic crops expressing stacked traits, for example cotton producing two *Bt* toxins (Cry2Ab and Cry1Ac) is now widely planted, thus control of *H. zea* by Cry2Ab would limit problems associated with resistance to Cry1Ac (Jackson et al. 2004). Considerable effort has been devoted to delaying the onset of evolution of resistance, e.g. the use of refugia has been required/recommended in most regions growing *Bt*-crops depending upon the country in question (Tabashnik and Carrière 2009). Gene-stacking and integrated pest management should be combined to help control this problem.

4.4 Environmental Impact of Insect Resistant Crops

Almost from the beginning of the production of transgenic crops there have been concerns over their use and introduction into the environment. There is international agreement that GM crops should be evaluated for their safety, including their environmental impact (Dale 2002). During the past 15–20 years, there have been extensive research programmes of risk assessment, with several areas of major concern identified.

4.4.1 Impact on Non-target Organisms

Assessing the consequences of pest control strategies on non-target organisms is an important precursor to their becoming adopted in agriculture. The expression of transgenes that confer enhanced levels of resistance to insect pests is of particular significance since it is aimed at manipulating the biology of organisms in a different trophic level to that of the plant. Potential risks to beneficial non-target arthropods exist. Those groups most at risk include: non-target Lepidoptera, beneficial insects (pollinators, natural enemies) and soil organisms.

Exposure of non-target Lepidoptera to insecticidal transgene products may occur through direct consumption of transgenic plant tissues including via consumption of transgenic pollen; many non-target Lepidoptera are rare butterflies having great conservation value. The case of the Monarch butterfly (*Danaus plexippus*), a conservation flagship species in the US, highlighted the need for ecological impact research. In a letter to Nature, Losey et al. (1999) claimed that both survival and consumption rates of Monarch larvae fed milkweed leaves (natural host) dusted with *Bt* pollen were significantly reduced, and that this would have profound implications for the conservation of this species. However, a series of ecologically based studies rigorously evaluated the impact of pollen from such crops on Monarchs and demonstrated that the commercial wide-scale growing of *Bt*-maize did not pose a significant risk to the Monarch population (Hellmich et al. 2001; Gatehouse et al. 2002). In fact, the initial experiments did not quantify the dose of pollen used, or indeed, if this was a realistic level likely to be encountered in the field, nevertheless this work highlighted the importance of studying non-target effects. In a separate field study Wraight et al. (2000) showed that *Papilio polyxenes* (black swallowtail) larvae were unaffected by pollen from *Bt* expressing maize event MON810 at distances up to 7 meters from the transgenic field edge, highlighting the need for a case-by-case study of organisms considered to be at risk. In addition to the potential direct impacts of *Bt* toxins on susceptible target insects, as in the case of the Morarch butterfly, some Lepidoptera have been shown to have a reduced sensitivity to the lepidopteran specific *Bt* toxins. For example, *Spodoptera littoralis* can survive on maize expressing Cry1Ab (Hilbeck et al. 1998) and thus present a route of exposure to the next trophic level. In the case of *Bt* Cry3Aa or Cry3Bb expressing potatoes or maize, some Lepidoptera may represent non-target secondary pests, and whilst not directly affected by the transgene product themselves, may again present a route of exposure to the next trophic level, as do other non-target herbivores. Organisms such as those belonging to the orders Homoptera, Hemiptera, Thysanoptera, and Tetranychidae are not targeted by *Bt* toxins expressed in transgenic plants, however they do utilize the *Bt* crop (Groot and Dicke 2002). The direct effect that this may have on these insects is dependent on the presence of *Bt* receptors in the first instance, and it is so far unclear whether such receptors are present in many non-target organisms (de Maagd et al. 2001). Furthermore, the fate of the toxin ingested by non-target herbivores is often unclear, since if it retains toxicity then this may have implications at the next trophic level.

The impacts of insect-resistant transgenic crops at higher trophic levels have also been considered, where there are concerns over the risks to beneficial arthropod biodiversity (Schuler et al. 1999; Bell et al. 2001). Of particular interest are the effects of such crops on predators and parasitoids, which play an important role in suppressing insect pest populations both in the field and under specialized cultivation systems (glasshouses). Natural enemies may ingest transgene products via feeding on herbivorous insects that have themselves ingested the toxin from the plant; such tritrophic interactions will be influenced by the susceptibility of the herbivore to the plant protection product. If, as in the case with *Bt* toxins, the prey item is susceptible to the toxin, then the predator would not come into contact with the toxin as the pest will effectively be controlled; the exception to this would be scavengers such as carabid beetles, which may well be exposed. In target insects the toxin is bound to receptors in the midgut epithelium that are structurally re-arranged and may lose their entomotoxicity (de Maagd et al. 2001), thus again reducing the potential for subsequent exposure, although it would be questionable as to whether all of the toxin would be bound. In non-target insects (and resistant insects) the toxins do not bind and may thus retain biological activity. However, the overwhelming weight of evidence from independent laboratory and field studies show that *Bt* toxins have a limited ability to affect the next trophic level (reviewed in Sanvido et al. 2007; Romeis et al. 2008; Ferry and Gatehouse 2010; Gatehouse et al. 2011).

Pollinators represent another group of important non-target organisms highlighted as at risk from *Bt* toxins in GM crops. The current generation of transgenic crops produce *Bt* toxin in the pollen as well as in the vegetative tissues. Several studies have been conducted to determine toxicity of *Bt* toxins to pollinators (Vandenberg 1990; Sims 1995, 1997; Arpaia 1997; Malone and Pham-Delegue 2001); generally they all conclude that neither the adults nor the larvae of bees were affected by *Bt* toxins. For a comprehensive review of the impact of transgenic crops on pollinators, the reader is referred to two recent reviews (Malone et al. 2008; Malone and Burgess 2009).

Finally non-target species may come into contact with *Bt* toxins via the environment. Several studies have shown that *Bt* toxins released from transgenic plants bind to soil particles (Palm et al. 1996; Crecchio and Stotzky 1998; Saxena et al. 1999). Soil-dwelling and epigeic insects such as Collembola and Carabidae may thus be exposed to the toxins. However several studies (Saxena and Stotzky 2001; Ferry et al. 2007) show no differences in mortality or body mass of bacteria, fungi, protozoa, nematodes and earthworms or carabid beetles exposed to *Bt*.

Exposure to the transgene products, however, does not necessarily imply a negative impact. Most studies to date have demonstrated that crops transformed for enhanced pest resistance have no deleterious effects on beneficial insects (reviewed in Ferry et al. 2003; Romeis et al. 2008; Ferry and Gatehouse 2010). Ultimately one must consider the impact of transgenic crops, and specifically *Bt* toxins, in comparison to other pest control strategies such as conventional crop protection using synthetic insecticides. While pesticides have undoubtedly brought vast yield improvements they have well documented undesirable non-target effects (Devine and Furlong 2007). It is worth remembering that whilst potential risks do exist to the environment from the cultivation of GM crops, their potential to decrease reliance on external inputs (less insecticide sprays) and to increase the availability of genetic resources to breeders, is great (Ferry et al. 2006).

4.5   Plant Defence Proteins

The concept of employing genes encoding *Bt* toxins to produce insect-resistant
transgenic plants arises from the successful use of *Bt*-based biopesticides. However,
a number of other strategies for protecting crops from insect pests actually exploit
endogenous resistance mechanisms (Harborne 1998; Gatehouse 2002). Genes
encoding constitutively expressed defence proteins are thus obvious candidates
for enhancing crop resistance to insect pests.

4.5.1   Enzyme Inhibitors

Interfering with digestion, and thus affecting the nutritional status of the insect, is
a strategy widely employed by plants for defense, and has been extensively
investigated as a means of producing insect resistant crops (Gatehouse 2002). Insect
digestive proteases fall into four mechanistic classes (serine, cysteine, aspartic or
metallo proteases - depending on the enzyme active site residue). Numerous studies
since the 1970s have confirmed the insecticidal properties of a broad range of protease
inhibitors from both plant and animal sources (Jouanin et al. 1998; Gatehouse 2002).
Proof of concept for exploiting such molecules for crop protection was first
demonstrated with expression of a serine protease inhibitor from cowpea (CpTI),
which was shown to significantly reduce insect growth and survival (Hilder et al.
1987). These studies were subsequently extended to include a greater range of target
pests including economically important pest species, particularly lepidopterans
(Gatehouse et al. 1994; Graham et al. 1995; Xu et al. 1996; Broadway 1997; De Leo
et al. 2001), and a broader range of inhibitors and plant species. Since, many
economically important coleopteran pests predominantly utilize cysteine proteases
for protein digestion, inhibitors for this class of enzyme (cystatins) have also been
investigated as a means for controlling pests from this order. Oryzacystatin, a cysteine
protease inhibitor isolated from rice seeds, is effective towards both coleopteran
insects and nematodes when expressed in transgenic plants (Leple et al. 1995;
Urwin et al. 1995; Pannetier et al. 1997). Similarly, the cysteine/aspartic protease
inhibitor equistatin, from sea anemone, is also toxic to several economically important
coleopteran pests, including the Colorado potato beetle (Outchkourov et al. 2003).
More recent studies have included the stacking of different families of inhibitors to
increase the spectrum of activity (Abdeen et al. 2005).

A major limitation, however, to this strategy for control of insect pests arises
from the ability of some lepidopteran and coleopteran species to respond and adapt
to ingestion of protease inhibitors by either overexpressing native gut proteases, or
producing novel proteases that are insensitive to inhibition (Bown et al. 1997;
Jongsma and Bolter 1997). Thus detailed knowledge about the enzyme-inhibitor
interactions, both at the molecular and biochemical levels, together with detailed
knowledge on the response of insects to exposure to such proteins is essential to

effectively exploit this strategy. The concept of inhibiting protein digestion as a means of controlling insect pests has been extended to inhibition of carbohydrate digestion. For example, inhibitors of α-amylase have been expressed in transgenic plants and shown to confer resistance to bruchid beetles (Shade et al. 1994; Schroeder et al. 1995).

4.5.2 Lectins

Lectins, found throughout the plant and animal kingdoms, form a large and diverse group of proteins identified by a common property of specific binding to carbohydrate residues, either as free sugars, or more commonly, as part of oligo- or polysaccharides. Many physiological roles have been attributed to plant lectins, including defense against pests and pathogens (Chrispeels and Raikhel 1991; Peumans and Vandamme 1995). Although some lectins are toxic to mammals, and are thus not suitable candidates for transfer to crops for enhanced levels of protection, this is by no means universal. Many lectins are not toxic to mammals, yet are effective against insects from several different orders (Gatehouse et al. 1995), including homopteran pests such as hoppers and aphids (Powell et al. 1995; Sauvion et al. 1996; Gatehouse et al. 1997). This finding has generated significant interest, not least since no *Bts* effective against this pest order have been identified to date. One such lectin is the snowdrop lectin (*Galanthus nivalis* agglutinin; GNA). Both constitutive and phloem specific (Rss1 promoter) expression of GNA in rice is an effective means of significantly reducing survival of rice brown plant hopper (*Nilaparvata lugens*), and green leafhopper (*Nephotettix virescens*), both serious economic pests of rice (Rao et al. 1998; Foissac et al. 2000; Tinjuangjun et al. 2000). GNA has been expressed in combination with other genes encoding insecticidal proteins, including the *cry* genes (Maqbool et al. 2001). Although lectins such as GNA, and ConA are not as effective against aphids as they are against hoppers, they nonetheless have significant effects on aphid fecundity when expressed in potato (Down et al. 1996; Gatehouse et al. 1997; Gatehouse and Gatehouse 1999) and wheat (Stoger et al. 1999).

The precise mode of action of lectins in insects is not fully understood although binding to gut epithelial cells appears to be a prerequisite for toxicity. In the case of rice brown planthopper, GNA not only binds to the luminal surface of the midgut epithelial cells, but also accumulates in the fat bodies, ovarioles and throughout the haemolymph, demonstrating that the lectin is able to cross the midgut epithelial barrier and pass into the insect's circulatory system, resulting in a systemic toxic effect (Powell et al. 1998; Du et al. 2000). More recently GNA has been used as an effective carrier for transporting peptide hormones and arthropod toxins across the insect midgut epithelial barrier for control of target insect pests, when expressed as a fusion protein (Fitches et al. 2004, 2010).

As with protease inhibitors, the levels of protection conferred by expression of lectins in transgenic plants are generally not sufficient to be considered commercially viable. However, the absence of genes with proven high insecticidal activity against homopteran pests may well mean that transgenic crops with partial resistance may still find acceptance in agriculture, especially if expressed with other genes that confer partial resistance, or if introduced into partially resistant genetic backgrounds.

## 4.6	Other Sources of Insecticidal Molecules

Generating insect-resistant transgenic crops harbouring genes from novel sources is an extremely active area, with amongst others, foreign genes from plants (e.g., enzymes inhibitors and novel lectins), animal sources including insects (e.g., biotin-binding proteins, neurohormones, venoms and enzyme inhibitors) and other microorganisms (in addition to *B. thuringiensis*) being a major focus (Ferry et al. 2006).

The development of second-generation transgenic plants with greater durable resistance has included the expression of multiple insecticidal genes such as the Vip (vegetative insecticidal proteins) produced by *B. thuringiensis* during its vegetative growth. The benefit of such an approach is a broader insect target range than conventional *Bt* proteins and the proposed expectation to control current *Bt* resistant pests due to the low levels of homology between the domains of the two proteins classes (Christou et al. 2006).With *Bt* toxins as the classical reference, toxins from other insect pathogens provide a potential repository of novel insecticidal compounds. *Photorhabdus* spp. are bacterial symbionts of entomopathogenic nematodes, which are lethal to a wide range of insects (Chattopadhyay et al. 2004; Ffrench Constant et al. 2007). *Photorhabdus* toxin expression in *Arabidopsis* caused significant insect mortality (see for review Ferry et al. 2006). Thus toxins from other insect pathogens are also opening up new routes to pest control using transgenic based strategies. Similar protection has also been achieved with insect peptide hormones (Tortiglione et al. 2003). Interestingly they replaced the tomato systemin peptide region of prosystemin (a plant signaling molecule) with the insect peptide and showed that this resulted in the production of biologically active insecticidal peptides.

Ultimately reliance on the expression of a single gene product for pest control is a relatively short-term strategy that parallels the use of exogenously applied chemical pesticides. Thus to obtain durable levels of resistance to insect pests in the field, a multimechanistic approach is required.

## 4.7	The Future

The deployment of a limited number of transgenes conferring insect resistance has proven to be highly successful in many crop species over recent years. Ultimately, this technology will be limited by the evolution of resistance in pest populations to these insecticidal molecules. There is an expectation that a systematic functional analysis of genes, their expression and interaction, commonly referred to as 'functional genomics', will reveal novel genes and gene control sequences conferring more complex traits involved in biotic stress tolerance. As a foundation for functional genomics, genome sequencing has been completed for several model plants, significantly the dicot model *Arabidopsis thaliana* and the monocot rice which is both a model and target (Upadhyaya et al. 2010).

Mapping and DNA sequencing of plant genomes and analysis of the information present in genomic sequences, commonly termed as 'genomics', has the potential to provide valuable insight into genes controlling complex traits. Such knowledge could be used effectively, not only in molecular marker-assisted breeding, but also in transgenic breeding strategies. This will extend well beyond *Arabidopsis*, and rice; genome sequencing efforts are underway/completed for several other food grain and tuber crops (barley, cassava, maize, mungbean, potato, sorghum and wheat), vegetable crops (tomato, cabbage and field mustard), fruit crops (grape, papaya, orange and apple), oil crops (rapeseed, Indian mustard, black mustard, soybean and castor), forage crops (barrel medic), biofuel crops (jatropha, miscanthus, switchgrass, pine, madhuca, arundo and pongamia) and other commercial crops such as tobacco and cotton (Upadhyaya et al. 2010).

This chapter will now focus on recent technological advances that have revealed detailed genetic information on crop plants and contributed significantly to our understanding of plant-insect interactions.

To fight the bug, we must understand the bug. Starship Troopers (film) 1997.

4.8 Functional Genomics

As proposed by Hieter and Boguski (1997), genomics can be broadly classified into two disciplines: 'structural genomics' and 'functional genomics'. Structural genomics corresponds to the initial phase of genome analysis resulting ultimately in the definition of the complete DNA sequence of an organism, while functional genomics makes use of the genome sequence to assess, on a large-scale, the functions of genes as well as their expression and interaction.

4.8.1 Gene Expression Studies at the Transcriptome Level

Although there could be more than 50,000 genes in a plant genome, not all of these are transcribed into RNA at any given time, in any given tissue or under any given environmental condition. Even some of the transcribed RNAs are suppressed, broken down or rendered nontranslatable. The characterization of all the transcribed genes, referred to as the 'transcriptome', is normally attempted by collecting large numbers of ESTs from diverse cDNA libraries. There are more than 17 million plant ESTs in the public database (http://www.ncbi.nlm.nih.gov/dbEST/). Genome-wide expression profiling (including differential expression) of genes in various crop species is being facilitated by high-throughput techniques, such as microarrays and recently ultra-deep sequencing (e.g., 454, Solexa and SOliD technologies). These procedures are typically used to compare two mRNA populations derived from tissues of different developmental stages or those subjected to different environmental stimuli, to yield information on the comparative changes in gene expression in each tissue. As such they provide valuable tools to better understand the response of plants to insect attack.

Studies by Baldwin and his group on the interaction between insect herbivores and tobacco (*Nicotiana attenuata*) have provided new insights into the molecular bases of plant defence. They estimate that approximately 500 mRNAs constitute the insect-responsive transcriptome in tobacco (Hermsmeier et al. 2001). However, many of these genes are of unknown function, and many changes in gene expression do not represent induction of defence-related proteins. Photosynthetic genes, for example, are down-regulated in tobacco plants in response to insect attack. Further microarray analysis (Hui et al. 2003) has demonstrated putative up-regulation of defence-associated and down-regulation of growth-associated transcripts. This analysis provided evidence for the simultaneous activation of salicylic acid, ethylene, cytokinin and jasmonic acid-regulated pathways during herbivore attack. Similar co-activation of numerous signalling cascades in response to various stresses has been found in *Arabidopsis* (Chen et al. 2002) and supports the idea of a network of interacting signal cascades. Microarray analysis also identified direct defensive responses in terms of dramatic increases in proteinase inhibitor (PI) transcripts, and increases in transcripts encoding putrescine *N*-methyl transferase (which catalyses the first committed step of biosynthesis of nicotine), as well as metabolic commitment to terpenoid based indirect defences.

In addition to work with chewing pests, interactions between plants and sap-sucking homopterans such as aphids are equally complex and various studies have shown that extensive gene reprogramming can occur when homopteran pests invade plants (Moran and Thompson 2001; Zhang et al. 2004; De Vos et al. 2005; Kaloshian and Walling 2005; Yuan et al. 2005; Wei et al. 2009). Moran and Thompson (2001) demonstrated that phloem feeding by the green peach aphid (*Myzus persicae*) on *Arabidopsis thaliana* induced expression of genes associated with salicylic acid (SA) responses to pathogens, as well as a gene involved in the jasmonic acid (JA) mediated response pathway. These results suggest stimulation of response pathways involved in both pathogen and herbivore responses. Microarray and macroarray data has identified genes involved in oxidative stress, calcium-dependent signalling, pathogenesis-related responses, and signalling as key components of the induced response (Moran et al. 2002). Similarly, Zhang *et al.* (2004) demonstrated that upon attack by a piercing-sucking insect, brown planthopper (*Nilaparvata lugens*), rice genes that were strongly regulated were grouped within the categories of signalling pathways, oxidative stress/apoptosis, wound-response, drought-inducible and pathogen-related proteins and also that unlike chewing pests, aphids have been shown to elicit responses in plants similar to those induced by pathogen attack.

De Vos et al. (2005) show a surprisingly complex set of transcriptional alterations in *A. thaliana* in response to aphid and pathogen attack. Analysis of global gene expression profiles revealed consistent changes induced by both pathogens and insects with attacker specific responses but considerable overlap. It may be that transgenic strategies that activate such signalling cascades could enhance plant resistance to these problematic pests. However, many changes in gene expression do not necessarily represent induction of defence-related proteins.

> Similar co-activation of numerous signalling cascades in response to various stresses has been found in *Arabidopsis* (Chen et al. 2002) and supports the idea of a network of interacting signal cascades. Microarray analysis also identified direct defensive responses in terms of dramatic increases in proteinase inhibitor (PI) transcripts, and increases in transcripts encoding putrescine *N*-methyl transferase (which catalyses the first committed step of biosynthesis of nicotine), as well as metabolic commitment to terpenoid based indirect defences

4.8.2 Gene Expression Studies at the Proteome Level

With recent advances in high-resolution two-dimensional polyacrylamide gel electrophoresis (2D-PAGE), staining, detection, peptide micro-sequencing and associated computer software, 'proteomics' is also emerging as a powerful functional genomics tool. Here, instead of looking at gene expression, an assessment is made on the gene product, i.e., the protein. With 2D-PAGE, intact proteins are separated and protein abundance is determined by the relative stain intensities of protein spots on the gel. The differential proteome is confirmed by image analysis. The identity of a specific protein is generally determined by mass-spectrometric (MS) analysis of peptides after proteolysis of the protein spot or by protein sequencing after blotting the gel to a membrane (Upadhyaya et al. 2010; Ferry et al. 2011). Comparison of the amino acid sequences of fragments of proteins with those predicted from DNA sequences greatly facilitates not only the validation of gene predictions but also provides insight into the cellular and developmental regulation of gene expression. Several public databases of 2D-PAGE-derived plant proteins are already available, such as WORLD-2DPAGE (http://expasy. org/ch2d/2d-index.html), Rice Membrane Protein Library (http://www.cbs.edu/ rice/) and the Rice Proteome Database (http://gene64.dna.affrc.go.jp/RPD/). In addition, progress is being made in detecting post-translational modifications such as glycosylation, lipid attachment, phosphorylation, methylation, disulfide bond formation and proteolytic cleavage.

There have been significantly fewer studies conducted on the plant proteome in response to insect feeding in comparison to studies at the transcriptome level. Recently a few papers have emerged (Giri et al. 2006; Lippert et al. 2007;

Wei et al. 2009; Maserti et al. 2011) using a wide range of methods to describe plant responses to various herbivores. The Case Study presented below describes the first attempt to study a major crop (wheat) response to a homopteran pest using differential proteomics.

4.8.3   A Case Study – The Aphid Responsive Proteome of Wheat

In a single study on wheat, Ferry et al. (2011) found that major re-programming of the proteome occurred in wheat leaves following feeding by the cereal aphid *Sitobion avenae*. The study was carried out to identify gene products involved in defence against insect herbivores, for use in directed strategies for wheat breeding. Although the wheat genome had not been completely sequenced at the time of publication, over 50% of the protein spots excised from gel could be putatively identified as showing significant similarity to known wheat genes using wheat EST databases combined with BLAST searches, or to similar genes in other cereals/plants. Of the proteins showing reproducible, statistically significant changes in expression, the majority could be grouped into functional classes on the basis of their similarity to plant genes of known function, as discussed below.

4.8.3.1   Looking for Aphid Resistance Genes

Resistance genes involved in gene-gene interactions with hemipteran insects have previously been identified in cereals (Smith et al. 1991; Teetes et al. 1999). Zhu-Salzman et al. 2004 identified an LRR (leucine-rich repeat)-containing glyco-protein sequence that is differentially expressed in leaves of sorghum, *Sorghum bicolour* (L.) infested by the greenbug *Schizaphis graminum*. Similarly Boyko et al. 2006 reported that a *Pto*-like serine/threonine kinase gene and a *Pti1* (*Pto* interactor) 1-like kinase gene are both up-regulated in resistant wheat plants infested by Russian wheat aphid, *Diuraphis noxia*. Homologues of these proteins were not observed. However, one protein with a direct role in insect defence was identified as down-regulated following aphid feeding, Hfr-2, a wheat cytolytic toxin-like gene with regions of similarity to agglutinins (Puthoff et al. 2005). The interactions of wheat with the Hessian fly (*Mayetiola destructor*) have been studied in some detail. This interaction is highly specialized, and involves a gene-for-gene interaction between pest and plant host (Sharma et al. 1997). This interaction is atypical of plant-insect interactions in general, and genes conferring resistance to Hessian fly are generally not effective against other pests. Hessian fly infestation results in the up-regulation of a range of genes. These include genes encoding xylanase inhibitors (Igawa et al. 2005), and cytolytic proteins (Puthoff et al. 2005) as well as other genes typical of responses to pathogens and/or stresses. Up-regulation of genes in the response to Hessian fly shows marked specificity, even within gene families. Some genes are only up-regulated by the specific plant-insect interaction, whereas others are up-regulated by a wide range of biotic and abiotic stresses, including wounding (Jang et al. 2005). For example, a gene designated *Hfr-1* responds specifically to Hessian fly feeding, whereas a similar gene, *Wci-1,* is generally responsive to stresses (Subramanyam et al. 2006). There is also a distinction between different insect pests; *Hfr-1* is not up-regulated by aphid feeding in wheat (Subramanyam et al. 2006), whereas the related Hessian fly-responsive gene *Hfr-2* is up-regulated by feeding of bird cherry-oat aphid (*Rhopalosiphum padi*) and methyl jasmonate treatment (Puthoff et al. 2005). This is in strong agreement with the work of Ferry et al. (2011) where no up-regulation was found in response to *S. avenae*, indicating a very insect specific response. Such proteins may represent targets for transgenic improvement of wheat lines.

As well as no observed induction of specific responses in wheat to the insect pest (cereal aphid) there was also little evidence for a significant wounding response; the change in the proteome on aphid infestation did not resemble that produced by treatment with methyl jasmonate, known to elicit a wound response (Gatehouse 2002). Although wheat recognises aphid-inflicted plant damage, in the case study reported here a more general stress response was triggered, analogous to the basal plant defence to phytopathogens (Smith and Boyko 2007). Changes in plant metabolism and gene expression induced by arthropod feeding have been shown to be multifaceted and to include a general stress response (Walling 2000; Moran and Thompson 2001). This suggests that transgenic enhancement of the wheat wound response may also increase insect resistance.

The aphid response observed in wheat was divided into several major categories, as described below:

4.8.3.2 Metabolic Re-Programming; Changes in Metabolism and Photosynthesis

Plants respond to herbivory by increased transcription of many "house-keeping" genes (Karban and Baldwin 1997) involved in photosynthesis, photorespiration, protein synthesis, and maintenance of cell homeostasis. Some of these genes may be involved in addressing changing source-sink relationships, such as those induced in plants affected by the removal of phloem during aphid feeding (Smith and Boyko 2007). For example, glyceraldehyde-3-phosphate dehydrogenase was initially observed to be down-regulated following aphid feeding in the Case Study, but was observed to be up-regulated at later time points, suggesting an initial metabolic drain following phloem removal, which may be compensated for subsequently. Several enzymes involved in the Calvin cycle, electron transport and ATP synthesis were observed to be differentially regulated suggesting a greater metabolic requirement in the plant following aphid infestation. Similar proteins associated with changes in metabolic processes also presented altered expression patterns in protein profiling experiments in both rice and barley following insect or pathogen infestation respectively (Wei et al. 2009; Geddes et al. 2008). Proteins directly associated with photosynthesis were also differentially regulated in wheat following cereal aphid infestation, including a key enzyme of the Calvin cycle, ribulose bisphosphate carboxylase (RuBisCO). Up-regulation in expression of photosynthetic or photorespiration genes have been observed in *M. persicae* feeding on leaves of celery, *D. noxia* feeding on wheat foliage, and *M. nicotianae* feeding on *N. attenuata* foliage, whereas photosynthesis genes were down-regulated after feeding by *M. nicotianae* or *S. graminum* (Smith and Boyko 2007. These changes may reflect the reallocation of plant metabolites from normal growth processes to defensive functions, after induction of plant responses by aphid feeding.

4.8.3.3   Changes in Protein Degradation

Plants rely on protein turnover by the ubiquitin/proteasome system (UPS) to respond to biotic and abiotic stresses (Pickart and Eddins 2004; Dreher and Callis 2007). Components of the UPS were observed to be altered in response to aphid feeding in the Case Study presented. This is in broad agreement with other studies showing that regulated proteolysis of endogenous proteins can contribute to multiple levels of plant defence (Feng et al. 2003).

4.8.3.4   Changes in Transcription and Translation

Changes in protein factors regulating transcription and translation are an essential part of plant responses to changed conditions. PIF4 (phytochrome-interacting factor 4) transcription factor was up-regulated in wheat by aphid feeding (Ferry et al. 2011). Recent studies suggest that such transcription factors play a central role in the crosstalk between environmental cues and hormone signalling (Lucyshyn and Wigge 2009; Koini et al. 2009; Casson et al. 2009). However, the molecular mechanisms of PIF function as transcription factors, and the identity of their targets have not yet been elucidated. Transcription factors like PIF4 can form flexible interaction networks (Toledo-Ortiz et al. 2003), and authors have suggested that particular combinations of PIFs may act to integrate environmental information, and pass on signals by specific interactions with other transcription factors, leading to positive or negative regulation in a finely tuned manner (Lucyshyn and Wigge 2009). The manipulation of transcription factors for specific activation of plant defenses should be viewed as a goal in transgenic crop research.

4.8.3.5   Signal Transduction

The proteomic analysis presented in the Case Study identified a number of potential signal transduction components as being differentially regulated by aphid infestation. Two proteins identified as up-regulated by aphid infestation showed similarity to a putative calmodulin binding protein, and a putative MAP4 kinase. Both were also up-regulated following salicylic acid treatment suggesting overlap between signalling responses to pathogen attack and aphid feeding. Similar up-regulation of a calcium binding protein after salicylic acid treatment was observed in barley (Glinwood et al. 2007).

Calcium is important in initiating oxidative stress cascades, such as that observed when *A. thaliana* leaves are infested with the aphid *M. persicae* (Bolwell and Wojtaszek 1997; Moran et al. 2002). The activity of calcium is exerted through binding to calmodulin, which in turn binds to calmodulin-binding proteins to continue the signalling pathway. Calmodulin-binding proteins were also identified as differentially regulated gene products associated with the water-stress response in rice via an EST approach (Gorantla et al. 2007).

Mitogen activated protein kinase (MAPK) pathways are signal transduction modules with layers of protein kinases having *c.* 120 genes in *Arabidopsis*, but only a few have been linked experimentally to functions (Menges et al. 2008). The mitogen-activated protein (MAP) kinase cascade is one of the well-characterized intracellular signalling modules, and it is highly conserved among eukaryotes (Hirt 1997; Kultz 1998). This phosphorylation cascade typically consists of three functionally interlinked protein kinases: a MAP kinase kinase kinase (MAPKKK), a MAP kinase kinase (MAPKK), and a MAP kinase (MAPK). In this phosphorylation module, a MAPKKK phosphorylates and activates a particular MAPKK, which in turn phosphorylates and activates a MAPK. MAPK is often immediately transported into the nucleus, where it phosphorylates and activates specific downstream signalling components such as transcription factors (Khokhlatchev et al. 1998). Most plant MAPKs are associated with the subgroup of extracellular signal-regulated kinases based on phylogeny (Kultz 1998). The mitogen-activated protein (MAP) kinase cascade is one of the well-characterized eukaryotic intracellular signalling modules (Hirt 1997; Kültz 1998) and plays an important role in plant responses to biotic and abiotic stresses. Activation of MAPKs have been observed in plants exposed to pathogens (Suzuki and Shinshi 1995; Ligterink et al. 1997; Zhang and Klessig 1998), cold (Jonak et al. 2002), salinity (Mikołajczyk et al. 2000), drought (Jonak et al. 2002), and wounding (Seo et al. 1995, 1999). Plant MAPKs can also be activated by fungal elicitors ((Suzuki and Shinshi 1995), salicylic acid (Zhang and Klessig 1997), jasmonic acid (Seo et al. 1999), and abscisic acid (Heimovaara-Dijkstra et al. 2000). The results presented by Ferry et al. (2011) show that this protein is highly responsive to aphid feeding, suggesting involvement in the plant response.

Recent studies have identified a number of components that may be involved in the reactive oxygen signal transduction of plants. These include the MAPKs, and calmodulin. In a hypothetical model depicting some of the players involved in this pathway (Fig. 4.2), a sensor that might be a two component Histidine-kinase, or a receptor-like protein kinase, is thought to sense H_2O_2. Calmodulin and a MAPK cascade are then activated resulting in the induction/activation/suppression of a number of transcription factors. These regulate the response of plants to oxidative stress. Cross-talk with the pathogen-response signal transduction pathway (gene-for-gene) also occurs and may involve interactions between different

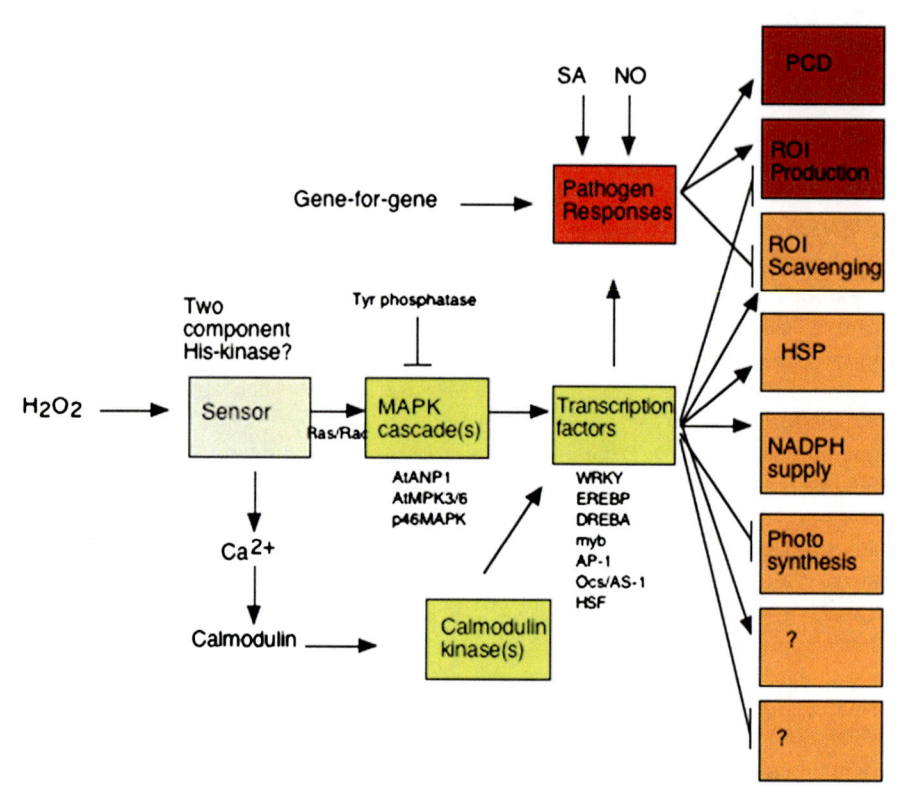

Fig. 4.2 A hypothetical model of the signaling pathway activated in plants in response to stress. *SA* salicylic acid, *NO* nitric oxide, *PCD* programmed cell death, *HSP* heat shock protein, *MAPK* mitogen-activated protein kinase, *Tyr* tyrosine, *HSF* heat shock transcription factor (Source: cas.unt.edu/~rmittler/oxistress.pdf)

MAPK pathways, feedback loops, and the action of salicylic acid and nitric oxide (Smith and Boyko 2007). The number of antioxidant proteins identified in the study by Ferry et al. (2011) (> 10% of differentially regulated proteins) further support the hypothesis that oxidative stress is induced following the initiation of aphid feeding.

In general, the results presented in the Case Study suggest that salicylic acid responses have greater similarity to aphid-induced responses in wheat than the herbivore defence signalling pathway in plants involving jasmonic acid. Interestingly, a protein was identified as having similarity to a cytochrome P450 74A (allene oxide synthase). In plants, an allene oxide is a precursor of jasmonic acid (Song et al. 1993), thus the observed down-regulation of allene oxide synthase observed following aphid feeding may represent a suppression of jasmonate-mediated responses. This may go some way to explain the lack of observation of any proteins associated with a classical wounding response, suggesting that wheat may lack/have a weak wounding response in response to aphids, again possibly suggesting a potential target for future control methods.

4.8.3.6   Antioxidants and Other Stress Responses

The oxidative burst, a rapid production of reactive oxygen species (ROS), is a well-documented early plant response to biotic stress (Apel and Hirt 2004). One of the consequences of ROS activity is oxidative damage of membrane integrity due to lipid peroxidation processes that may result in the generation of highly cytotoxic compounds. In order to maintain homeostasis under stress conditions, plants need to fortify the resistance mechanisms, such as ROS scavenging and cell defence. The number of antioxidant proteins identified in the Case Study ($>10\%$ of differentially regulated proteins) supports the hypothesis that oxidative stress is induced following the initiation of aphid feeding. Both catalase and a protein with greatest similarity to a GST from *Arabidopsis,* which are typical antioxidants, were strongly up-regulated by aphid infestation and were not induced by treatment with signalling molecules. Enhancement of GST expression has become a marker for plant response to stress and the activation of antioxidant networks has previously been described in both dicots and cereals in response to homopteran insect feeding (Walling 2000; DeVos et al. 2005; Wei et al. 2009). At present it is unclear as to whether enhancement of this initial response would be beneficial in protecting crops against insects. ROS need to be kept in fine balance to prevent direct damage to plant tissues and few successful attempts to enhance these mechanisms have proved successful. However manipulation of ROS generation has provided enhanced drought tolerance in *Arabidopsis* (Chu et al. 2010).

Data presented in the Case Study above provide no evidence for a popular commercial wheat line being able to provide direct resistance to the cereal aphid *Sitobium avenae*, thus suggesting that a transgenic approach may be needed to improve many of our elite crop lines.

4.9   Metabolomics

Metabolomics is the comprehensive analysis of low-molecular-weight compounds in biological samples and is emerging as a biochemical phenotyping tool along with transcriptomics and proteomics in functional genomics (Tarpley and Roessner 2007). Technologies used in metabolomics are normally based on the chromatographic separation of complex compound mixtures, using either liquid or gas chromatography and mass-spectrometric detection. Nuclear magnetic resonance (NMR) spectroscopy is also playing a major role in metabolomic approaches. Fourier-transform ion cyclotron mass spectrometry (FT-ICR-MS) can mass-resolve metabolites with a mass accuracy of <1 ppm (Dunn et al. 2005), and thus provides a high-throughput method for metabolite fingerprinting (Upadhyaya et al. 2010).

A metabolomics approach is useful in discovering unexpected bioactive compounds involved in ecological interactions between plants and their herbivores. Plants possess a wealth of structural and chemical mechanisms to defend themselves against a wide range of attackers (Bennett and Wallsgrove 1994;

Hanley et al. 2007). It is estimated that plants may produce 200,000 metabolites, of which only ca. 10% has been identified to date (Bino et al. 2004). A plant metabolomic study by Jansen et al. (2009) expertly demonstrates the potential of this technology.

In recent years, large-scale analyses of gene sequences (genomics), expression (transcriptomics), protein patterns (proteomics), metabolite profiles (metabolomics) and lipid compounds (lipidomics) have been performed and remain the focus of many studies. The tremendous amount of accumulated data has led to new insights into how plants respond to biotic or abiotic stress and how a growth and developmental programme takes place in the plant life cycle. Ultimately crop resistance may be improved using either transgenic or molecular breeding approaches. Studies profiling crop responses to pests will provide new leads for crop improvement.

4.10 Summary

To return to the title of this chapter, if we are to stay one step ahead of the insects and move away from reliance on single resistance genes locking us into a perpetual co-evolutionary arms race, then multi-mechanistic resistance must be breed/engineered into our crops. The pressures on food production resulting from human population growth dictates that crop productivity must increase. The authors see transgenic crops as an essential and inevitable technology towards ensuring sustainable food security. However, this should not be viewed as a 'stand alone' technology but as part of IPM programmes where molecular breeding strategies are also used to develop pest resistant crops. As discussed above, detailed knowledge of how plants respond to biotic stress, and in particular to insect attack, at the molecular levels should enable us to identify suitable genes/suites of genes or markers (QTLs) for subsequent breeding programmes. If such an approach is taken, we may finally see high levels of insect control with low/minimal environmental impacts.

References

Abdeen A, Virgos A, Olivella E, Villanueva J, Aviles X, Gabarra R, Prat S (2005) Multiple insect resistance in transgenic tomato plants over-expressing two families of plant proteinase inhibitors. Plant Mol Biol 57:189–202

Amman K (2010) Biodiversity and genetically modified crops. In: Ferry N, Gatehouse AMR (eds) Environmental impact of genetically modified/novel crops. CAB Int, Oxford, pp 240–265

Ananieva KI, Ananiev ED (1999) Effect of methyl ester of jasmonic acid and benzylaminopurine on growth and protein profile of excised cotyledons of Cucurbita pepo L. (zucchini). Biol Plant 42:549–557

Apel K, Hirt H (2004) Reactive oxygen species: metabolism, oxidative stress, and signal transduction. Annu Rev Plant Biol 55:373–399

Arimura G, Tashiro K, Kuhara S, Nishioka T, Ozawa R, Takabayashi J (2000) Gene responses in bean leaves induced by herbivory and by herbivore-induced volatiles. Biochem Biophys Res Commun 277:305–310

Armstrong CL, Parker GB, Pershing JC, Brown SM, Sanders PR, Duncan DR, Stone T, Dean DA, DeBoer DL, Hart J, Howe AR, Morrish FM, Pajeau ME, Petersen WL, Reich BJ, Rodriguez R, Santino CG, Sato SJ, Schuler W, Sims SR, Stehling S, Tarochione LJ, Fromm ME (1995) Field evaluation of European corn borer control in progeny of 173 transgenic corn evens expressing an insecticidal protein from *Bacillus thuringiensis*. Crop Sci 35:550–557

Arpaia S (1997) Ecological impact of Bt-transgenic plants: 1. Assessing possible effects of CryIIIB toxin on honey bee (*Apis mellifera*) colonies. J Genet Breed 50:315–319

Baldwin IT, Halitschke R, Kessler A, Schittko U (2001) Merging molecular and ecological approaches in plant-insect interactions. Curr Opin Plant Biol 4(4):351–358

Bauer LS (1990) Response of the cottonwood leaf beetle (Coleoptera: Chrysomelidae) to *Bacillus thuringiensis* var. *San Diego*. Environ Entomol 19:428–431

Bell HA, Fitches EC, Marris GC, Bell J, Edwards JP, Gatehouse JA, Gatehouse AMR (2001) Transgenic crop enhances beneficial biocontrol agent performance. Transgen Res 10:35–42

Bennett RN, Wallsgrove RM (1994) Secondary metabolites in plant defense-mechanisms. New Phytol 127:617–633

Bino RJ, Hall RD, Fiehn O, Kopka J, Saito K, Draper J et al (2004) Potential of metabolomics as a functional genomics tool. Trends Plant Sci 9:418–425

Bolwell GP, Wojtaszek P (1997) Mechanisms for the generation of reactive oxygen species in plant defence – a broad perspective. Physiol Mol Plant Pathol 51:347–366

Bown DP, Wilkinson HS, Gatehouse JA (1997) Differentially regulated inhibitor-sensitive and insensitive protease genes from the phytophagous insect pest, *Helicoverpa armigara*, are members of complex multigene families. Insect Biochem Mol Biol 27:625–638

Boyko EV, Smith CM, Thara VK, Bruno JM, Deng Y, Starkey SR, Klaahsen DL (2006) Molecular basis of plant gene expression during aphid invasion: wheat Pto- and Pti-like sequences are involved in interactions between wheat and Russian wheat aphid (Homoptera: Aphididae). J Econ Entomol 99:1430–1445

Broadway RM (1997) Dietary regulation of serine proteinases that are resistant to serine proteinase inhibitors. J Insect Physiol 43(9):855–874

Brooks EM, Hines ER (1999) Viral biopesticides for heliothine control- fact or fiction. Today's Life Sci: 38–44

Brotman Y, Silberstein L, Kovalski I, Perin C, Dogimont C, Pitrat M, Klingler J, Thompson GA, Perl-Treves R (2002) Resistance gene homologues in melon are linked to genetic loci conferring disease and pest resistance. Theor Appl Genet 104:1055–1063

Carson R (1962) Silent spring. Houghton Mifflin, Boston

Casson SA, Franklin KA, Gray JE, Grierson CS, Whitelam GC, Hetherington AM (2009) Phytochrome B and PIF4 regulate stomatal development in response to light quantity. Curr Biol 19:229–234

Chattopadhyay A, Bhatnagar NB, Bhatnagar R (2004) Bacterial insecticidal toxins. Crit Rev Microbiol 30:33–54

Chen WQ, Provart NJ, Glazebrook J, Katagiri F, Chang HS, Eulgem T, Mauch F, Luan S, Zou GZ, Whitham SA, Budworth PR, Tao Y, Xie ZY, Chen X, Lam S, Kreps JA, Harper JF, Si-Ammour A, Mauch-Mani B, Heinlein M, Kobayashi K, Hohn T, Dangl JL, Wang X, Zhu T (2002) Expression profile matrix of Arabidopsis transcription factor genes suggests their putative functions in response to environmental stresses. Plant Cell 14:559–574

Chrispeels MJ, Raikhel NV (1991) Lectins, lectin genes, and their role in plant defense. Plant Cell 3(1):1–9

Chrispeels M, Sadava D (2003) Plants, genes and crop biotechnology. ASPB/Jones and Bartlett Publ, Boston

Christou P, Capell T, Kohli A, Gatehouse JA, Gatehouse AMR (2006) Recent developments and future prospects in insect pest control in transgenic crops. Trends Plant Sci 11:302–308

Chu SH, Noh H, Kim S, Kim KH, Hong SW, Lee H (2010) Enhanced drought tolerance in *Arabidopsis* via genetic manipulation aimed at the reduction of glucosamine-induced ROS generation. Plant Mol Biol 74:493–502

Crecchio C, Stotzky G (1998) Insecticidal activity and biodegradation of the toxin from *Bacillus thuringiensis* subspecies *kurstaki* bound to humic acids from soil. Soil Biol Biochem 30:463–470

Dale PJ (2002) The environmental impact of genetically modified (GM) crops: a review. J Agric Sci 138:245–248

De Leo F, Bonade-Bottino M, Ceci LR, Gallerani R, Jouanin L (2001) Effects of a mustard trypsin inhibitor expressed in different plants on three lepidopteran pests. Insect Biochem Mol Biol 31:593–602

de Maagd RA, Bosch D, Stiekema W (1999) *Bacillus thuringiensis* toxin-mediated insect resistance in plants. Trends Plant Sci 4:9–13

de Maagd RA, Bravo A, Crickmore N (2001) How *Bacillus thuringiensis* has evolved specific toxins to colonize the insect world. Trends Genet 17:193–1999

De Vos M, Van Oosten VR, Van Poecke RMP, Van Pelt JA, Pozo MJ, Mueller MJ, Buchala AJ, Metraux JP, Van Loon LC, Dicke M, Pieterse CMJ (2005) Signal signature and transcriptome changes of Arabidopsis during pathogen and insect attack. Mol Plant-Microbe Interact 18:923–937

Degenhardt J, Gershenzon J, Baldwin IT, Kessler A (2003) Attracting friends to feast on foes: engineering terpene emission to make crop plants more attractive to herbivore enemies. Curr Opin Biotechnol 14:169–176

Delaney TP, Uknes S, Vernooij B, Friedrich L, Weymann K, Negrotto D, Gaffney T, Gut-Rella M, Kessmann H, Ward E, Ryals J (1994) A central role of salicylic acid in plant disease resistance. Science 266:1247–1250

Denolf P (1996) Isolation, cloning and characterisation of *Bacillus thuringiensis* delta-endotoxin receptors in Lepidoptera. PhD, University of Gent, Belgium

Devine GJ, Furlong MJ (2007) Insecticide use: contexts and ecological consequences. Agric Hum Values 24:281–306

Dicke M, Agrawal AA, Bruin J (2003) Plants talk, but are they deaf? Trends Plant Sci 8(9):403–405

Down RE, Gatehouse AMR, Hamilton WDO, Gatehouse JA (1996) Snowdrop lectin inhibits development and decreases fecundity of the glasshouse potato aphid (*Aulacorthum solani*) when administered *in vitro* and via transgenic plants both in laboratory and glasshouse trials. J Insect Physiol 42:1035–1045

Dreher KA, Callis J (2007) Ubiquitin, hormones and biotic stress in plants. Ann Bot 9:787–822

Du JP, Foissac X, Carss A, Gatehouse AMR, Gatehouse JA (2000) Ferritin acts as the most abundant binding protein for snowdrop lectin in the midgut of rice brown planthoppers (Nilaparvata lugens). Insect Biochem Mol Biol 30:297–305

Dunn WB, Bailey NJ, Johnson HE (2005) Measuring the metabolome: current analytical technologies. Analyst 130:606–625

Feng W, Shi Y, Li M, Zhang M (2003) Tandem PDZ repeats in glutamate receptor-interacting proteins have a novel mode of PDZ domain-mediated target binding. Nat Struct Biol 10:972–978

Ferré J, Van Rie J (2002) Biochemistry and genetics of insect resistance to *Bacillus thuringiensis*. Annu Rev Entomol 47:501–533

Ferro DN, Gerlernter WD (1989) Toxicity of a new strain of *Bacillus thuringiensis* to Colorado potato beetle (Coleoptera: Chrysomelidae). J Econ Entomol 82:750–755

Ferry N, Gatehouse AMR (2010) Transgenic crop plants for resistance to biotic stress. In: Kole C, Michler CH, Abbott AG, Hall TC (eds) Transgen crop plants: vol 2. Utilization and biosafety. Spinger, Berlin/Heidelberg

Ferry N, Edwards MG, Mulligan EA, Emami K, Petrova A, Frantescu M, Davison GM, Gatehouse AMR (2003) Engineering resistance to insect pests. In: Christou P, Klee H (eds) Handbook of plant biotechnology. Wiley, New York, pp 373–394

Ferry N, Edwards MG, Gatehouse JA, Capell T, Christou P, Gatehouse AMR (2006) Transgenic plants for insect pest control: a forward looking scientific perspective. Transgen Res 15:13–19

Ferry N, Mulligan EA, Majerus MEN, Gatehouse AMR (2007) Bitrophic and tritrophic effects of Bt Cry3A transgenic potato on beneficial, non-target, beetles. Transgen Res 16:795–812

Ferry N, Stavroulakis S, Guan W, Davison GM, Gatehouse JA, Gatehouse AMR (2011) Molecular interactions between wheat and the insect herbivore *Sitobion avenae (cereal aphid)*; analysis of changes to the wheat proteome. Proteomics 11:1985–2002

Feyereisen R (1995) Molecular biology of insecticide resistance. Toxicol Lett 82(83):83–90

Ffrench-Constant RH, Daborn PJ, Le Goff G (2004) The genetics and genomics of insecticide resistance. Trends Genet 20:163–170

Ffrench-Constant RH, Dowling A, Waterfiled NR (2007) Insecticidal toxins from *Photorhabdus* bacteria and their potential use in agriculture. Toxicon 49:436–451

Fitches E, Edwards MJ, Mee C, Grishan E, Gatehouse AMR, Edwards JP, Gatehouse JA (2004) Fusion proteins containing neurotoxins as insect control agents: snowdrop lectin delivers fused insecticidal spider venom neurotoxin to insect haemolymph following oral ingestion. J Insect Physiol 50:61–71

Fitches EC, Bell HA, Powell ME, Back E, Sargiotti C, Weaver RJ, Gatehouse JA (2010) Insecticidal activity of scorpion toxin (ButaIT) and snowdrop lectin (GNA) containing fusion proteins towards pest species of different orders. Pest Manage Sci 66:74–83

Fletcher, MR, Hunter K, Barnett EA, Sharp EA (2000) Pesticide poisoning of animals 1998: Investigations of suspected incidents in the United Kingdom. A Report of the Environment Panel of the Advisory Committee on Pesticides. London, UK, MAFF, (PB4786), 54 p

Flor HH (1955) Host-parasite interaction in flax rust – its genetics and other implications. Phytopathology 45:680–685

Foissac X, Loc NT, Christou P, Gatehouse AMR, Gatehouse JA (2000) Resistance to green leafhopper (*Nephotettix virescens*) and brown planthopper (*Nilaparvata lugens*) in transgenic rice expressing snowdrop lectin (*Galanthus nivalis agglutinin*; GNA). J Insect Physiol 46:573–583

Forcada C, Alcacer E, Garcera MD, Tato A, Martinez R (1999) Resistance to *Bacillus thuringiensis* Cry1Ac toxin in three strains of *Heliothis virescens*: proteolytic and SEM study of the larval midgut. Arch Insect Biochem Physiol 42:51–63

Frey M, Stettner C, Pare PW, Schmelz EA, Tumlinson JH, Gierl A (2000) An herbivore elicitor activates the gene for indole emission in maize. Proc Natl Acad Sci USA 97:14801–14806

Gahan LJ, Gould F, Heckel DG (2001) Identification of a gene associated with Bt resistance in *Heliothis virescens*. Science 293:857–860

Gatehouse JA (2002) Plant resistance towards insect herbivores: a dynamic interaction. Tansley review No 140. New Phytol 156:145–169

Gatehouse JA, Gatehouse AMR (1999) Genetic engineering of plants for insect resistance. In: Rechcigl JE, Reichcigl NA (eds) Biological and biotechnological control of insect pests. CRC Press LLC, Boca Raton, pp 211–241

Gatehouse AMR, Hilder VA, Powell KS, Wang M, Davison GM, Gatehouse LN, Down RE, Edmonds HS, Boulter D, Newell CA, Merryweather A, Hamilton WDO, Gatehouse JA (1994) Insect-resistant transgenic plants – choosing the gene to do the job. Biochem Soc Trans 22:944–949

Gatehouse AMR, Powell K, Peumans W, Damme EV, Gatehouse JA (1995) Insecticidal properties of plant lectins: their potential in plant protection. In: Pusztai A, Bardocz S (eds) Lectins biomedical perspectives. Taylor & Francis, London, pp 35–57

Gatehouse AMR, Davison GM, Newell CA, Merryweather A, Hamilton WDO, Burgess EPJ, Gilbert RJC, Gatehouse JA (1997) Transgenic potato plants with enhanced resistance to the tomato moth, *Lacanobia oleracea*: growth room trials. Mol Breed 3:49–63

Gatehouse AMR, Ferry N, Raemaekers RJM (2002) The case of the Monarch butterfly: a verdict is returned. Trends Genet 18:249–251

Gatehouse AMR, Ferry N, Edwards MG, Bell HA (2011) Insect-resistant biotech crops and their impacts on beneficial arthropods. Philos Trans R Soc B 366:1438–1452

Geddes J, Eudes F, Laroche A, Selinger LB (2008) Differential expression of proteins in response to the interaction between the pathogen *Fusarium graminearum* and its host, *Hordeum vulgare*. Proteomics 8:545–554

Gill SS, Cowles EA, Francis V (1995) Identification, isolation, and cloning of a *Bacillus thuringiensis* CryIAc toxin-binding protein from the midgut of the Lepidopteran insect *Heliothis virescens*. J Biol Chem 270:27277–27282

Giri AP, Wünsche H, Mitra S, Zavala JA, Muck A, Svatoš A, Baldwin IT (2006) Molecular interactions between the specialist herbivore *Manduca sexta* (Lepidoptera, Sphingidae) and its natural host *Nicotiana attenuata*. VII. Changes in the plant's proteome. Plant Physiol 142:1621–1641

Glinwood RT, Gradin T, Karpinska B, Ahmed E, Jonsson LMV, Ninkovic V (2007) Aphid acceptance of barley exposed to volatile phytochemicals differs between plants exposed in daylight and darkness. Plant Signal Behav 2:205–210

Gorantla M, Babu PR, Lachagari VB, Reddy AM, Wusirika R, Bennetzen JL, Reddy AR (2007) Identification of stress-responsive genes in an *indica* rice (*Oryza sativa* L.) using ESTs generated from drought-stressed seedlings. J Exp Bot 58:253–265

Graham J, McNicol RJ, Greig K (1995) Towards genetic based insect resistance in strawberry using the Cowpea trypsin inhibitor gene. Ann Appl Biol 127:163–173

Groot AT, Dicke M (2002) Insect-resistant transgenic plants in a multi-trophic context. Plant J 31:387–406

Gurr GM, Kvedaras OL (2010) Synergizing biological control: scope for sterile insect technique, induced plant defences and cultural techniques to enhance natural enemy impact. Biol Control 52:198–207

Hanley ME, Lamont BB, Fairbanks MM, Rafferty CM (2007) Plant structural traits and their role in anti-herbivore defence. Perspect Plant Ecol Evol Syst 8:157–178

Harborne J (1998) Introduction to ecological chemistry. Academic, London

He KJ, Wang ZY, Zhang YJ (2009) Monitoring Bt resistance in the field: China as a case study. In: Ferry N, Gatehouse AMR (eds) Environmental impact of genetically modified/novel crops. CAB Int, Oxford, pp 344–360

Heimovaara-Dijkstra S, Testerink C, Wang M (2000) Mitogen-activated protein kinase and abscisic acid signal transduction. Results Probl Cell Differ 27:131–144

Hellmich RL, Siegfried BD, Sears MK, Stanley-Horn DE, Daniels MJ, Mattila HR, Spencer T (2001) Monarch larvae sensitivity to *Bacillus thuringiensis*-purified proteins and pollen. Proc Natl Acad Sci USA 98:11925–11930

Hermsmeier D, Schittko U, Baldwin IT (2001) Molecular interactions between the specialist herbivore *Manduca sexta* (Lepidoptera, Sphingidae) and its natural host *Nicotiana attenuata*. I. Large-scale changes in the accumulation of growth- and defence-related plant mRNAs. Plant Physiol 125:683–700

Hieter P, Boguski M (1997) Functional genomics: it's all how you read it. Science 278:601–602

Hilbeck A, Baumgartner M, Fried PM, Bigler F (1998) Effects of transgenic *Bacillus thuringiensis* corn-fed prey on mortality and development of immature *Chrysoperla carnea* (Neuroptera: Chrysopidae). Environ Entomol 27:480–487

Hilder VA, Gatehouse AMR, Sheerman SE, Barker RF, Boulter D (1987) A novel mechanism of insect resistance engineered into tobacco. Nature 330:160–163

Hilker M, Meiners T (2002) Induction of plant responses to oviposition and feeding by herbivorous arthropods: a comparison. Entomol Exp Appl 104:181–192

Hirt H (1997) Multiple roles of MAP kinases in plant signal transduction. Trends Plant Sci 2:11–15

Hui DQ, Iqbal J, Lehmann K, Gase K, Saluz HP, Baldwin IT (2003) Molecular interactions between the specialist herbivore Manduca sexta (Lepidoptera, Sphingidae) and its natural host Nicotiana attenuata: V. Microarray analysis and further characterization of large-scale changes in herbivore-induced mRNAs. Plant Physiol 131:1877–1893

Igawa T, Tokai T, Kudo T, Yamaguchi I et al (2005) A wheat xylanase inhibitor gene, Xip-I, but not Taxi-I, is significantly induced by biotic and abiotic signals that trigger plant defense. Biosci Biotechnol Biochem 69:1058–1063

Jackson RE, Bradley JR Jr, Van Duyn JW (2004) Performance of feral and Cry1Ac-selected *Helicoverpa zea* (Lepidoptera: Noctuidae) strains on transgenic cottons expressing either one or two *Bacillus thuringiensis* ssp. *kurstaki* proteins under greenhouse conditions. J Entomol Sci 39:46–55

James C (2007) Global status of commercialized biotech/GM crops. ISAAA briefs 37. ISAAA, Ithaca

James C (2010) Global status of commercialized biotech/GM crops: 2010. ISAAA brief no. 42. ISAAA, Ithaca

Jang CS, Johnson JW, Seo YW (2005) Differential expression of TaLTP3 and TaCOMT1 induced by Hessian fly larval infestation in a wheat line possessing H21 resistance gene. Plant Sci 168:1319–1326

Jansen JJ, Allwood JW, Marsden-Edwards E, van der Putten WH, Goodacre R, van Dam NM (2009) Metabolomic analysis of the interaction between plants and herbivores. Metabolomics 5:150–161

Jonak C, Okresz L, Bögre L, Hirt H (2002) Complexity, cross talk and integration of plant MAP kinase signaling. Curr Opin Plant Biol 5:415–424

Jongsma MA, Bolter C (1997) The adaptation of insects to plant protease inhibitors. J Insect Physiol 43:885–895

Jouanin L, Bonade-Bottino M, Girard C, Morrot G, Giband M (1998) Transgenic plants for insect resistance. Plant Sci 131:1–11

Kaloshian I, Walling LL (2005) Hemipterans as plant pathogens. Annu Rev Plant Biol 43:491–521

Karban R, Baldwin IT (1997) Induced responses to herbivory. The University of Chicago Press, Chicago

Kessler A, Baldwin IT (2002) Plant responses to insect herbivory: the emerging molecular analysis. Annu Rev Plant Biol 53:299–328

Khokhlatchev AV, Canagarajah B, Wilsbacher J, Robinson M, Atkinson M, Goldsmith E, Cobb MH (1998) Phosphorylation of the MAP kinase ERK2 promotes its homodimerization and nuclear translocation. Cell 93:605–615

Knight PJK, Crickmore N, Ellar DJ (1994) The receptor for *Bacillus thuringiensis* Cryla(C) delta-endotoxin in the brush-border membrane of the Lepidopteran *Manduca sexta* is aminopepti-dase-N. Mol Microbiol 11:429–436

Knoester M, van Loon LC, van den Heuvel J, Hennig J, BolJF LHJM (1998) Ethylene-insensitive tobacco lacks nonhost resistance against soil-borne fungi. Proc Natl Acad Sci USA 95:1933–1937

Koini MA, Alvey L, Allen T, Tilley CA, Harberd NP, Whitelam GC, Franklin KA (2009) High temperature-mediated adaptations in plant architecture require the phytochrome-interacting bHLH factor PIF4. Curr Biol 19:408–413

Krieg A, Hugner AM, Lagenbruch GA, Schnetter W (1983) *Bacillus thuringiensis* var. *tenebrionis:* Ein neuer gegenuber Larven von Coleopteran wirksamer Pathotyp. Z Angew Entomol 96:500–508

Kültz D (1998) Phylogenetic and functional classification of mitogen- and stress-activated protein kinases. J Mol Evol 46:571–588

Leple JC, Bonadebottino M, Augustin S, Pilate G, Letan VD, Delplanque A, Cornu D, Jouanin L (1995) Toxicity to *Chrysomela tremulae* (Coleoptera, Chrysomelidae) of transgenic poplars expressing a cysteine proteinase-inhibitor. Mol Breed 1:319–328

Li XC, Berenbaum MR, Schuler MA (2000) Molecular cloning and expression of CYP6B8: a xanthotoxin-inducible cytochrome P450 cDNA from *Helicoverpa zea*. Insect Biochem Mol Biol 30:75–84

Li XC, Schuler MA, Berenbaum MR (2002) Jasmonate and salicylate induce expression of herbivore cytochrome P450 genes. Nature 419(6908):712–715

Ligterink W, Kroj T, zur Nieden U, Hirt H, Scheel D (1997) Receptor-mediated activation of a MAP kinase in pathogen defense of plants. Science 276:2054–2057

Lippert D, Chowrira S, Ralph SG, Zhuang J, Aeschliman D, Ritland C, Ritland K, Bohlmann J (2007) Conifer defense against insects: proteome analysis of Sitka spruce (*Picea sitchensis*) bark induced by mechanical wounding or feeding by white pine weevils (*Pissodes strobi*). Proteomics 7:248–270

Losey JE, Rayor LS, Carter ME (1999) Transgenic pollen harms monarch larvae. Nature 399:214–214

Lucyshyn D, Wigge PA (2009) Plant development: PIF4 integrates diverse environmental signals. Curr Biol 19:265–266

Luo K, Sangadala S, Masson L, Mazza A, Brousseau R, Adang MJ (1997) The *Heliothis virescens* 170 kDa aminopeptidase functions as "receptor A" by mediating specific *Bacillus thuringiensis* Cry1A delta-endotoxin binding and pore formation. Insect Biochem Mol Biol 27:735–743

Ma G, Roberts H, Sarjan M, Featherstone N, Lahnstein J, Akhurst R, Schmidt O, Ma G, Roberts H, Sarjan M, Featherstone N, Lahnstein J, Akhurst R, Schmidt O (2005) Is the mature endotoxin Cry1Ac from *Bacillus thuringiensis* inactivated by a coagulation reaction in the gut lumen of resistant *Helicoverpa armigera* larvae? Insect Biochem Mol Biol 35:729–739

MacIntosh SC, Stone TB, Sims SR, Hunst PL, Green-plate JT, Marrone PG, Perlak FJ, Fischhoff DA, Fuchs RL (1990) Specificity and efficacy of purified *Bacillus thuringiensis* proteins against agronomically important insects. J Invertebr Pathol 56:258–266

Malone LA, Burgess EPJ (2009) Impact of genetically modified crops on pollinators. In: Ferry N, Gatehouse AMR (eds) Environmental impact of genetically modified/novel crops. CAB Int, Oxford, pp 199–225

Malone LA, Pham-Delegue M-H (2001) Effects of transgene products on honey bees (*Apis mellifera*) and bumblebees (*Bombus* sp.). Adidologie 32:287–304

Malone LA, Gatehouse AMR, Barratt BIP (2008) Beyond Bt: alternative strategies for insect-resistant crops. In: Romeis J, Shelton T, Kennedy G (eds) Integration of insect-resistant genetically modified crops within integrated pest management programs. Series on progress in biological control. Springer, Berlin

Maqbool SB, Riazuddin S, Loc NT, Gatehouse AMR, Gatehouse JA, Christou P (2001) Expression of multiple insecticidal genes confers broad resistance against a range of different rice pests. Mol Breed 7:85–93

Maserti BE, Del Carratore R, Della Croce CM, Podda A, Migheli Q, Froelicher Y, Luro F, Morillon R, Ollitrault P, Talon M, Rossignol M (2011) Comparative analysis of proteome changes induced by the two spotted spider mite *Tetranychus urticae* and methyl jasmonate in citrus leaves. J Plant Physiol 168:392–402

Menges M, Doczi R, Okresz L, Morandini P, Mizzi L, Soloviev M, Murray JAH, Bogre L (2008) Comprehensive gene expression atlas for the Arabidopsis MAP kinase signaling pathways. New Phytol 179:643–666

Mikołajczyk M, Awotunde OS, Muszyńska G, Klessig DF, Dobrowolska G (2000) Osmotic stress induces rapid activation of a salicylic acid-induced protein kinase and a homolog of protein kinase ASK1 in tobacco cells. Plant Cell 121:165–178

Milligan SB, Bodeau J, Yaghoobi J, Kaloshian I, Zabel P, Williamson VM (1998) The root knot nematode resistance gene Mi from tomato is a member of the leucine zipper, nucleotide binding, leucine-rich repeat family of plant genes. Plant Cell 10:1307–1319

Moran PJ, Thompson GA (2001) Molecular responses to aphid feeding in Arabidopsis in relation to plant defence pathways. Plant Physiol 125:1074–1085

Moran PJ, Cheng Y, Cassell JL, Thompson GA (2002) Gene expression profiling of *Arabidopsis thaliana* in compatible plant-aphid interactions. Arch Insect Biochem Physiol 51:182–203

Nagamatsu Y, Toda S, Koike T, Miyoshi Y, Shigematsu S, Kogure M (1998) Cloning, sequencing, and expression of the *Bombyx mori* receptor for *Bacillus thuringiensis* insecticidal CryIA(a) toxin. Biosci Biotechnol Biochem 62:727–734

Nauen R, Ebbinghaus-Kintscher U, Elbert A, Jeschke P, Tietjen K (2001) Acetylcholine receptors as sites for developing neonicotinoid insecticides. In: Ishaaya I (ed) Biochemical sites important in insecticide action and resistance. Springer, Berlin/Heidelberg, pp 77–105

Nicholson GM (2007) Fighting the global pest problem: preface to the special Toxicon issue on insecticidal toxins and their potential for insect pest control. Toxicon 49:413–422

Novotny V, Basset Y, Miller SE, Weiblen GD, Bremer B, Cizek L, Drozd P (2002) Low host specificity of herbivorous insects in a tropical forest. Nature 416:841–844

Oppert B, Kramer KJ, Johnson DE, MacIntosh SC, McGaughey WH (1994) Altered protoxin activation by midgut enzymes from a *Bacillus thuringiensis* resistant strain of *Plodia interpunctella*. Biochem Biophys Res Commun 198:940–947

Ortego F, Pons X, Albajes R, Castañera P (2010) European commercial GM plantings and field trials. In: Ferry N, Gatehouse AMR (eds) Environmental impact of genetically modified/novel crops. CAB Int, Oxford, pp 327–344

Ostlie KR, Hutchison WD, Hellmich RL (1997) Bt Corn and European Corn Borer (NCR Publ 602). Univ of Minnesota, St. Paul, MN, USA

Ostlie K (2001) Crafting crop resistance to corn rootworms. Nat Biotechnol 19:624–625

Outchkourov NS, Rogelj B, Strukelj B, Jongsma MA (2003) Expression of sea anemone equistatin in potato. Effects of plant proteases on heterologous protein production. Plant Physiol 133:379–390

Palm CJ, Schaller DL, Donegan KK, Seidler RJ (1996) Persistence in soil of transgenic plant produced *Bacillus thuringiensis* var. *kurstaki* δ-endotoxin. Can J Microbiol 42:1258–1262

Pannetier C, Giband M, Couzi P, LeTan V, Mazier M, Tourneur J, Hau B (1997) Introduction of new traits into cotton through genetic engineering: insect resistance as example. Euphytica 96:163–166

Peumans WJ, Vandamme EJM (1995) Lectins as plant defense proteins. Plant Physiol 109:347–352

Phipps RH, Park JR (2002) Environmental benefits of genetically modified crops: global and European perspectives on their ability to reduce pesticide use. J Anim Feed Sci 11:1–8

Pickart CM, Eddins MJ (2004) Ubiquitin: structures, functions, mechanisms. Biochim Biophys Acta 1695:55–72

Pickett JA, Wadhams LJ, Woodcock CM (1997) Developing sustainable pest control from chemical ecology. Agr Ecosyst Environ 64:149–156

Powell KS, Gatehouse AMR, Hilder VA, Gatehouse JA (1995) Antifeedant effects of plant-lectins and an enzyme on the adult stage of the rice brown planthopper, *Nilaparvata lugens*. Entomol Exp Appl 75:51–59

Powell KS, Spence J, Bharathi M, Gatehouse JA, Gatehouse AMR (1998) Immuno histo chemical and developmental studies to elucidate the mechanism of action of the snowdrop lectin on the rice brown planthopper, *Nilaparvata lugens* (Stal). J Insect Physiol 44:529–539

Puterka GJ, Peters DC (1989) Inheritance of greenbug, *Schizaphis graminum* (Rondani), virulence to Gb2 and Gb3 resistance genes in wheat. Genome 32:109–114

Puthoff DP, Sardesai N, Subramanyam S, Nemacheck JA et al (2005) Hfr-2, a wheat cytolytic toxin-like gene, is up-regulated by virulent Hessian fly larval feeding. Mol Plant Pathol 6:411–423

Rahman MM, Roberts HL, Sarjan M, Asgari S, Schmidt O (2004) Induction and transmission of *Bacillus thuringiensis* tolerance in the flour moth *Ephestia kuehniella*. Proc Natl Acad Sci USA 101:2696–2699

Rao KV, Rathore KS, Hodges TK, Fu X, Stoger E, Sudhakar D, Williams S, Christou P, Bharathi M, Bown DP, Powell KS, Spence J, Gatehouse AMR, Gatehouse JA (1998) Expression of snowdrop lectin (GNA) in transgenic rice plants confers resistance to rice brown planthopper. Plant J 15:469–477

Raymond-Delpech V, Matsuda K, Sattelle BM, Rauh JJ, Sattelle DB (2005) Ion channels: molecular targets of neuroactive insecticides. Invert Neurosci 5:119–133

Reymond P, Farmer EE (1998) Jasmonate and salicylate as global signals for defense gene expression. Curr Opin Plant Biol 1:404–411

Roberts PA (1995) Conceptual and practical aspects of variability in root-knot nematodes related to host-plant resistance. Annu Rev Phytopathol 33:199–221

Rojo E, Solano R, Sanchez-Serrano JJ (2003) Interactions between signaling compounds involved in plant defence. J Plant Growth Regul 22:82–98

Romeis J, Bartsch D, BiglerF CMP, Gielkens MMC, Hartley SE, Hellmich RL, Huesing JE, Jepson PC, Layton R, Quemada H, Raybould A, Rose RI, Schiemann J, Sears MK, Shelton AM, Sweet J, Vaituzis Z, Wolt JD (2008) Assessment of risk of insect-resistant transgenic crops to nontarget arthropods. Nat Biotechnol 26:203–208

Ross H (1986) Potato breeding – problems and perspectives: advances in plant breeding, J Plant Breed (Suppl 13). Paul Parey, Berlin/Hamburg, 132 p

Rossi M, Goggin FL, Milligan SB, Kaloshian I, Ullman DE, Williamson VM (1998) The nematode resistance gene Mi of tomato confers resistance against the potato aphid. Proc Natl Acad Sci USA 95:9750–9754

Sangadala S, Walters FS, English L, Adang MJA (1994) Mixture of *Manduca sexta* aminopeptidase and phosphatase enhances *Bacillus thuringiensis* insecticidal Cryia(C) toxin binding and (Rb + −K+)-Rb-86 efflux in vitro. J Biol Chem 269:10088–10092

Sanvido O, Romeis J, Bigler F (2007) Ecological impacts of genetically modified crops: ten years of field research and commercial cultivation. Green Gene Technol: Res Area Soc Confl 107:235–278

Sardesai N, Rajyashri KR, Behura SK, Nair S, Mohan M (2001) Genetic, physiological and molecular interactions of rice and its major Dipteran pest, gall midge. Plant Cell Tissue Organ 64:115–131

Sauvion N, Rahbe Y, Peumans WJ, Van Damme EJM, Gatehouse JA, Gatehouse AMR (1996) Effects of GNA and other mannose binding lectins on development and fecundity of the peach-potato aphid *Myzus persicae*. Entomol Exp Appl 79:285–293

Saxena D, Stotzky G (2001) *Bacillus thuringiensis* (*Bt*) toxin released from root exudates and biomass of *Bt* corn has no apparent effect on earthworms, nematodes, protozoa, bacteria, and fungi in soil. Soil Biol Biochem 33:1225–1230

Saxena D, Flores S, Stotzky G (1999) Insecticidal toxin in root exudates from *Bacillus thuringiensis* corn. Nature 402:480

Sayyed A, Gatsi R, Kouskoura T, Wright DJ, Crickmore N (2001) Susceptibility of a field-derived, *Bacillus thuringiensis*-resistant strain of diamondback moth to in vitro-activated Cry1Ac toxin. Appl Environ Microbiol 67:4372–4373

Schenk PM, Kazan K, Wilson I, Anderson JP, Richmond T, Somerville SC, Manners JM (2000) Coordinated plant defence responses in Arabidopsis revealed by microarray analysis. Proc Natl Acad Sci USA 97:11655–11660

Schroeder HE, Gollasch S, Moore A, Tabe LM, Craig S, Hardie DC, Chrispeels MJ, Spencer D, Higgins TJV (1995) Bean alpha-amylase inhibitor confers resistance to the pea weevil (*Bruchus pisorum*) in transgenic peas (*Pisum sativum* L). Plant Physiol 107:1233–1239

Schuler TH, Potting RPJ, Denholm I, Poppy GM (1999) Parasitoid behaviour and *Bacillus thuringiensis* plants. Nature 400:825–826

Seo S, Okamoto M, Seto H, Ishizuka K, Sano H, Ohashi Y (1995) Tobacco MAP kinase: a possible mediator in wound signal transduction pathways. Science 270:1988–1992

Seo S, Sano H, Ohashi Y (1999) Jasmonate-based wound signal transduction requires activation of WIPK, a tobacco mitogen-activated protein kinase. Plant Cell 11:289–298

Shade RE, Schroeder HE, Pueyo JJ, Tabe LM, Murdock LL, Higgins TJV, Chrispeels MJ (1994) Transgenic pea seeds expressing the alpha-amylase inhibitor of the common bean are resistant to bruchid beetles. Biotechnology 12:793–796

Sharma HC, Ohm HW, Patterson FL, Benlhabib O et al (1997) Genetics of resistance to Hessian fly (*Mayetiola destructor*) [Diptera: Cecidomyiidae] biotype L in diploid wheats. Phytoprotection 78:61–65

Shen BZ, Zheng ZW, Dooner HK (2000) A maize sesquiterpene cyclase gene induced by insect herbivory and volicitin: Characterization of wild-type and mutant alleles. Proc Natl Acad Sci USA 97:14807–14812

Sims SR (1995) *Bacillus thuringiensis* var. *kurstaki* [Cry1A(c)] protein expressed in transgenic cotton: effects on beneficial and other non-target insects. Southwest Entomol 20:493–500

Sims SR (1997) Host activity spectrum of the CryIIA *Bacillus thuringiensis* subsp. *kurstaki* protein: effects on Lepidoptera, Diptera, and non-target arthropods. Southwest Entomol 22:395–404

Smith CM, Boyko EV (2007) The molecular bases of plant resistance and defense responses to aphid feeding: current status. Entomol Exp Appl 122:1–16

Smith CM, Schotzko D, Zemetra RS, Souza EJ, Schroeder-Teeter S (1991) Identification of Russian wheat aphid (Homoptera: Aphididae) resistance in wheat. J Econ Entomol 84:328–332

Song WC, Funk CD, Brash AR (1993) Molecular cloning of an allene oxide synthase: a cytochrome P450 specialized for the metabolism of fatty acid hydroperoxides. Proc Natl Acad Sci USA 90:8519–8523

Sticher L, Mauch-Mani B, Métraux JP (1997) Systemic acquired resistance. Annu Rev Phytopathol 35:235–270

Stoger E, Williams S, Christou P, Down RE, Gatehouse JA (1999) Expression of the insecticidal lectin from snowdrop (*Galanthus nivalis* agglutinin; GNA) in transgenic wheat plants: effects on predation by the grain aphid *Sitobion avenae*. Mol Breed 5:65–73

Stone TB, Sims SR, Marrone PG (1989) Selection of tobacco budworm for resistance to a genetically engineered *Pseudomonas fluorescens* containing the δ-endotoxin of *Bacillus thuringiensis* subsp. *Kurstaki*. J Invertebr Pathol 53:228–234

Stuart JJ, Schulte SJ, Hall PS, Mayer KM (1998) Genetic mapping of Hessian fly avirulence gene *Vh6* using bulked segregant analysis. Genome 41:702–708

Subramanyam S, Sardesai N, Puthoff DP, Meyer JM et al (2006) Expression of two wheat defense-response genes, Hfr-1 and Wci-1, under biotic and abiotic stresses. Plant Sci 170:90–103

Suzuki K, Shinshi H (1995) Transient activation and tyrosine phosphorylation of a protein kinase in tobacco cells treated with a fungal elicitor. Plant Cell 7:639–647

Tabashnik BE, Carrière Y (2009) Insect resistance to genetically modified crops. In: Ferry N, Gatehouse AMR (eds) Environmental impact of genetically modified/novel crops. CAB Int, Oxford, pp 74–101

Tabashnik BE, Gassmann AJ, Crowder DW, Carriére Y (2008) Insect resistance to *Bt* crops: evidence versus theory. Nat Biotechnol 26:199–202

Tarpley L, Roessner U (2007) Metabolomics: enabling systems-level phenotyping in rice functional genomics. In: Upadhyaya NM (ed) Rice functional genomics – challenges, progress and prospects. Springer, New York, pp 91–107

Teetes GL, Peterson GC, Nwanze KF, Pendleton BB (1999) Genetic diversity of sorghum a source of insect resistant germplasm. In: Clement SL, Quisenberry SS (eds) Global plant genetic resources for insect-resistant crops. CRC Press, Boca Raton, pp 63–82

Tinjuangjun P, Loc NT, Gatehouse AMR, Gatehouse JA, Christou P (2000) Enhanced insect resistance in Thai rice varieties generated by particle bombardment. Mol Breed 6(4):391–399

Toledo-Ortiz G, Huq E, Quail PH (2003) The *Arabidopsis* basic/helix-loop-helix transcription factor family. Plant Cell 15:1749–1770

Tortiglione C, Fogliano V, Ferracane R, Fanti P, Pennacchio F, Monti LM, Rao R (2003) An insect peptide engineered into the tomato prosystemin gene is released in transgenic tobacco plants and exerts biological activity. Plant Mol Biol 53:891–902

Upadhyaya NM, Pereira A, Watson JM (2010) Transgenic crops and functional genomics. In: Kole C, Michler CH, Abbott AG, Hall TC (eds) Transgenic crop plants: vol 2. Utilization and biosafety. Spinger, New York

Urwin PE, Atkinson HJ, Waller DA, McPherson MJ (1995) Engineered oryzacystatin-I expressed in transgenic hairy roots confers resistance to *Globodera pallida*. Plant J 8:121–131

Vadlamudi RK, Weber E, Ji IH, Ji TH, Bulla LA (1995) Cloning and expression of a receptor for an insecticidal toxin of *Bacillus thuringiensis*. J Biol Chem 270:5490–5494

Vaeck M, Reynaerts A, Hofte H, Jansens S, Debeuckeleer M, Dean C, Zabeau M, Vanmontagu M, Leemans J (1987) Transgenic plants protected from insect attack. Nature 328:33–37

Van Loon LC (1997) Induced resistance in plants and the role of pathogenesis related proteins. Eur J Plant Pathol 103:753–765

Vancanneyt G, Sanz C, Farmaki T, Paneque M, Ortego F, Castanera P, Sanchez-Serrano JJ (2001) Hydroperoxide lyase depletion in transgenic potato plants leads to an increase in aphid performance. Proc Natl Acad Sci USA 98:8139–8144

Vandenberg JD (1990) Safety of four entomopathogens for cages adult honey bees (Hymenoptera: Apidae*)*. J Econ Entomol 83:755–759

Wäckers F, van Rijn P, Bruin J (2005) Plant-provided food for carnivorous insects – a protective mutualism and its applications. Cambridge University Press, Cambridge

Walling LL (2000) The myriad plant responses to herbivores. J Plant Growth Regul 19:195–216

Wei Z, Hu W, Lin QS, Cheng XY, Tong MJ, Zhu LL, Chen RZ, He GC (2009) Understanding rice plant resistance to the brown planthopper (*Nilaparvata lugens*): a proteomic approach. Proteomics 9:2798–2808

Williams WP, Davis FM (1997) Maize germplasm with resistance to south-western corn borer and fall armyworm. In: Mihm JA (ed) Insect resistant maize: recent advances and utilization. Proceedings of the international symposium, 27 Nov–3 Dec 1994. CIMMYT, Mexico, pp 226–229

Wraight CL, Zangerl AR, Carroll MJ, Berenbaum MR (2000) Absence of toxicity of *Bacillus thuringiensis* pollen to black swallowtails under field conditions. Proc Natl Acad Sci USA 14:7700–7703

Xu DP, Xue QZ, McElroy D, Mawal Y, Hilder VA, Wu R (1996) Constitutive expression of a cowpea trypsin inhibitor gene, CpTi, in transgenic rice plants confers resistance to two major rice insect pests. Mol Breed 2:167–173

Yuan H, Chen X, Zhu L, He G (2005) Identification of genes responsive to brown planthopper *Nilaparvata lugens* Stål (Homoptera: Delphacidae) feeding in rice. Planta 221:105–112

Zantoko L, Shukle RH (1997) Genetics of virulence in the Hessian fly to resistance gene H13 in wheat. J Hered 88:120–123

Zhang S, Klessig DF (1997) Salicylic acid activates a 48-kD MAP kinase in tobacco. Plant Cell 9:809–824

Zhang S, Klessig DF (1998) Resistance gene N-mediated de novo synthesis and activation of a tobacco mitogen-activated protein kinase by tobacco mosaic virus infection. Proc Natl Acad Sci USA 95:7433–7438

Zhang F, Zhu L, He G (2004) Differential gene expression in response to brown planthopper feeding in rice. J Plant Physiol 161:53–62

Zhu-Salzman K, Salzman RA, Ahn JE, Koiwa H (2004) Transcriptional regulation of sorghum defense determinants against a phloem-feeding aphid. Plant Physiol 134:420–431

Chapter 5
Multitrophic Interactions: The Entomovector Technology

Guy Smagghe, Veerle Mommaerts, Heikki Hokkanen, and Ingeborg Menzler-Hokkanen

5.1 Introduction

The entomovector technology (Hokkanen and Menzler-Hokkanen 2007; Mommaerts and Smagghe 2011) utilizes insects as vectors of biological control agents for targeted precision biocontrol of plant pests and diseases, providing an intriguing example of multitrophic interactions. As the insect vector normally is a pollinator of the crop plant, it adds a further dimension to these interactions. The technology depends on bee management, manipulation of bee behaviour, components of the cropping system, and on the plant-pathogen-vector-antagonist-system. We investigate in this chapter how to exploit and support the natural ecological functions of biocontrol and pollination, and enhance these via innovative management. Recent systematic developments of the entomovector technology are described, with focus on the component technologies such as the dispensers and carrier substances (see Mommaerts and Smagghe 2011; Mommaerts et al. 2011; Hokkanen et al. 2012). With functioning dispensers and improved, new microbiological control agents (MCA) available, excellent results have been obtained, and will be described in two case studies. The first

G. Smagghe (✉)
Department of Crop Protection, Faculty of Bioscience Engineering, Ghent University, Coupure Links 653, 9000 Ghent, Belgium
e-mail: guy.smagghe@ugent.be

V. Mommaerts
Department of Crop Protection, Faculty of Bioscience Engineering, Ghent University, Coupure Links 653, 9000 Ghent, Belgium

Department of Biology, Faculty of Science and Bio-engineering Sciences, Vrije Universiteit Brussel, Pleinlaan 2, 1050 Brussels, Belgium

H. Hokkanen • I. Menzler-Hokkanen
Department of Agricultural Sciences, University of Helsinki, Latokartanonkaari 5, Box 27, 00014 Helsinki, Finland

G. Smagghe and I. Diaz (eds.), *Arthropod-Plant Interactions: Novel Insights and Approaches for IPM*, Progress in Biological Control 14, DOI 10.1007/978-94-007-3873-7_5, © Springer Science+Business Media B.V. 2012

involves open field studies conducted in Finland with honey bees (*Apis mellifera* Linnaeus (Hymenoptera: Apidae)) as the vector of "Prestop-Mix", containing *Gliocladium catenulatum* J1446 (Hypocreales, Bionectriaceae), to control *Botrytis cinerea* Pers.: Fr. (Helotiales: Sclerotiniaceae) in strawberries, and the second describes the efficiency of bumble bees (*Bombus terrestris* Linnaeus (Hymenoptera: Apidae)) to vector the commercial product "Prestop-Mix" to control *B. cinerea* in strawberries in the greenhouse.

In this chapter the key components of the entomovector technology are described in detail. We also give space to the different dispenser types and their capacity to load the insect vector (also called as acquisition), the selection criteria for the carrier materials with respect to MCA stability (Hjeljord et al. 2000), and vector safety (Israel and Boland 1993; Pettis et al. 2004). The acquisition of a powdery MCA product is not only affected by the formulation type, but also by the insect body characteristics and behaviour of the insect and by the dispenser type. The already 20-years of entomovectoring studies resulted in the development of two dispenser types: namely the one-way type and the two-way type. For honey bees it was shown that the use of two-way dispensers resulted in a higher loading of the bees (Bilu et al. 2004). Similarly, Mommaerts et al. (2010) developed a new two-way dispenser for *B. terrestris*, realizing a 10-fold higher loading of bumble bees compared to earlier dispensers, and also without affecting the foraging intensity of the workers. Important criteria crucial to the success of the entomovector approach are that an optimal dispenser should (a) load the vector with a sufficient amount of the powdery product, (b) not interfere with the foraging behaviour, and that (c) the dispenser should have long refilling intervals (>1 day) (Kevan et al. 2008; Mommaerts and Smagghe 2011).

The choice of the most efficient pollinator species as vector is crucial for maximizing pollination and disease control at the same time. Due to their availability, three pollinators have been used as vectors: honey bees, bumble bees, and mason bees. For instance, Maccagnani et al. (2005, 2006) worked with solitary bees (*Osmia cornuta* Latreille (Hymenoptera: Megachilidae)) and honey bees in delivering MCA for the control of fire blight, an important apple and pear disease caused by *Erwinia amylovora* (Burrill) Winslow (Enterobacteriales: Enterobacteriaceae).

The successful control of a disease or pest by the entomovector technology, either in the greenhouse or in the open field, depends on a web of criteria, ranging from the dispenser design, the selection of the vector, and the transport of the control agent, to the safety of the control agent to the environment and humans. The chapter deals with the complex interactions and provides the reader at the end with a future perspective.

5.1.1 Biological Control in Modern Horticulture

The adoption of biological control in horticulture is driven by the lack of suitable chemical pesticide options in many countries, failures in controlling pests and diseases with classic pesticide applications (spray and/or irrigation), as well as by consumer demand for products free of pesticide residues, and by the health risks to workers applying pesticides in enclosed structures such as greenhouses. Additional benefits to growers include that there are no phytotoxic effects on plants associated with biological control, no problems with residues, and no loss of control efficacy due to resistance in the pests with biological control agents (van Lenteren 2008).

Biological control has rapidly become an essential component of modern horticultural production systems. The use of biological control is in many cases attractive both from an environmental and economic perspective (van Lenteren 2008), and consequently, there has been recently a large increase in the uptake of biological control of vegetable pests as well as in the production of ornamental crops in Europe. Most pests and some diseases of greenhouse vegetables can now be controlled with biological control agents (van Lenteren 2008). The most striking development has taken place in Spain, where 20,000 ha of greenhouses are now routinely using biological control (Pilkington et al. 2010). Until 2006 biological control was only used on a small scale in Spain, but by the growing season of 2007–2008 already more than 75% of the 8,000 ha of sweet pepper in Almeria started to implement biological control (Van der Blom et al. 2009). The market size for biological control agents in Almeria is estimated to amount to 30 million Euros, which is more than the total market for the rest of Europe.

Currently more than 150 species of natural enemies are commercially available to growers around the world, which is already much more than the number of active ingredients for insect control by chemical pesticides (van Lenteren 2008). About 80% of the overall commercial value of biological control (excluding *Bacillus thuringiensis* Berliner (Bacillales: Bacillaceae) (*Bt*)) has been estimated to be generated by their use in greenhouses (van Lenteren 2008; Pilkington et al. 2010).

In contrast with arthropod control in the greenhouses, which has been the spearhead of commercial biological control, the control of plant diseases with biological antagonists has clearly lagged behind. Plant pathologists have discovered many suitable control agents, but their commercial development has been slow and only relatively few products are available to growers (Christensen 2006; Strømeng 2008). The difficult and expensive registration process required for MCAs is the likely explanation for this situation (Ravensberg 2011).

Biological control on outdoors horticultural crops is far less developed than that in protected crops. Also here the control of insect and mite pests by biological means is much more advanced than the biocontrol of plant diseases. The most widespread current practice in outdoor horticultural crops in Europe is the application of predatory mites against mite pests in vineyards: about 40,000 ha are treated annually (Sigsgaard 2006). Of the European orchard area, some 30,000 ha is applying biological control, with the use of predatory mites for spider mite and rust mite control being the most widespread practice. Biocontrol agents are also released against Lepidopteran and Homopteran pests. On strawberries, phytoseiid predatory mites are applied against spider mites on approximately 20,000 ha annually (Sigsgaard 2006). Biological control of some major diseases of horticultural crops, notably of the grey mould fungus (*B. cinerea*) on strawberries, has repeatedly been attempted with several antagonistic organisms, but with inconsistent and usually poor results (e.g., Hjeljord et al. 2000; Prokkola and Kivijärvi 2007; Strømeng 2008).

5.1.2   Efficacy of MCAs as Disease and Pest Management Agents

Efficacy of MCAs in controlling pests and diseases in practical pest management is a key component of their commercial success. For any biological control product, however, several conditions must be met at the same time in order for them to be successful (Gelernter and Lomer 2000):

(a) technical efficacy: the product must control the target pest or disease satisfactorily;
(b) practical efficacy: target pest control must be achieved with a high degree of predictability, and the application procedure of the MCA must be simple and should fit into the grower's other routines;
(c) commercial viability: procedures required to bring the product into the market, and the economics of its production, pricing, and returns to the producer as well as to the grower must be satisfactory, and compare favourably with competing control options;
(d) sustainability: the use of the product must be economically sustainable to all parties, and it should be ecologically sustainable so that it can be applied into the foreseeable future;
(e) provision of public benefit: if the use of a MCA, in addition to meeting the other criteria, also brings along tangible public benefits (such as improvement in the quality of life, improved pollination services, etc.), it will improve the likelihood of success of the product in question.

If the product fails in any of the 'must'-criteria, it is unlikely to become a truly successful MCA. A large majority of the unsuccessful MCA product leads probably did not meet some of the above criteria, and has resulted in their far lower share than anticipated of pest and disease control markets: it still is not more than about 2%, instead of 10% or 20%, as it was predicted to be (Frost and Sullivan 2001). Indeed, far less than 5% of the initial leads developed at the universities ever result in a commercially viable biopesticide (Törmälä 1995). One simple reason most often cited for the lack of commercial success is the poor and erratic field performance of the tested products.

Encouragingly, however, there are an increasing number of examples of successful MCA. The most important family of products are the *Bt*-based bioinsecticides, which alone account for over one-half of all MCA sales worldwide. They are widely used against pests in high-value crops such as vegetables and fruits, but also in corn and cotton. For example, over 50% of cabbage, celery, eggplant and raspberry area in the USA are treated annually with *Bt*. Additionally *Bt*-based products are very important in forest protection, accounting for about 25% of total insecticide use in forest systems, and in mosquito control for example in Italy, France and Germany.

Over the past two decades a large number of fungal MCA have been registered for insect control, including at least five species in Europe and four in North-America, and several more fungal products for plant disease and nematode control. Among the dozens of insect virus-based MCA, worldwide, the most successful and striking example is the NPV of the velvetbean caterpillar *Anticarsia gemmatalis* Hübner (Lepidoptera: Noctuidae) on soybean in Brazil (Moscardi 1999). It is currently applied annually over one million hectares every year, and it gives over 80% control of the pest at a lower cost than that for chemical insecticides.

Despite some successes, an important question remains: why has the expected market potential of MCA yet to be realised? To understand this more fully, we need to consider the most important selection criteria of the MCA from the point of view of the grower. These include the:

- Speed and degree of activity
- Persistence of activity
- Selectivity of activity
- Economic performance

It is clear that a satisfactory control of the target pest or disease must be obtained. The problem for MCAs usually is that their mode of action is different from that of the chemical pesticides: the growers expect to see immediate results (e.g., dead insects after the treatment), which is hardly ever the case with MCAs. From the crop protection point of view, the speed of kill is not usually critical, because infected insects soon stop feeding, but remain on the plant and do not 'drop dead'. Therefore they do not cause any further damage, but still appear on the plant, causing the grower to lose trust in the efficacy of the product. For MCA-bioinsecticides the *Bt*-based products and nematodes usually have an adequate speed of kill, while other MCA need more consideration in this respect. Developments on two fronts are needed here: (1) investigations on how to increase the speed of kill, and/or (2) abandoning the chemical pesticide paradigm requiring fast results. Approaches utilised so far to increase the speed of kill include:

(i) selection, and even engineering, of more virulent strains of the MCA (e.g., viruses, nematodes);
(ii) increasing the effective dose (e.g., fungi, whose degree of activity often is dose-dependent);
(iii) combining the MCA with other agents or 'enhancers' for a synergistic action; the other agents utilized include low doses of pesticides, oils, optical brighteners, and specific proteins (e.g., 'enhancin') (Moscardi 1999).

In an educational process to train the growers to abandon the chemical pesticide paradigm, emphasis should be on the behaviour of the target organisms (e.g., cessation of feeding) as well as on the persistence of activity and long-term impacts. MCAs are all living organisms with their own population dynamics and interaction with the environment, usually providing a long-term control for example via secondary cycling and/or horizontal transfer of the MCA. In addition, the persistence of the MCA can be enhanced via rapidly developing and improved formulation technology, including microencapsulation of *Bt*-based toxin crystals, using UV-protectants, oils, substrate matrixes, etc. for extended activity of the MCA. There is much scope for improvement in this area, and rapid advances can be expected if adequate research emphasis is focused on these questions.

Cohen et al. (1999) discussed very interestingly the barriers to progress in using biocontrol agents, asking whether these are biological, engineering, economic, or cultural. They pointed out that the biological constraints have largely been solved, or are intensively being studied and can be expected to be solved soon. On the other hand, the engineering bottlenecks (e.g., automated mass-production of MCAs) have received less attention, but are equally important to success. Likely the most important obstacle is the "can't-do-culture", which limits the willingness to invest, and even to think along these new lines, even when it is realistic to expect that the biological and engineering bottlenecks can be solved. Many beneficial organisms could already, or in the near future, be produced and sold at a fraction of their current cost, making them very competitive and significant in overall crop protection, in practically any crop. The question still stands: what prevents this from happening?

5.2   Historical Perspective of the Entomovector Technology

The term 'entomovector technology' was first used by Hokkanen and Menzler-Hokkanen (2007), and the approach incorporates different ecological components such as pollinators, biocontrol agents and plant pathogens/insect pests (Kevan et al. 2008). However, its success is based on mutual and suited interactions between the appropriate components of vector, control agent, formulation and dispenser, and it needs to be safe for the environment and the human health (Fig. 5.1).

5.2.1   *Pollinators as Vector: A Choice Determined by the Vector Species*

Within the entomovector technology the selection of the pollinator vector is a crucial step towards success. To date many crops are being pollinated (see for an overview Goulson 2010), but not all used pollinators are classified as 'good pollinators'. Thus, the main criterion to maximize MCA deposition on the target is to choose a vector which will visit the crop efficiently. Therefore it

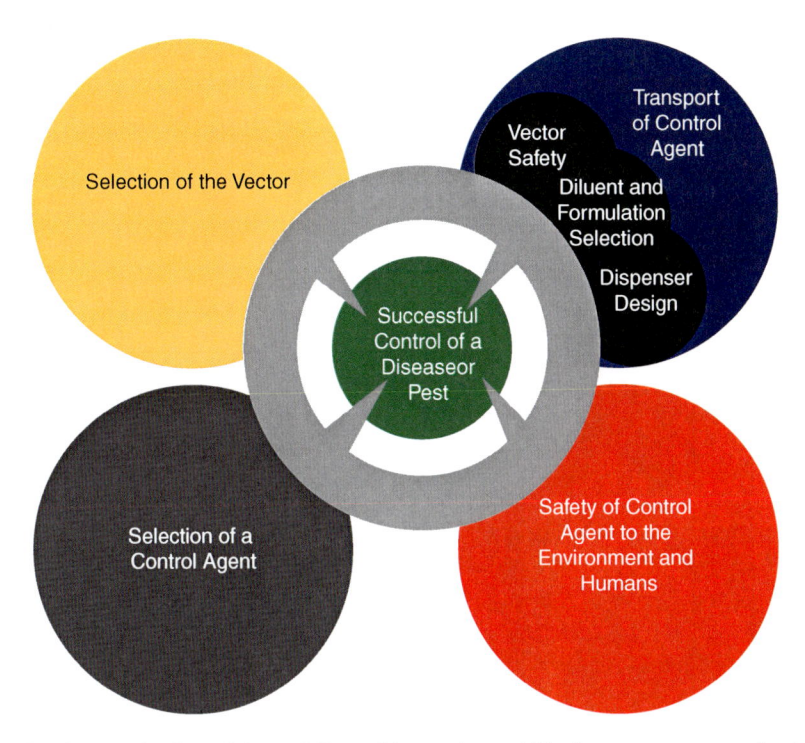

Fig. 5.1 Schematic view of the multifaceted interactions within the entomovector technology. (Adapted from Kevan et al. 2008)

is important to explore our knowledge on vector-plant interactions, and on the influence of environmental conditions on vector activity.

Until now due to their availability, three pollinator genera have been reported to vector MCAs onto crops: honey bees (*A. mellifera*), bumble bees (*Bombus impatiens* Cresson (Hymenoptera: Apidae) and *B. terrestris*) and the mason bee (*O. cornuta*). As listed in Table 5.1, the first studies were conducted in orchards with honey bees, and more recently also with the solitary mason bee, while bumble bees disseminated MCAs under both field and greenhouse conditions. This vector choice can be explained by their behaviour: bumble bees tolerate high temperature fluctuations and are bad weather foragers (Guerra-Sanz 2008; Goulson 2010); the mason bee has so far only been used in orchard crops (see for review Bosch and Kemp 2002), and honey bees are less appropriate to pollinate greenhouse long-blooming crops (Cribb and Hand 1993; Guerra-Sanz 2008). But for the latter also the foraging behaviour was shown to be sensitive to weather conditions (temperature, cloudiness and rain) (Goulson 2010), which can affect the capacity of the bees to disseminate the MCA onto flowers (Vanneste 1996; Maccagnani et al. 1999). In addition, upon selection of a vector also their foraging ranges need to be considered, as they need to be able to transport the MCA into the flowers. Honey bees have a larger foraging range (up to 3 km) compared to both the mason bee (100–200 m) and bumble bees (for *B. terrestris*: 800–1500 m) (Vicens and Bosch 2000; Osborne et al. 2008; Wolf and Moritz 2008).

Table 5.1 Overview of the entomovector technology studies so far conducted with honey bees, bumble bees and a mason bee species

Vector	Active ingredient (CFU/g)	Crop	Dispenser	Bee load direct (CFU/bee)	Flower load (CFU/flower)	Target	Control	Reference
A. mellifera	Gliocladium roseum (5×10^8)	Strawberry F* F	Peng	6.3×10^4 5.7×10^5	$1.6\text{–}27 \times 10^3$ $2.0\text{–}114 \times 10^2$	Botrytis cinerea	S NS	Peng et al. (1992)
A. mellifera	Pseudomonas fluorescens A506 (1.7×10^9)	Apple and pear	Antles	$5.0\text{–}8.0 \times 10^3$	$0\text{–}1 \times 10^3$	NC	NC	Thomson et al. (1992)
	Erwinia herbicola Eh318 (3.1×10^8)	F		2.5×10^3	$0\text{–}5.4 \times 10^3$			
A. mellifera B. impatiens	Gliocladium roseum (1.2×10^9)	Raspberry F* F* G	Peng Yu & Sutton	H:$1.3\text{–}81 \times 10^4$ B: $0.3\text{–}128 \times 10^4$	H: $0\text{–}8000$ B: $0\text{–}20666$ B: 945 ± 1255	Botrytis cinerea	S*	Yu and Sutton (1997)
A. mellifera	Trichoderma harzianum $(1.0\text{–}4.9 \times 10^8)$	Strawberry F*	NI	$2.7\text{–}45 \times 10^5$	$0.4 \pm 0.3 \times 10^5$ $4.1 \pm 0.5 \times 10^5$	Botrytis cinerea	NS	Maccagnani et al. (1999)
A. mellifera B. impatiens	Trichoderma harzianum 1295-22 $(1 \times 10^8\text{–}10^{10})$	Strawberry F	H: Tray B: Tube	H: 1.0×10^5 B: NC	NC	Botrytis cinerea	S	Kovach et al. (2000)
A. mellifera	Trichoderma harzianum T39 (1.0×10^9)	Strawberry F	Triwaks	1.5×10^5	$2.2 \pm 0.5 \times 10^4$	Botrytis cinerea	S*	Shafir et al. (2006)
A. mellifera	Bacillus subtilis QRD132	Blueberry F*	Gross	$5.1\text{–}6.4 \times 10^5$	5.1×10^3	Monilinia vaccinii-corymbosi	S	Dedej et al. (2004)

A. mellifera	Trichoderma koningii, Trichoderma aureoviride Trichoderma longibrachiatum, Trichoderma harzianum sp. (13×10^9)	Sunflower F*	NI F	NC NC	$1 \times 10^7-10^9$ S*	Sclerotinia sclerotiorum	S	Escande et al. (2002)
A. mellifera	Bacillus subtilis BS-F4[rif] $(1 \times 10^{10}-10^{11})$	Pear F* F*	A: NI	A: 1×10^4	A: $1-2 \times 10^4$ (=1st–3rd flower) O: $4-14 \times 10^4$ (=1st–3rd flower)	NC	NC	Maccagnani et al. (2006)
O. cornuta		F	O: Maccagnani	O:$1 \times 10^4-10^7$	O: $1 \times 10^2-10^6$			
B. impatiens	Clonostachys rosea + Beauveria bassiana GHA $(1.4 \times 10^7 + 6.3 \times 10^{10})$	[1]Tomato [2]Sweet pepper G	NI	[1]2.6×10^4 [2]5.5×10^5 [1]5.0×10^4 [2]8.0×10^5	[1]$4.3 \pm 0.3 \times 10^3$ [2]1.6×10^4 [1]$4.8 \pm 0.2 \times 10^3$ [2]1.0×10^4	Botrytis cinerea Trialeurodes vaporariorum Lygus lineolaris	S M	Kapongo et al. (2008a)
B. impatiens	Beauveria bassiana GHA $(9 \times 10^9 -6.2 \times 10^{10}-2 \times 10^{11})$	[1]Tomato [2]Sweet pepper G	Modified Yu & Sutton	[1]$1.7 \times 10^3 - 3.1 \times 10^5$ [2]$1.1-7.4 \times 10^5$	[1]$0.4-3.2 \times 10^3$ [2]$0.3-2.4 \times 10^3$	Trialeurodes vaporariorum Lygus lineolaris Myzus persicae	M	Kapongo et al. (2008b)
A .mellifera B. impatiens	Trichoderma harzianum T22 (9.8×10^6)	Strawberry H: F B: G	Houle	H: $3.9 \pm 1.8 \times 10^3$ B: $7.2 \pm 2.2 \times 10^4$	H: 26 ± 90 B: $1.3 \pm 0.9 \times 10^3$	NC	NC	Albano et al. (2009)

(continued)

Table 5.1 (continued)

Vector	Active ingredient (CFU/g)	Crop	Dispenser	Bee load direct (CFU/bee)	Flower load (CFU/flower)	Target	Control	Reference
B. terrestris	*Trichoderma harzianum* + *Gliocladium virens* (1×10^8) *Trichoderma harzianum* (13/3 RBDPH1,15/ 2RBD5 and 4/ 18RBD1) (3.5×10^9)	Tomato G	SSP OP	$1.3 \pm 4.0 \times 10^2$ $4.3 \pm 4.2 \times 10^4$	$0.7 \pm 0.3 \times 10^2$ $1.3 \pm 1.1 \times 10^2$	NC	NC	Maccagnani et al. (2005)
B. terrestris	*Trichoderma atroviride* + *Hypocrea parapilulifera* (7.0×10^7)	NC	Newly developed	$4.3 \pm 0.2 \times 10^5$	NC	NC	NC	Mommaerts et al. (2010)
A. mellifera	*Heliothis* nuclear polyhedrosis virus $(106 \times 10^{12}$ IU/g)	Crimson clover F	Gross	NC	NC	*Helicoverpa zea* larvae	M	Gross et al. (1994)
A. mellifera	*Metarhizium anisopliae*	Oil seed rape F*	Peng	NC	NC	*Meligethes aeneus*	M	Butt et al. (1998)
A. mellifera	*Bacillus thuringiensis var kurstaki*	Sunflower F	Modified Gross	NC	NC	*Cochylis hospes* larvae	M	Jyoti and Brewer (1999)
A. mellifera	*Beauveria bassiana* GHA (1×10^9)	Canola G F* F	NI	NC 4.5–6.0×10^5 NC	NC 2.7–3.7×10^4 NC	*Lygus lineolaris*	M	Al-mazra'awi et al. (2006a)

B. impatiens	Beauveria bassiana GHA (1×10^9)	Sweet pepper		$4.5 \pm 0.5 \times 10^5$	$1.8 \pm 0.5 \times 10^4$	Lygus lineolaris		Al-mazra'awi et al. (2006b)
		G	NI	$3.8 \pm 0.8 \times 10^5$	$3.2 \pm 1.0 \times 10^4$	Frankliniella occidentalis	M	
A. mellifera	Metarhizium anisopliae	Canola				Meligethes aeneus		Carreck et al. (2007)
						Ceutorhynchys assimilis	M	

F field, F* covered field, G greenhouse, M significant mortality of the pest, NC not considered, NI no information, NS no suppression, S suppression, S* suppression at low and medium disease level but not at high level

To transport the MCA into the flowers of the crops, vectors need to become attracted to the flowers. Studies investigating the foraging behaviour of bees identified two factors affecting crop visitation rate: the population levels of other bees in the vicinity (Vanneste 1996), and floral-related cues (innate and learned preferences which include flower size, colour, odour, temperature and reward) (Spaethe et al. 2001; Stout and Goulson 2002; Farina et al. 2007; Raine and Chittka 2007; Rands and Whitney 2008; Whitney et al. 2008; Forrest and Thomson 2009; Lunau et al. 2009; Molet et al. 2009; Gil 2010). Although these factors are difficult to control, several entomovectoring studies assured vector visitation in crops under open field conditions either by spraying a bee-attractant (Bee-Scent®) (Peng et al. 1992), by increasing the colony size up to 50,000 bees/hive (Escande et al. 2002; Shafir et al. 2006), or by using up to four plant cultivars per field plot (Yu and Sutton 1997; Kovach et al. 2000; Escande et al. 2002). In addition, and based on Ngugi et al. (2002), also the timing of vectoring was shown to be crucial and should be started before most of the flowers are open, and thus before the presence of the plant pathogen (Wilson et al. 1992; Johnson et al. 1993a, b; Wilson and Lindow 1993; Alexandrova et al. 2002).

Next to transport also deposition of the MCA on the target organ, such as the stigma of flowers, was found necessary for the successful control of the plant pathogen *E. amylovora* (Scherm et al. 2004). For the different vector species the successes so far obtained using the entomovectoring technology in connection with the crop considered, are summarized in Table 5.1.

5.2.2   Dissemination of MCAs into Flowers of Target Crops

The next step towards the success of the entomovector technology is to guarantee that vectors acquire a sufficient amount of MCA, which allows optimal transport to the flowers. Most of the commercial powdery MCA formulations are not developed for vectoring and thus need to be improved (see Sect. 5.2.2.2). But next to the formulation of the MCA-based product, acquisition on the vector body is also determined by the dispenser type (see Sect. 5.2.2.1).

5.2.2.1   The Loading Capacity of the Different Dispenser Types So Far Developed

Dispensers must be designed in a way that they are safe for the vector, that they load the vector with a sufficient amount of MCA, and that they have refilling intervals of >1 day. Over the 20-years of entomovectoring studies multiple dispensers have been developed (for review see Mommaerts and Smagghe 2011), representing two basic approaches: one-way dispensers and two-way dispensers (Fig. 5.2). One-way dispensers were shown less suitable, as their use resulted in low MCA acquisition on the vector, affected the foraging behaviour, and demanded a daily refill (Thomson et al. 1992; Johnson et al. 1993a; Dag et al. 2000; Bilu et al. 2004; Maccagnani et al. 2005; Mommaerts et al. 2010). There is one exception: a high loading of $>10^4$ CFU

Dispenser entrance via hive

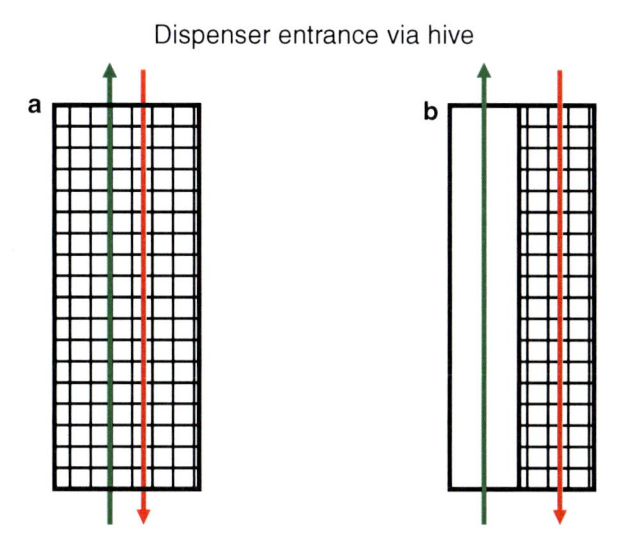

Dispenser exit to field or greenhouse

Fig. 5.2 Schematic view of (**a**) one-way type dispensers where the chamber through which the bees enter or leave the dispenser is the same (or is not completely separated), and (**b**) two-way dispensers where the chamber (with control agent) through which bees leave the dispenser is separated from the chamber (without control agent) via which they enter the dispenser. (*red arrow* = outgoing bees, *green arrow* = incoming bees, and □ = MCA powder formulation)

per bee, a threshold of efficiency as determined by Bilu et al. (2004), was obtained with the over-and-under one-way dispenser developed for the bumble bee *B. impatiens* by Yu and Sutton (1997), and its modified version (Kapongo et al. 2008b).

Using the same over-and-under design but within a two-way dispenser system such as the Tray-, Peng-, Triwaks-, Gross- and Houle-dispenser, multiple authors reported on an increased vector loading of $>10^5$ CFU per honey bee and/or obtained good control of the targeted disease/pest (Gross et al. 1994; Jyoto et al. 1999; Kovach et al. 2000; Bilu et al. 2004; Dedej et al. 2004; Albano et al. 2009; Hokkanen et al. 2012). This suggests that the two-way type dispensers, in total six, developed for honey bees are satisfactory, and actually to date one commercial dispenser, namely the "BeeTreat", is available on the market (Fig. 5.3) (Hokkanen et al. 2012). The latter dispenser type has not been patented, and its specifications are freely available on the internet (www.aasatek.fi).

Other vectors such as bumble bees and the mason bee differ in their behaviour compared to honey bees, and thus require an appropriate two-way dispenser. Nonetheless, reports on the development of two-way dispenser designs are limited to only a few studies. For bumble bees two-way dispensers comprise the overlapping-passageway-dispenser (Maccagnani et al. 2005), the Houle-dispenser (Albano et al. 2009) and the Mommaerts-dispenser (Mommaerts et al. 2010). Next to a satisfactory loading ($>10^4$ per bee) only the latter dispenser showed no adverse effects on the foraging activity and refilling intervals of >1 day (Fig. 5.4) (Mommaerts et al. 2010); this system is currently under investigation under large

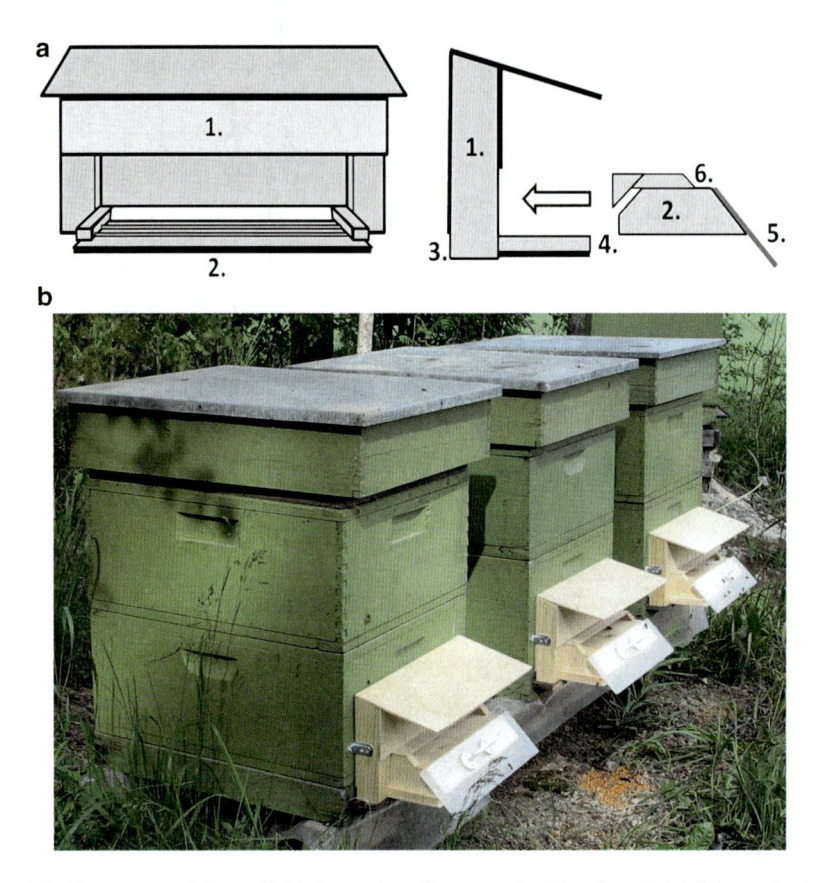

Fig. 5.3 The commercially available honey bee dispenser, the "Bee-Treat". (**a**) Schematic view: *1* the dispenser; *2* detachable steering part, to be inserted into the dispenser; *3* entrance to the dispenser; *4* exit of the dispenser; *5* landing platform for the incoming bees made of transparent plexiglass; *6* entrance corridor for bees to return to the hive (crawl over the solid block 2 to access opening 3). The powder MCA is loaded in an area between 3 and 4. (**b**) Photograph of a "Bee-Treat" connected to a honey bee hive (Drawings and photo: Heikki Hokkanen)

greenhouse conditions. Also, to date one two-way dispenser has been developed for *O. cornuta* showing a good average load (10^4–10^7 CFU/bee) (Maccagnani et al. 2006). Further validation in practice is needed because of observed avoidance behaviour towards the powder immediately after filling of the dispenser.

5.2.2.2 The Role of Dilutions and Formulations in MCA Acquisition and Transport by the Vector

For transport by bees, MCAs need to be formulated as a powder (Fig. 5.5). In early studies the self-prepared mixtures consisted of MCAs and pollen (Thomson et al. 1992), but thereafter entomovectoring studies added carriers such as flours to MCAs or to commercially available MCA formulations. These carriers increase

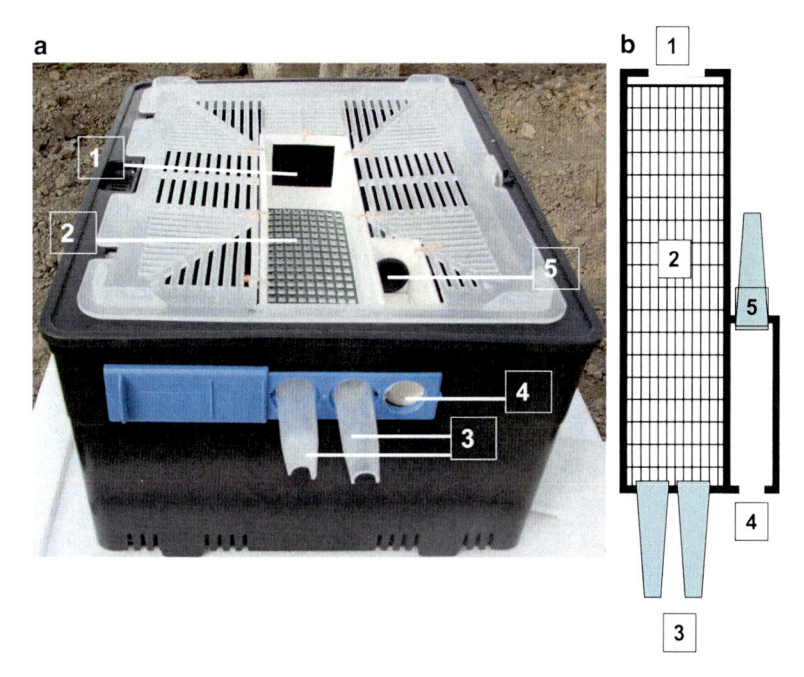

Fig. 5.4 The two-way Mommaerts-dispenser. (**a**) Photograph from the front without cover lid, and (**b**) schematic drawing top view. *1* connection of the exit compartment with the bumble bee hive; *2* exit compartment with a grid at the bottom which contains the powder MCA formulation; *3* exit holes with bumble bee-in-closer; *4* entrance hole; *5* bumblebee-in-closer, connecting the entrance compartment with the bumble bee hive. Dispenser length = 20 cm

Fig. 5.5 Acquisition of a powder MCA formulation: (**a**) Honey bee disseminating "Prestop-Mix" to a strawberry flower (Photo by Heikki Hokkanen). (**b**) SEM picture showing the presence of conidia and the carrier corn meal of the powder formulation of *G. roseum* on the femoral setae of a honey bee. (Bar = 70 μm) (Picture adapted from Peng et al. (1992))

the acquisition and transport efficacy by reducing loss during flight (Kevan et al. 2008; Mommaerts et al. 2011). However, carriers of MCA formulations must show MCA stability and vector safety. For example the mineral carrier talc adversely affected MCA growth (Hjeljord et al. 2000) and honey bee brood (Pettis et al. 2004), and induced honey bee grooming behaviour due to irritation (Israel and Boland 1993), while flours as carrier decreased grooming by 50% (Kevan et al. 2008). Potential carriers comprise corn flour (Peng et al. 1992; Al-mazra'awi et al. 2006b), bentonite (Kevan et al. 2008), "Maizena-Plus" (corn starch) (Mommaerts et al. 2011), and polystyrene beads (Butt et al. 1998). The latter carrier was too expensive for commercial formulations, compared with flours and meals which are inexpensive, safe and food grade qualified. Unfortunately, 20 years after the first entomovectoring study was conducted there is still inadequate information on the potential of different carriers and their role in vector acquisition. A study by Al-mazra'awi et al. (2007) found a negative correlation between honey bee loading and carrier particle size and moisture content. For example carriers with a particle diameter above 300 μm were not acquired on the honey bee body, while smaller particles in the range of 1–150 μm resulted in a good loading. This confirms the need for future studies.

5.2.3 The Reliability of Vectored MCAs

Many of the MCAs isolated from soils, leaves or insects have been tested for their control capacity, but only a few have been registered for agricultural use. In the European Union the active substances of microbial pesticides that are included in Directive 91/414/CEE comprise 14 fungicides and 5 insecticides. Regulations require that the MCA formulation is safe for the environment and humans. However, they are not all suitable for use in the entomovector system, and to date only two microbial pesticides are registered for this purpose, namely "Binab-T-vector" that is based on a combination of two antagonistic fungi, *Trichoderma atroviride* P.Karst (Hypocreales, Hypocreaceae) and *Hypocrea parapilulifera* B.S.Lu, Druzhin & Samuels (Hypocreales, Hypocreaceae), and "Prestop-Mix" that is a preparation of *G. catenultum* J1446.

5.2.3.1 MCAs Against Plant Pathogens

Table 5.1 summarizes the different successes already obtained with the entomovector technology for the dissemination of MCAs against several economically important plant pathogens of orchard fruits as apple and pear, strawberry, raspberry, blueberry and sunflowers.

B. *cinerea,* the grey mould fungus, is a pathogen which is difficult to control because of its high genetic variability and its capacity to grow on every plant part (Mertley et al. 2002; Beever and Weeds 2004; Williamson et al. 2007).

Suppression was shown after vectoring with honey bees or bumble bees of three MCAs, namely *Clonostachys rosea* (Link.: Fr.) Schroers, Samuels, Seifert & Gams (formerly *Gliocladium roseum* Bainier) (Hypocreales: Bionectriaceae), *Trichoderma harzianum* T39 and *G. catenulatum* J1446. In strawberry, the incidence of *B. cinerea* could be reduced in the stamens from 53% to 35%, in the petals from 18% to 15%, and also in the fruits (Peng et al. 1992; Shafir et al. 2006). Similar results were also shown for covered areas and greenhouses vectoring the same MCAs in raspberries, in strawberries, and in tomato and sweet pepper (Peng et al. 1992; Yu and Sutton 1997; Kapongo et al. 2008a; Mommaerts et al. 2011). To date, all entomovector studies reported a mean deposition of 10^3–10^4 CFU per flower, an amount shown by Elad and Freeman (2002) to be sufficient to suppress the plant pathogen *B. cinerea*. Moreover, based on the obtained results it can be concluded that vectoring by bumble bees suppressed *B. cinerea* as efficiently as when vectored by honey bees, although the colony size is only 1/60 of that of the honey bees.

Sclerotinia sclerotiorum (Lib.) de Bary (Helotiales: Sclerotiniaceae) on sunflower can not be efficiently controlled chemically because the treatment is not economically feasible. *Trichoderma* spp. have been recognized as suitable fungal antagonists. The control by *Trichoderma* spp. was also confirmed when the MCA was disseminated by a high density of honey bees, vectoring into the flowers of sunflowers, which were protected until 31 days after application of the pathogen (Escande et al. 2002).

Monilinia vaccinii-corymbosi (JM Reade) Honey (Helotiales: Sclerotiniaceae), a disease of blueberry, has two infection phases but flowers are only infected in the second phase via the gynoecial pathway (Shinners and Olson 1996; Ngugi et al. 2002). As blueberries are a crop being typically pollinated by honey bees, vectoring studies have also been conducted with this vector. So far only one study evaluated the capacity to control this plant pathogen with the entomovector technology. Dedej et al. (2004) showed that vectoring of *Bacillus subtilis* (Ehrenberg) Cohn (Bacillales: Bacilaceae) QRD132 significantly suppressed the infection level of this plant pathogen from 21–67% to 7–44%.

E. amylovora, a disease of mainly Rosaceae such as apple, pear and raspberry, infects host plants primarily through nectarthodes in flowers (Oh and Beer 2005). Epiphytic bacteria, such as *Erwinia herbicola* (Löhnis) Dye and *Pseudomonas fluorescens* (Trevison) Migula, have been vectored by honey bees in pear and apple orchards (Thomson et al. 1992). Analyses of the flowers showed the capacity of the bees to disseminate the MCA, but cold temperatures reduced the numbers of bacterium per flower, and did not allow to evaluate the biocontrol capacity. The early blooming of orchard crops might pose a problem for entomovector studies using honey bees, because of their sensitivity to bad weather conditions. But next to honey bees also the mason bee is a known orchard pollinator. In one study by Maccagnani et al. (2006) the mason bee was identified as a better vector to disseminate powdery MCA into pear flowers due to a higher CFU per flower (10^4 versus 10^4–10^7) and a higher deposition of CFU/flower up to the 6th consecutively visited flower as compared to honey bees. However, firm conclusions will be drawn in future studies when biocontrol capacity will be investigated.

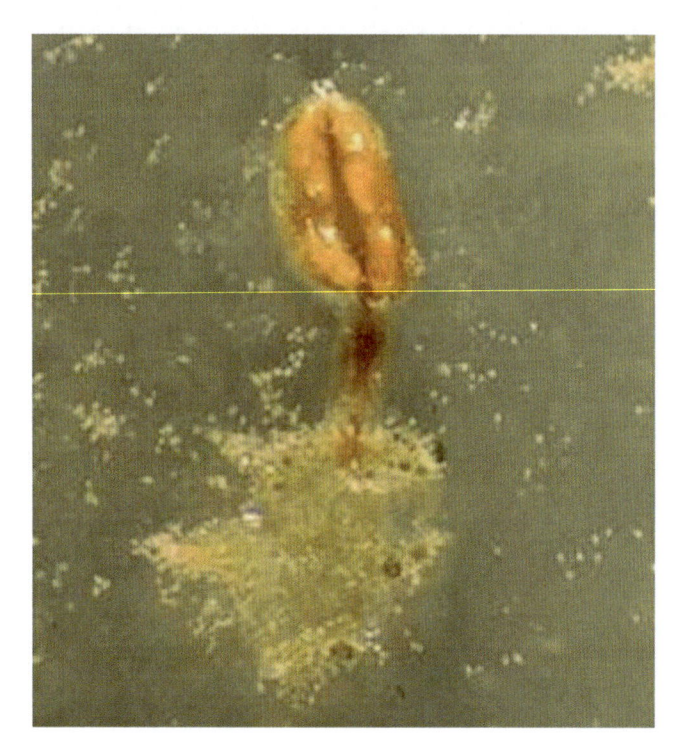

Fig. 5.6 Colonization of a strawberry anther by *Gliocladium catenulatum* J1446 ("Prestop-Mix") (Photo by Marja-Leena Lahdenperä (Verdera))

In conclusion, entomovector studies conclude that the success of plant pathogen control depends on the frequency of the visits by the vector to the crop. Thus when a satisfactory level of the MCA into flowers is realized, a good control level will be achieved. Several authors reported that the suppression capacity was variable, and Shafir et al. (2006) even found a loss of control at high disease pressure. Indeed, bees will visit flower crops while foraging but some will be visited more than others and so a high variability in numbers of CFU per flower (ranging between 0 and 10^4 CFU/flower) is not unlikely (Thomson et al. 1992; Kovach et al. 2000; Shafir et al. 2006; Albano et al. 2009). It should be noted that there is less loss of viability of the MCA by vectoring, in comparison to a spray application (Yu and Sutton 1997; Kovach et al. 2000). Under optimal conditions these viable spores can colonize the flower giving protection towards infectious diseases (Fig. 5.6).

It is of interest that, in addition to its presence in the flowers, disseminated MCA has also been recovered on the leaves. Kapongo et al. (2008a) showed that 90% of the sampled tomato leaves and 76% of the sampled sweet pepper leaves contained *C. rosea*, vectored by *B. impatiens*. Therefore, it is likely that the entomovector technology could be of help in protecting also plant structures other than flowers, and also to control foliar diseases such as powdery mildews.

5.2.3.2 MCAs of Pest Insects

Pollinators (honey bees and bumble bees) have also been shown to be useful in the control of pest insects which feed on, or inhabit, the flowers of oil seed rape, canola, sunflowers, tomato and sweet pepper (Table 5.1). However, to date studies have been conducted in field cages or in small greenhouses, while information on open field is lacking. The first MCA disseminated in this context was *Heliothis* nuclear polyhedrosis virus (HNPV) which killed 74–87% of the collected *Helicoverpa zea* Boddy (Lepidoptera: Noctuidae) larvae from crimson clover (Gross et al. 1994). Later successes were also reported for several other entomopathogenic fungi and one bacterium: *Metharizium anisopliae* (Metchnikoff) Sorokin (Hypocrreales: Clavicipitaceae) against larvae/adults of *Meligethes aeneus* Fabricius (Coleoptera: Nitidulidae) and *Ceuthorhynchus assimilis* Dejean (Coleoptera: Curculionidae) (Butt et al. 1998; Carreck et al. 2007); *B. thuringiensis var kurstaki* against *Cochylis hospes* (Jyoti and Brewer 1999), and *Beauveria bassiana* (Balsamo) Vuillemin GHA (Hypocrreales: Cordycipitaceae) against *Lygus lineolaris* (Palisot de Beauvois) (Hemiptera: Miridae) (Al-mazra'awi et al. 2006a). For the latter MCA also several greenhouse pest populations (thrips, tarnished plant bug, whiteflies and aphids) were controlled when bumble bees of *B. impatiens* vectored 10^3–10^4 CFU of *B. bassiana* into the flowers (Al-mazra'awi et al. 2006b; Kapongo et al. 2008a, b). Previously *B. bassiana* and *M. anisopliae* were often inefficient due to the difficulty of suspending the conidia in water because of their hydrophobic cell walls (Noma and Strickler 2000), and the adverse effects on the viability of the conidia of a spray application (Nilsson and Gripwall 1999) but the entomovector technology opens future perspectives.

As with MCAs against plant diseases, MCAs of pest insects have also been reported on the leaves of vectored crops. For example *B. bassiana* GHA was recovered at high amounts on the sampled leaves: 92% for tomato, 87–92% for sweet pepper, and 70–82% for canola (Al-mazra'awi et al. 2006a; Kapongo et al. 2008b). It can therefore be concluded, as above for plant pathogens, that the entomovector technology is not limited to targeting insect pests of flowers only.

5.3 The Entomovector Technology as Tool Against an Important Plant Pathogen of Strawberries: Case Studies in Open Field and Greenhouse

This section describes an innovative approach to solve one of the most difficult disease problems in strawberry production. The EU is the biggest producer of strawberries in the world, and of the single member countries, Spain is the number two producer after the USA. Turkey is the third most important strawberry producer in the world. In total, the strawberry area in the EU was 111,801 ha in 2008 (FAO 2011). In terms of economic importance, strawberry is in Finland the

12th most valuable agricultural commodity (after a long list of top-ranking animal-based products such as meat, milk, eggs, etc.), and ranks similarly among top-20 agricultural commodities in countries like Germany (15th), Estonia (15th), and Belgium (16th) (FAO 2011). Besides, organic strawberry growing has rapidly expanded in Europe, but organic berry and fruit production suffers heavily from the lack of effective disease and pest management tools, and occasionally from inadequate insect pollination. As a consequence, the expanding demand on organic berries cannot be filled today.

Grey mould (*B. cinerea*) is the most important biotic threat to strawberry growing, and conventional growing uses more fungicides on strawberry than on any other crop, usually 3–8 treatments per season. The industry is concerned about the slow progress in the development of biological control methods (biofungicides) against *Botrytis* (AAFC 2009), as the chemical fungicides rapidly lose their ability to control the disease. Currently organic strawberry growers have no means of preventing grey mould on their crop, and consequently, they occasionally lose the harvest almost entirely. Conventional growers suffer 10–20% pre-harvest crop losses to grey mould on the average (Strømeng 2008), even up to 25–35% (IPMCenters 2011) despite the numerous fungicide treatments.

The entomovector approach represents the only significant breakthrough in sight for providing control of this problem disease, to improve the pollination of straw-berry crops, and to significantly improve the yield and quality of berry production and thus, farm economics. The entomovector technology contributes to improved resource use and efficiency in production, and enhances local biodiversity unlike most other plant protection systems.

5.3.1 Open Field Studies Conducted in Finland with Honey Bees of Apis mellifera as Vector of "Prestop-Mix", Containing G. catenulatum J1446, to Control B. cinerea in Strawberries

Comprehensive on-farm research has been carried out in Finland on the use of honey bee-disseminated, targeted biocontrol of the grey mould *B. cinerea*, with the antagonistic MCA *G. catenulatum* J1446. Research started in 2006 with three commercial farms (one of them organic) in the northern Savo region of the country, and expanded in 2007 to five farms (including two organic farms), on which intensive research and monitoring was carried out. In addition, about 20 commer-cial strawberry farms participated in extensive trials in different parts of Finland (Fig. 5.7). Intensive research was continued on four farms in the years 2008 and 2009, while some 20 additional farms joined annually in the extensive trials in all parts of the country. Based on the data generated during the first two study years, the method was officially approved in spring 2008 by the Finnish Food Safety Authority to be used for grey mould control by all growers. Two-way dispensers

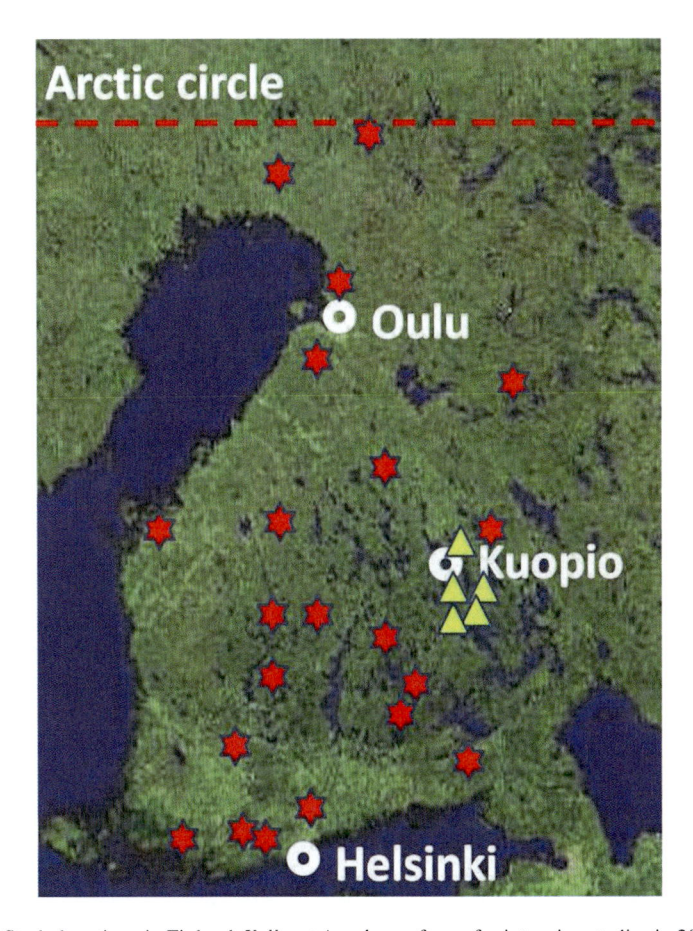

Fig. 5.7 Study locations in Finland. *Yellow triangles* = farms for intensive studies in 2006–2009; *red stars* = farms for extensive studies in 2007–2009. Map dimensions ca. 500 km × 900 km (Background map is Eniro satellite image)

must be used and the recommended rate of the MCA is 400 g/ha (but no restrictions on the dose are set).

In the first preliminary field trials in 2005, some of the available dispenser types were tested, but they proved unreliable or not suitable for practical use under the variable Finnish weather conditions. We therefore developed our own model, namely the "Bee-Treat" dispenser (Fig. 5.3), which is now commercially produced and the only type available in Finland (also used in Estonia) (www.aasatek.fi).

During the 4 years of the intensive studies we compared : (i) bee-disseminated biocontrol with (ii) standard chemical fungicide treatments, (iii) combined treatments, and with (iv) untreated controls, all on strawberry cultivar'Polka'. The fungicide treatments varied according to the farmer's practice and the year, but normally included 3–5 spray treatments against the grey mould; in addition, standard other pesticide treatments were carried out as usual (including insecticides and herbicides).

The total number of pesticide treatments varied between 3 and 9 per season in the conventionally grown strawberry (no pesticide treatments on the organic farms).

Bee dissemination of the antagonistic MCA was started at the onset of strawberry flowering, and was continued at each site until the end of flowering (about 3 weeks). The powdery MCA product was added in the dispenser by the grower daily in the morning before flight activity of the bees (about 5–10 g product at a time), and in total about 300–500 g of the MCA product was dispensed per ha. On rainy days MCA was added only if there was a clear break in the rain, and the bees were flying out.

Two strong beehives/ha were employed in the dissemination, placed at the edge of the study field (Fig. 5.3). Very light exclusion cages with large mesh size were used to keep bees from entering the control plots, and plots with pesticide treatment only, during the flowering. The cages were removed after flowering, and in a separate assay it was determined that the cages did not have a statistically significant influence on the level of grey mould in the enclosed area.

Four replicates for each treatment were included at each study location (farm). Assessments of grey mould incidence were carried out in 2006 and in 2007 by counting 100 berries (minimum diameter 1 cm) in each replicate, and determining its health status (marketable or mouldy). In 2008 and 2009 all the ripe berries were picked at normal picking time (every 2 days) from 1 m of the strawberry row (marked in the field), divided into two baskets (marketable and mouldy), and weighed. Additionally, in 2007, an assay of storage durability was carried out with berries picked from each treatment.

Technically all the experiments worked out very well. Bees accustomed themselves rapidly to the dispensers, and after 1–2 days exited and entered the hive through the dispenser without hesitation. Bee visits on the strawberry flowers were monitored throughout the flowering season in 2007, and they were found to visit each flower on the average ten times every day, throughout the season, carrying a load of the biocontrol fungus to the flower each time.

Consistently, all control treatments were highly effective regardless the weather conditions during the four study years: overall, bee-vectored biocontrol alone decreased disease incidence on average by 50%, chemical control by 65%, and both methods together by 80%. However, when the marketable yield was measured (instead of disease incidence), biocontrol alone proved to be as effective as, or more effective than the two other methods. Remarkably the detailed results obtained in 2008 do show for chemical control a high degree of disease control, but only marginal increase in marketable yield over untreated control in the absence of enhanced pollination by bees (Fig. 5.8a). In contrast, biocontrol alone provided in 2008 the highest marketable yields, with 90% overall increase over untreated controls. Surprisingly, combining biocontrol with chemical control did not increase the total marketable yield, despite providing superior disease control (Fig. 5.8a).

The pattern displayed by treatments involving chemical fungicide sprays can only be explained by a likely impact of the sprays on the yield potential of the strawberry plants. Based on the total berry yield produced by the strawberry plants (marketable and mouldy berries together), the lowering of the yield potential by the

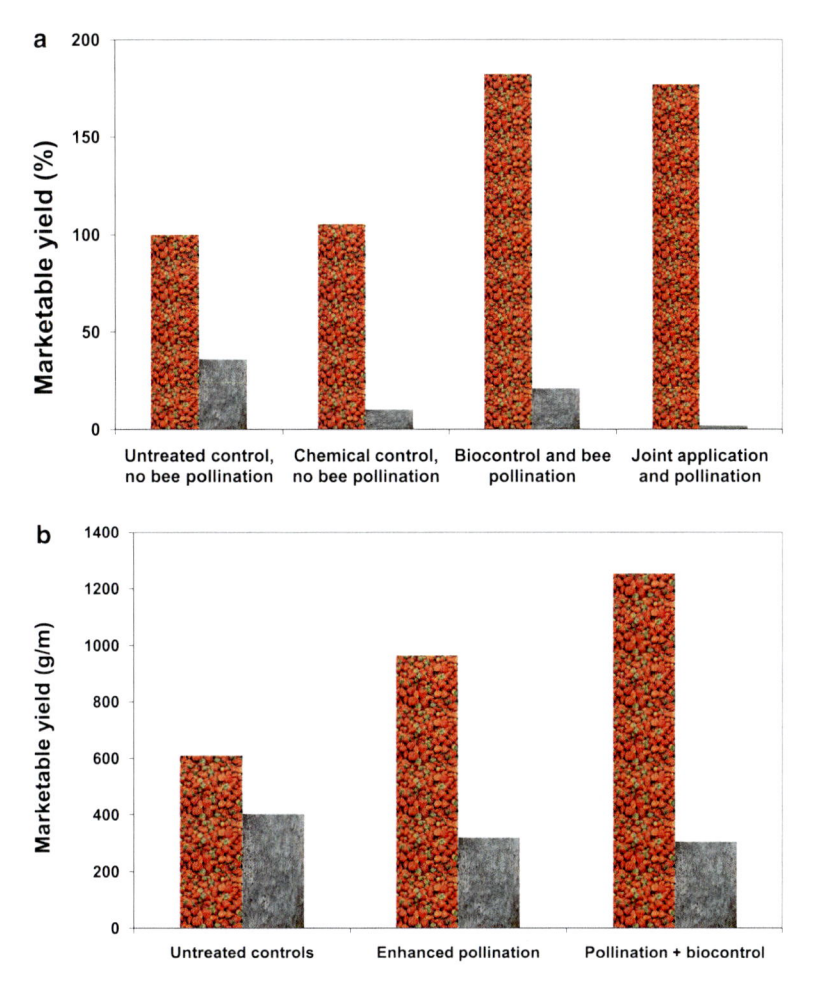

Fig. 5.8 Overview of the yield of marketable (healthy = *red bars*) and mouldy berries (= *grey bars*) per 1 m of strawberry row obtained in case study 1. (**a**) Comparison of marketable yield from the four different treatments, relative to untreated control (= 100%) [Data from 2008 (4 farms, each with 4 replicates)]. (**b**) Yields on an organic strawberry farm in 2008, comparing untreated controls (no active disease control, natural pollination), enhanced pollination by honey bees but without active disease control, and enhanced pollination combined with targeted honey bee disseminated biocontrol (Hokkanen et al. 2012)

chemical sprays was 17% in the treatments without enhanced pollination (first two columns in Fig. 5.8a), and 18% in the bee pollinated treatments (last two columns in Fig. 5.8a).

A significant part of the yield increases, as shown in Fig. 5.8a, are due to improved pollination of the flowers, inherent in the concept of using bees to vector the MCA. This was clearly demonstrated by the detailed data from an organic farm, where the pollination effect of bees was studied (Fig. 5.8b): marketable berry yield per 1 m of the strawberry row was 610 g in the untreated control, but 965 g when abundant bee pollination was provided (58% yield increase). Adding the grey mould control by *C. catenulatum* further increased the marketable berry yield to 1,253 g (105% yield increase over the untreated controls).

All grey mould control treatments improved significantly the shelf-life of strawberries, approximately doubling their durability. Combined use of biocontrol and fungicides extended the shelf-life of the berries more than either method alone.

In conclusion, the experiences of dozens of growers in Finland using the bee-disseminated biological control have been very positive (Hokkanen and Menzler-Hokkanen 2007, 2009). The method is technically easy to handle – not a single grower has complained that the system would be difficult to manage. Despite the frequent filling of the dispensers, the amount of work required is far less than for chemical treatments: filling a dispenser takes less than 2 min in the morning. Besides the facts that the use of biocontrol saved time, work and equipment for the growers, its main advantage may be that it saves money: while the costs of chemical control range at about 500–1,000 €/ha, biocontrol cost is only about 300 €/ha.

One berry grower – and beekeeper – who has participated in this research from its very beginning, summarises his experiences after 4 years as follows: *"When I started growing strawberries 20 years ago, yield levels were typically about 5,000 kg/ha. After I started to keep bees ten years ago, the yields rose up to about 6,000-9,000 kg/ha. During the past four years, the yields never have been below 10,000 kg/ha"*.

5.3.2 The Efficiency of Bumble Bees of Bombus terrestris to Vector the Commercial Product "Prestop-Mix", Containing G. catenulatum J1446, in Greenhouse Strawberries to Control B. cinerea in Belgium

In this series of experiments, we report on the potential of *B. terrestris* to suppress the plant pathogen, *B. cinerea* under the controlled conditions of a greenhouse as greenhouse strawberries are currently pollinated by *B. terrestris*. As depicted in Fig. 5.9, four different treatments were investigated (Mommaerts et al. 2011):

- T1: "control" = manual infection with *B. cinerea* and no pollination by bumblebees;
- T2: "Maizena-Plus" = manual infection with *B. cinerea* and dissemination of "Maizena-Plus" by bumblebees via the Mommaerts-dispenser;
- T3: "Prestop-Mix" = manual infection with *B. cinerea* and dissemination of "Prestop-Mix" by bumblebees via the Mommaerts-dispenser, and
- T4: "Prestop-Mix + Maizena-Plus" = manual infection with *B. cinerea* and dissemination of mixture of "Prestop-Mix" and "Maizena-Plus" (1/1, w/w) by bumblebees via the Mommaerts-dispenser.

In these experiments the Mommaerts-dispenser was used as developed and described in Mommaerts et al. (2010) and presented here in Fig. 5.4.

Fig. 5.9 Experimental set up of case study 2 with bumble bees of *Bombus terrestris* to control *Botrytis cinerea* in greenhouse strawberries with use of the Mommaerts-dispenser, "Prestop-Mix" and "Maizena-Plus". *T1–T4* represent the four different treatments

Table 5.2 Overview of the pre- and post-harvest yield percentages and total efficacy of vectoring to suppress *B. cinerea* for the four different treatments (T1–T4)

Treatment	% Yield (pre-harvest)	% Yield (post-harvest)	% Total yield (pre- and post-harvest)[a]
T1	54 ± 21	43 ± 13	24 ± 14 a
T2	51 ± 9	50 ± 10	25 ± 8 a
T3	72 ± 17	67 ± 13	47 ± 10 b
T4	71 ± 9	79 ± 17	56 ± 10 b

[a]Analyses with one-way ANOVA resulted for the % total yield in two groups (F = 8.291; df = 15; p = 0.003). The values per treatment followed by a different letter (a–b) are significantly different after a *post hoc* Duncan test with $\alpha = 0.05$

The biocontrol efficacy was determined at two different time points, namely at picking of the red berries (pre-harvest yield), and after 2 days of incubation in the laboratory after picking (post-harvest yield). The strawberry fruits were scored according to a binary system of 0 and 1, where '0' stands for fruits without symptoms of damage, and '1' for infected fruit.

As depicted in Table 5.2, for the pre-harvest yield, yields were higher in the treatments including "Prestop-Mix" as compared to the treatments without MCA. This increase of control was obtained when flowers were inoculated with a mean number of 640 CFU of *B. cinerea* per flower, resembling a medium disease pressure. Analyses of the flowers confirmed the capacity of manual inoculation *B. cinerea* to grow on the petals as mycelium was found after 24 h. Also the greenhouse conditions with temperature of 18.5–20.6°C and relative humidity of 88–97% were in the optimal range for growth of the plant pathogen during the entire experimental period.

For the post-harvest yield, the same trend was observed. Strawberries were better protected in T3 and T4 (including "Presop-Mix") because the numbers of picked berries that remained free of *B. cinerea* after a 2-day-incubation was higher than in T1 and T2 (without MCA) (Table 5.2).

Finally, when considering the total yield (pre-harvest yield x post-harvest yield), vectoring of "Prestop-Mix + Maizena-Plus" (T4) or "Prestop-Mix" alone (T3) resulted in the highest yield, while the total yield percentages were significantly lower ($p < 0.05$) for T1 and T2 (Table 5.2).

In conclusion, the total yield in the treatments including biocontrol was 2–2.5 times higher than in the controls. In addition, it is of interest that addition of the carrier "Maizena-Plus" to "Prestop-Mix" at 1:1 (w/w) resulted in a similar yield to that of "Prestop-Mix" used alone. This observation might be interesting for future studies investigating the formulation of MCA products.

Next to the biocontrol efficiency, the present study also confirmed the safety of the MCA for the vector. Over the 4-week blooming period during which bumble bees disseminated "Prestop-Mix", the amount of worker mortality in T3 and T4 was not higher than in T2. In addition, monitoring of the foraging activity of the bumble bee workers did not show any adverse effects by the MCA.

5.4 Conclusions and Future Perspectives

Biological control in open field is a challenge. In general berry and fruit production currently suffers heavily from the lack of effective disease and pest management tools, as well as from inadequate insect pollination. The present chapter provides evidence that these limitations can be overcome via development of targeted precision biocontrol and pollination enhancement involving honey bees, bumble bees, and solitary bees. Also the case study conducted in open field strawberry farms in Finland confirmed the potential of this technology for specific applications, where classic control methods are not possible. Similar to open field, several research groups have reported promising results on plant pathogen and/or pest control under greenhouse conditions using *Bombus* bumble bees. Until today, research by several groups on the entomovector context have resulted in advances in:

(i) the development of suitable dispensers;
(ii) exploring the knowledge of pollinator efficiency;
(iii) the identification of potential carriers;
(iv) several commercially available MCAs.

Here it is important to mention that to date the entomovector technology is already recommended and officially approved since spring 2008 by the Finnish Food Safety Authority for practice in open field strawberries in Finland. As future perspective, we expect that the use of the entomovector technology in agriculture and horticulture would benefit from further research on aspects as follows:

• First, in order to minimize yield loss it is crucial that MCA efficacy can be guaranteed towards growers under diverse environmental conditions. In this context it would be interesting to evaluate combinations of different MCA strains and/or mixtures of MCAs with low risk chemical pesticides. For example, case study 1 indicated a potential for the combined use of "Prestop-Mix" within the entomovector technology and the chemical control strategy currently applied by cultivators in open field strawberry.

- Second, so far several entomovector studies reported a high loss of the MCA product during vector flight. We are therefore convinced that there is still a need to further fine-tuning of commercial MCA formulations. The final goal of these adaptations would be to optimize MCA product deposition in the flowers to improve control capacity under high disease or pest pressure. So far several potential carriers have been reported, but their usefulness within the entomovector technology will strongly depend on their safety towards the vectors. For example, case study 2 of this chapter demonstrated the efficacy of corn starch ("Maizena-Plus") when mixed with "Prestop-Mix", along with its safety towards bumble bees.
- Third, the entomovector technology has confirmed to control different plant pathogens and pest insects, particularly *B. cinerea* in strawberries. However, it should be said that this technology can be of use in more and other crops and in the control of more and other diseases and pests. Together with this growing development, we believe it opens the search for new antagonists, and in addition for investigations into depth on antagonist-plant pathogen/pest-plant interactions.
- Fourth, entomovector studies reported so far only reported on the use of one vector per study to disseminate the MCA product. Here we envisage that to guarantee pollination during the whole day or during the whole blooming season, it might be useful to investigate the efficiency when a combination of two or more vectors species is used. For this selection, information on the behaviour of the pollinator(s) is crucial as well as the guarantee that enough floral resources are present to avoid over-pollination and/or competition.

In summary, the entomovector technology is a "win-win" situation because the use of pollinators does not only lead to improved pollination but also to a reduction of the pest/disease pressure. We hope therefore that further research will contribute to its success and to its implementation in current IPM programs as a sustainable agriculture solution for crop protection.

Acknowledgements The authors acknowledge support for their research by the Fund for Scientific Research-Flanders (FWO-Vlaanderen), the Flemish agency for Innovation by Science and Technology (IWT-Vlaanderen), the Special Research Funds of Ghent University and of VUB, NordForsk grant 70066 (Entomovector technology), NordForsk Project no. 45941 (BICOPOLL-NET), and CORE-Organic II project "BICOPOLL" in an ERA-NET funded by the European Commission's 7th Framework Program.

References

AAFC (2009) Crop profile for strawberry in Canada. Agriculture and Agri-Food Canada. Available at http://dsp-psd.pwgsc.gc.ca/collection_2009/agr/A118-10-17-2005E.pdf. Accessed on 10 Feb 2011

Albano S, Chagon M, de Oliveira D, Houle E, Thibodeau PO, Mexia A (2009) Effectiveness of *Apis mellifera* and *Bombus impatiens* as dispensers of the Rootshield® biofungicide (*Trichoderma harzianum*, strain T-22) in a strawberry crop. Hell Plant Prot J 2:57–66

Alexandrova M, Bazzi C, Lameri P (2002) *Bacillus subtilis* strain BS-F3: colonisation of pear organs and its action as a biocontrol agent. Acta Hortic 590:291–297

Al-mazra'awi MS, Shipp JL, Broadbent AB, Kevan PG (2006a) Dissemination of *Beauveria bassiana* by honey bees (Hymenoptera: Apidae) for control tarnished plant bug (Hemiptera: Miridae) on canola. Biol Control 35:1569–1577

Al-mazra'awi MS, Shipp JL, Broadbent AB, Kevan PG (2006b) Biological control of *Lygus lineolaris* (Hemiptera: Miridae) and *Frankiniella occidentalis* (Thysanoptera: Thripidae) by *Bombus impatiens* (Hymenoptera: Apidae) vectored *Beauveria bassiana* in greenhouse sweet pepper. Biol Control 37:89–97

Al-mazra'awi MS, Kevan PG, Shipp L (2007) Development of *Beauveria bassiana* dry formulation for vectoring by honey bees *Apis mellifera* (Hymenoptera: Apidae) to the flowers of crops for pest control. Biocontrol Sci Technol 17:733–741

Beever RE, Weeds PL (2004) Taxonomy and genetic variation of *Botrytis* and *Botryotinia*. In: Elad Y, Williamson B, Tudzynski B, Delen N (eds) *Botrytis*: biology, pathology and control. Kluwer Academic Publishers, Dordrecht/Boston/London, pp 29–52

Bilu A, Dag A, Elad Y, Shafir S (2004) Honey bee dispersal of biocontrol agents: an evaluation of dispensing devices. Biocontrol Sci Technol 14:607–617

Bosch J, Kemp WP (2002) Developing and establishing bee species as crop pollinators: the example of *Osmia* spp. (Hymenoptera: Megachilidae) and fruit trees. Bull Entomol Res 92:3–16

Butt TM, Carreck NL, Ibrahim L, Williams IH (1998) Honey bee-mediated infection of pollen beetle (*Meligethes aeneus* Fab.) by the insect-pathogenic fungus, *Metarhizium anisopliae*. Biocontrol Sci Technol 8:533–538

Carreck NL, Butt TM, Clark SJ, Ibrahim L, Isger EA, Pell JK, Williams IH (2007) Honey bees can disseminate a microbial control agent to more than one inflorescence pest of oilseed rape. Biocontrol Sci Technol 17:179–191

Christensen L (2006) Practical use of biological control of pest and diseases in Danish glasshouses - bottlnecks and challenges. In: Hansen LS, Enkegaard A, Steenberg T, Ravnskov S, Larsen J (eds) Proceedings of the International Workshop "Implementation of Biocontrol in Practice in Temperate Regions - Present and Near Future". DIAS report Plant Production 119:169–171

Cohen AC, Nordlund DA, Smith RA (1999) Mass rearing of entomophagous insects and predaceous mites: are the bottlenecks biological, engineering, economic, or cultural? Biocontrol News Info 20(3):85N–90N

Cribb DM, Hand DW (1993) A comparative study of the effects of using the honeybee as a pollinating agent of glasshouse tomato. J Hortic Sci 68:79–88

Dag A, Weinbaum SA, Thorp R, Eiskowitch D (2000) Evaluation of pollen dispensers ('inserts') effect on fruit set and yield in almond. J Apic Res 39:117–123

Dedej S, Delaplane KS, Scherm H (2004) Effectiveness of honey bees in delivering the biocontrol agent *Bacillus subtilis* to blueberry flowers to suppress mummy berry disease. Biol Control 31:422–427

Elad Y, Freeman S (2002) Biological control of fungal plant pathogens. In: Kempken F (ed) The Mycota, a comprehensive treatise on fungi as experimental systems for basic and applied research. Springer, Heidelberg

Escande AR, Laich FS, Pedraza MV (2002) Field testing of honeybee-dispersed *Trichoderma* spp. to manage sunflower head rot (*Sclerotinia sclerotiorum*). Plant Pathol 51:346–351

FAO (2011) FAOSTAT agricultural production statistics. Available at http://faostat.fao.org/site/567/ and at http://faostat.fao.org/site/339/. Accessed on 10 Feb 2011

Farina WM, Gruter C, Acosta L, Cabe SMC (2007) Honeybees learn floral odors while receiving nectar from foragers within the hive. Naturwissensch 94:55–60

Forrest J, Thomson JD (2009) Background complexity affects colour preference in bumblebees. Naturwissensch 96:921–925

Frost & Sullivan (2001) European biopesticides market. Available at http://www.frost.com. Accessed 15 Apr 2005

Gelernter WD, Lomer CJ (2000) Success in biological control of above-ground insects by pathogens. In: Gurr G, Wratten SD (eds) Biological control: measures of success. Kluwer Academic Publishers, Dordrecht

Gil M (2010) Reward expectations in honeybees. Commun Integr Biol 3:95–100

Goulson D (2010) Bumblebees behaviour, ecology and conservation. Oxford University Press, New York, pp 317

Gross HR, Hamm JJ, Carpenter JE (1994) Design and application of a hive-mounted device that uses honey bees (Hymenoptera: Apidae) to disseminate *Heliothis* nuclear polyhedrosis virus. Biol Control 23:492–501

Guerra-Sanz JM (2008) Crop pollination in greenhouses. In: James RR, Pitts-Singer T (eds) Bee pollination in agriculture ecosystems. Oxford University Press, New York

Hjeljord LG, Stensvand A, Tronsmo A (2000) Effect of temperature and nutrient stress on the capacity of commercial *Trichoderma* products to control *Botrytis cinerea* and *Mucor piriformis* in greenhouse strawberries. Biol Control 19:149–160

Hokkanen HMT, Menzler-Hokkanen I (2007) Use of honeybees in the biological control of plant diseases. Entomol Res 37(suppl 1):A62–A63

Hokkanen HMT, Menzler-Hokkanen I (2009) Successful use of honey bees for grey mould biocontrol on strawberries and raspberries in Finland. Apidologie 40:659

Hokkanen HMT, Menzler-Hokkanen I, Mustalahti A-M (2012) Honey bees (*Apis mellifera*) for precision biocontrol of grey mould (*Botrytis cinerea*) with *Gliocladium catenulatum* on strawberries and raspberries in Finland. Arthropod-Plant Interactions (submitted)

IPMCenters (2011) Crop profile for strawberries in Louisiana. Available at http://www. ipmcenters.org/cropprofiles/docs/LAstrawberries.pdf. Accessed on 10 Feb 2011

Israel MS, Boland GJ (1993) Influence of formulation on efficacy of honey bees to transmit biological controls for management of *Sclerotinia* stem rot of canola. Can J Plant Pathol 14:244

Johnson KB, Stockwell VO, Burgett DM, Sugar D, Loper JE (1993a) Dispersal of *Erwinia amylovora* and *Pseudomonas fluorescens* by honeybees from hives to apple and pear blossoms. Phytopathology 83:478–484

Johnson KB, Stockwell VO, Mclaughlin RJ (1993b) Effect of antagonistic bacteria on establishment of honey bee-dispersed *Erwinia amylovora* in pear blossoms and on fire blight control. Phytopathology 83:995–1002

Jyoti JL, Brewer GJ (1999) Honeybees (Hymenoptera: Apidae) as vector of *Bacillus thuringiensis* for control of branded sunflower moth (Lepidoptera: Tortricidae). Environ Entomol 28:1172–1176

Kapongo JP, Shipp L, Kevan P (2008a) Optimal concentration of *Beauveria bassiana* vectored by bumble bees in relation to pest and bee mortality in greenhouse tomato and sweet pepper. Biocontrol 53:797–812

Kapongo JP, Shipp L, Kevan P, Sutton JC (2008b) Co-vectoring of *Beauveria bassiana* and *Clonostachys rosea* by bumblebees (*Bombus impatiens*) for control of insect pests and suppression of grey mould in greenhouse tomato and sweet pepper. Biol Control 46:508–514

Kevan PG, Kapongo J-P, Al-mazra'awi M, Shipp L (2008) Honey bees, bumble bees and biocontrol. In: James RR, Pitts-Singer T (eds) Bee pollination in agriculture ecosystems. Oxford University Press, New York

Kovach J, Petzoldt R, Harman GE (2000) Use of honeybees and bumble bees to disseminate *Trichoderma harzianum* 1295–22 to strawberries for *Botrytis* control. Biol Control 18:235–242

Lunau K, Unseld K, Wolter F (2009) Visual detection of diminutive floral guides in the bumblebee *Bombus terrestris* and in the honeybee *Apis mellifera*. J Comp Physiol 195A:1121–1130

Maccagnani B, Mocioni M, Gullino ML, Ladurner E (1999) Application of *Trichoderma harzianum* by using *Apis mellifera* as a vector for the control of grey mold of strawberry: first results. IOBC Bull 22:161–164

Maccagnani B, Mocioni M, Ladurner E, Gullino ML, Maini S (2005) Investigation of hive-mounted devices for the dissemination of microbiological preparations by *Bombus terrestris*. Bull Insectol 58:3–8

Maccagnani BBC, Biondi E, Tesoriero D, Maini S (2006) Potential of *Osmia cornuta* as a carrier of antagonist bacteria in biological control of fire blight: a comparison with *Apis mellifera*. Acta Hort (ISHS) 704:379–386

Mertley JC, Mackenzie SJ, Legard DE (2002) Timing of fungicide applications for *Botrytis cinerea* based on development stage of strawberry flowers and fruit. Plant Dis 86:1019–1024

Molet M, Chittka L, Raine NE (2009) How floral odours are learned inside the bumblebee (*Bombus terrestris*) nest? Naturwissensch 96:213–219

Mommaerts V, Smagghe G (2011) Entomovectoring in plant protection. Arthropod-Plant Interact 5:81–95

Mommaerts V, Put K, Vandeven J, Jans K, Sterk G, Hoffmann L, Smagghe G (2010) Development of a new dispenser for bumblebees and evaluation to disseminate microbiological control agents in strawberry in the greenhouse. Pest Manag Sci 66:1199–1207

Mommaerts V, Kurt P, Smagghe G (2011) *Bombus terrestris* as pollinator-and-vector to suppress *Botrytis cinerea* in greenhouse strawberry. Pest Manag Sci 67:1069–1075

Moscardi F (1999) Assessment of the application of baculoviruses for control of Lepidoptera. Annu Rev Entomol 44:257–289

Ngugi HK, Scherm H, Lehman JS (2002) Relationship between blueberry flower age, pollination and conidial infection by *Monilinia vaccinii-corymbosi*. Ecol Popul Biol 92:1104–1109

Nilsson U, Gripwall E (1999) Influence of application technique on the viability of the biological control agents *Verticillium lecanii* and *Stenernema feltiae*. Crop Prot 18:53–59

Noma T, Strickler K (2000) Effects of *Beauveria bassiana* on *Lygus hesperus* (Hemiptera: Miridae) feeding and oviposition. Environ Entomol 29:394–402

Oh CS, Beer SV (2005) Molecular genetics of *Erwinia amylovora* involved in the development of fire blight. FEMS Microbiol Lett 253:185–192

Osborne JL, Martin AP, Carreck NL, Swain JL, Knight ME, Goulson D, Hale RJ, Sanderson RA (2008) Bumblebee flight distances in relation to the forage landscape. J Anim Ecol 77:401–415

Peng G, Sutton JC, Kevan PG (1992) Effectiveness of honeybees for applying the biocontrol agent *Gliocladium rosea* to strawberry flowers to suppress *Botrytis cinerea*. Can J Plant Pathol 14:117–129

Pettis JS, Kochansky J, Feldlaufer MF (2004) Larval *Apis mellifera* L. (Hymenoptera: Apidae) mortality after topical application of antibiotics and dusts. J Econ Entomol 97:171–176

Pilkington LJ, Messelink G, van Lenteren JC, Le Mottee K (2010) Protected biological control – biological pest management in the greenhouse industry. Biol Control 52:216–220

Prokkola S, Kivijärvi P (2007) Effect of biological sprays on the incidence of grey mould, fruit yield and fruit quality in organic strawberry production. Agric Food Sci 16:25–33

Raine NE, Chittka L (2007) The adaptive significance of sensory bias in a foraging context: floral colour preferences in the bumblebee *Bombus terrestris*. PLoS One 2(6):e556. doi:10.1371/journal.pone.0000556

Rands SA, Whitney HM (2008) Floral temperature and optimal foraging: is heat a feasible floral reward for pollinators? PLoS One 3(4):e2007. doi:10.1371/journal.pone.0002007

Ravensberg W (2011) A roadmap to the successful development and commercialization of microbial pest control products for control of arthropods, vol 10, Progress in biological control. Springer, Zoetermeer, 383 p

Scherm H, Ngugi HK, Savelle AT, Edwards JR (2004) Biological control of infection of blueberry flowers caused by *Monilinia vaccinii-corymbosi*. Biol Control 29:199–206

Shafir S, Dag A, Bilu A, Abu-Toamy M, Elad Y (2006) Honeybee dispersal of the biocontrol agent and *Trichoderma harzianum* T39: effectiveness in suppressing *Botrytis cinerea* on strawberry under field conditions. Eur J Plant Pathol 116:119–128

Shinners TC, Olson AR (1996) The gynoecial infection pathway of *Monilinia vaccinii-corymbosi* in lowbush blueberry (*Vaccinium angustifolium*). Can J Plant Sci 76:493–497

Sigsgaard L (2006) Biological control of arthropod pests in outdoor crops – the new challenge. DIAS report Plant Production 119, pp 153–168

Spaethe J, Tautz J, Chittka L (2001) Visual constraints in foraging bumblebees: flower size and color affect search time and flight behaviour. Proc Natl Acad Sci USA 98:3898–3903

Stout JC, Goulson D (2002) The influence of nectar secretion rates on the responses of bumblebees (*Bombus* spp.) to previously visited flowers. Behav Ecol Sociobiol 52:239–246

Strømeng GM (2008) Aspects of the biology of *Botrytis cinerea* in strawberry (*Fragaria* x *ananassa*) and alternative methods for disease control. Philosophiae Doctor (PhD) thesis 2008, pp 56

Thomson SV, Hansen DR, Flint KM, Vandenberg JD (1992) Dissemination of bacteria antagonistic to *Erwinia amylovora* by honey bees. Plant Dis 76:1052–1056

Törmälä T (1995) Economics of biocontrol agents: an industrial view. In: Hokkanen HMT, Lynch JM (eds) Biological control: benefits and risks. Cambridge University Press, Cambridge, pp 277–282

van der Blom J, Robledo A, Torres S, Sánchez JA (2009) Consequences of the wide scale implementation of biological control in greenhouse horticulture in Almeria. Spain. IOBC/WPRS Bull 49:9–13

van Lenteren JC (ed) (2008) Internet book of biological control, 5th edn. IOBC, Wageningen, www.IOBC-Global.org

Vanneste JL (1996) Honey bees and epiphytic bacteria to control fire blight, a bacterial disease of apple and pear. Biocont News Info 17:67N–78N

Vicens N, Bosch J (2000) Pollinating efficacy of *Osmia cornuta* and *Apis mellifera* (Hymenoptera: Megachilidae, Apidae) on 'Red Delicious' apple. Environ Entomol 29:235–240

Whitney HM, Dyer A, Chittka L, Rands SA, Glover BJ (2008) The interaction of temperature and sucrose concentration on foraging preferences in bumblebees. Naturwissensch 95:845–850

Williamson B, Tudzynski B, Tudzynski P, Van Kan JAL (2007) *Botrytis cinerea*: the cause of grey mold disease. Mol Plant Pathol 8:561–580

Wilson M, Lindow SE (1993) Interactions between the biological control agent *Pseudomonas fluorescens* A506 and *Erwinia amylovora* in pear blossoms. Phytopathology 83:117–123

Wilson M, Epton HAS, Sigee DC (1992) Interactions between *Erwinia herbicola* and *E. amylovora* on the stigma of hawthorn blossoms. Phytopathology 82:914–918

Wolf S, Moritz RFA (2008) Foraging distance in *Bombus terrestris* L. (Hymenoptera: Apidae). Apidologie 39:419–427

Yu H, Sutton JC (1997) Effectiveness of bumblebees and honeybees for delivering inoculum of *Gliocladium roseum* to raspberry flowers to control *Botrytis cinerea*. Biol Control 10:113–122

Chapter 6
Biotechnological Approaches to Combat Phytophagous Arthropods

Isabel Diaz and M. Estrella Santamaria

New biotechnological approaches, based on the use of defence genes and the beneficial effects of the novel technology systems in terms of sustained agriculture are being explored to be integrated in pest resistance management. The development of insect-resistant crops have drastically increased since the commercial release of the first Bt-plant generation expressing a single *Bacillus thuringiensis* (Bt) toxin, 15 years ago (James 2010). These modified crops, successfully applied for agricultural use, triggered an important reduction of pesticide usage. However, pest resistance is still discussed (Carriere et al. 2010) and some phytophagous arthropods, particularly aphids and mites are moderate or insensitive to toxins encoded by Bt genes (Lawo et al. 2009; Li and Romeis 2010). Therefore, a great effort to search for alternative strategies of protecting crops from pest has been made. The development of plant genetic transformation has provided tools for transferring multiple pest resistance traits into agronomic important crop plants (Hilder and Boulter 1999; Christou et al. 2006; Ferry et al. 2006; Gatehouse 2008). This technology has been mainly focussed on the use of plant-derived genes with insecticidal and/or acaricidal properties and molecules or toxins from multiple sources that when expressed in a variety of plants resulted in enhanced resistance towards a wide spectrum of pests in laboratory assays. Currently, new molecular strategies for sustainable pest resistance in genetically enhanced crops are being provided from the understanding of endogenous resistant mechanisms developed by the plants as a result of the plant-herbivore interaction.

This chapter reviews the potential of current biotechnological prospects for pest control to be integrated in pest management programs for reducing crop losses and pesticide uses and for avoiding pest resistance and emergence of new pest.

I. Diaz (✉) • M.E. Santamaria
Centro de Biotecnología y Genómica de Plantas (UPM-INIA), Campus Montegancedo,
Universidad Politécnica de Madrid, Autovía M40 (km 38), 28223 Pozuelo de Alarcón,
Madrid, Spain
e-mail: i.diaz@upm.es

6.1   Potential Use of Defence Proteins

A huge number of publications have reported the potential of plant proteins synthesized in response to plant-herbivore interaction. Probably, the hydrolase inhibitor group is the most studied although the gene expression of its members in transgenic plants confer only a partial protection towards arthropod pests. Hydrolytic enzyme inhibitors, including protease and amylase inhibitors, act as pseudo-substrates of target digestive enzymes, interfere with the digestive herbivore process and bring about disruption of dietary compound assimilation with the consequent reduction in the pest growth and a delay in their development. The best characterized are the protease inhibitors (PIs), grouped in serine-, cysteine- aspartyl- and metallo-proteases, according to their specificity. PIs act as effective antidigestive compounds by interacting with their target proteases which can no longer cleave peptide bonds (Schluter et al. 2010). The first successful PI gene shown to improve resistance against larvae of *Heliothis virescens* when expressed in transgenic tobacco was CPTI, the cowpea trypsin inhibitor gene (Hilder et al. 1987). Then, the CPTI gene was integrated in the genome of cotton, rice, cabbage, strawberry, sweet potato, potato and pigeonpea, thus enhancing the resistance to different lepidopteran species (Hilder et al. 1989; Boulter et al. 1990; Hoffman et al. 1992; Gatehouse et al. 1993; Xu et al. 1996; Fang et al. 1997; Gatehouse et al. 1997; Golmirizaie et al. 1997; Graham et al. 1997; Hao and Ao 1997; Sane et al. 1997; Li et al. 1998; Bell et al. 2001; Lawrence and Koundal 2001). Subsequently, a number of serine-protease inhibitor genes were expressed in tomato, tobacco, rice, sweet potato, poplar, arabidopsis, oilseed rape, cauliflower, sugarcane, wheat, pea, apple and caoba to confer resistance against lepidopteran (Shulke and Murdock 1983; Johnson et al. 1989; McManus et al. 1994, 1999; Jongsma et al. 1995; Duan et al. 1996; Heath et al. 1997; Wu et al. 1997; Yeh et al. 1997; Confalonieri et al. 1998; De Leo et al. 1998, 2001; Ding et al. 1998; Altpeter et al. 1999; Cipriani et al. 1999; Charity et al. 1999; Gatehouse et al. 1999; Lee et al. 1999; Mochizuki et al. 1999; Nandi et al. 1999; Lara et al. 2000; Marchetti et al. 2000; Winterer and Bergelson 2001; De Leo and Gallerani 2002; Vila et al. 2005; Maheswaran et al. 2007; Da Silveira et al. 2009; Khadeeva et al. 2009; Luo et al. 2009; Srinvasan et al. 2009), coleopteran (Klopfenstein et al. 1997; Nutt et al. 1999; Alfonso-Rubí et al. 2003), homopteran (Hesler et al. 2005) and Acari (Castagnoli et al. 2003). See Table 6.1 and references herein.

Transgenic plants harbouring plant cysteine-protease inhibitors, phytocystatins, have been generated in several plants species (Table 6.2). Cystatins from rice, Arabidopsis, potato and barley expressed in mono- and di-cotyledoneous species have conferred resistance to coleopteran (Leple et al. 1995; Gatehouse et al. 1996; Irie et al. 1996; Girard et al. 1998a, b; Jouanin et al. 1998; Bonade-Bottino et al. 1999; Lecardonnel et al. 1999; Cloutier et al. 1999, 2000; Delledonne et al. 2001; Bouchard et al. 2003a, b; Ninković et al. 2007), lepidopteran (Ferry et al. 2003; Alvarez-Alfageme et al. 2007); homopteran (Schuler et al. 2001; Cowgill et al. 2004; Rahbe et al. 2003; Ribeiro et al. 2006; Carrillo et al. 2011a), thysanopteran

Table 6.1 Transgenic plants over-expressing protease inhibitors and target arthropod pests

Serine-protease inhibitor	Transgenic plant	Target pest species	References
CPTI	Tobacco	*Heliothis virescens*	Hilder et al. (1987)
			Hilder et al. (1989)
			Boulter et al. (1990)
			Gatehouse et al. (1993)
		Manduca sexta	Hilder et al. (1989)
	Cotton	*Helicoverpa zea*	Hilder et al. (1989)
			Hoffman et al. (1992)
		Spodoptera littoralis	Hilder et al. (1989)
		Autographa gamma	
		Spodoptera litura	Sane et al. (1997)
		Helicoverpa armigera	Li et al. (1998)
	Rice	*Chilo suppressalis*	Xu et al. (1996)
		Sesamia inferens	
		Spodoptera infestans	
	Cabbage	*Pieris rapae*	Fang et al. (1997)
		Helicoverpa armigera	
		Pieris rapae	Hao and Ao (1997)
		Heliothis armigera	
	Strawberry	*Otiorhynchus sulcatus*	Graham et al. (1997)
	Sweet potato	*Euscepes postfaciatus*	Golmirizaie et al. (1997)
	Potato	*Lacanobia oleracea*	Gatehouse et al. (1997)
			Bell et al. (2001)
	Pigeonpea	*Helicoverpa armigera*	Lawrence and Koundal (2001)
SKTI	Tobacco	*Spodoptera litura*	McManus et al. (1999)
		Helicoverpa armigera	Nandi et al. (1999)
		Spodoptera littoralis	Marchetti et al. (2000)
	Potato	*Spodoptera littoralis*	Marchetti et al. (2000)
		Lacanobia oleracea	Gatehouse et al. (1999)
	Poplar	*Lymantria dispar*	Confalonieri et al. (1998)
		Clostera anastomosis	
	Rice	*Nilaparvata lugens*	Lee et al. (1999)
SKTi-3	Tomato	*Tetranychus urticae*	Castagnoli et al. (2003)
SKTI-4	Sweet potato	*Cylas spp.*	Cipriani et al. (1999)
SKC-II	Tobacco/potato	*Manduca sexta*	Shulke and Murdock (1983)
		Spodoptera littoralis	Marchetti et al. (2000)
SKPI-IV	Tobacco/potato	*Spodoptera littoralis*	Marchetti et al. (2000)
MTI-2	Tobacco	*Spodoptera littoralis*	De Leo et al. (1998)
			De Leo and Gallerani (2002)
		Mamestra brassicae	De Leo et al. (2001)
	Arabidopsis	*Spodoptera littoralis*	De Leo et al. (1998)
		Mamestra brassicae	De Leo et al. (2001)
		Plutella xylostella	
	Oilseed rape	*Plutella xylostella*	De Leo et al. (2001)
		Mamestra brassicae	
		Spodoptera littoralis	

(continued)

Table 6.1 (continued)

Serine-protease inhibitor	Transgenic plant	Target pest species	References
SpTi-1	Tobacco	*Spodoptera litura*	Yeh et al. (1997)
	Cauliflower	*Spodoptera litura*	Ding et al. (1998)
		Plutella xylostella	
TPI-I and II	Tobacco	*Manduca sexta*	Johnson et al. (1989)
PPI-II	Rice	*Chilo suppressalis*	Duan et al. (1996)
		Sesamia inferens	
	Sugarcane	*Antitrogus consanguineus*	Nutt et al. (1999)
	Tobacco	*Manduca sexta*	Johnson et al. (1989)
		Thysanoplusia orichalcea	
		Spodoptera litura	McManus et al. (1994)
		Chrysodeixis eriosoma	
		Thysanoplusia orichalcea	
		Spodoptera exigua	Jongsma et al. (1995)
		Helicoverpa armigera	Wu et al. (1997)
	Poplar	*Plagiodera versicolor*	Klopfenstein et al. (1997)
	Oilseed rape	*Plutella xylostella*	Winterer and Bergelson (2001)
	Wheat	*Rhopalosiphum padi*	Hesler et al. (2005)
Mpi	Rice	*Chilo suppressalis*	Vila et al. (2005)
WTI-1B	Rice	*Chilo suppressalis*	Mochizuki et al. (1999)
GPTI	Tobacco	*Helicoverpa* armigera	Wu et al. (1997)
BTI-CMe	Wheat	*Sitotroga cerealella*	Altpeter et al. (1999)
	Tobacco	*Spodoptera exigua*	Lara et al. (2000)
	Rice	*Sitophilus oryzae*	Alfonso-Rubí et al. (2003)
NaPI-II	Tobacco	*Helicoverpa punctigera*	Heath et al. (1997)
		Helicoverpa armigera	Charity et al. (1999)
	Pea	*Helicoverpa armigera*	Charity et al. (1999)
	Apple	*Epiphyas postvittana*	Maheswaran et al. (2007)
BPTI	Sugarcane	*Scirpophaga excerptalis*	Christy et al. (2009)
PFTI	Caoba	*Anagasta kuehniella*	Da Silveira et al. (2009)
TTPI	Tobacco	*Spodotera litura*	Srinvasan et al. (2009)
		Helicoverpa armigera	
BSPI	Tobacco	*Trialeurodes vaporariorum*	Khadeeva et al. (2009)
SAPI	Tobacco	*Helicoverpa armigera*	Luo et al. 2009
		Spodoptera litura	

Serine-protease inhibitors: Cowpea trypsin inhibitor (CpTI); Soybean Kunitz trypsin inhibitors (SKTI and SKTI-IV); Soybean kunitz serine proteinase inhibitors (C-II and PI-IV); Mustard trypsin inhibitor 2 (MTI-2); Sweet potato trypsin inhibitor (*spTi-1*) or sporamin; Tomato protease inhibitors I and II (TPI-I and TPI-II); Potato proteinase inhibitor II (PPI-II); Maize proteinase inhibitor (Mpi); Winged bean trypsin inhibitor (WTI-1B); Giant taro proteinase inhibitor (GPTI); Barley trypsin inhibitor (BTI-CMe); *Nicotiana alata* multidomain trypsin and chymotrypsin inhibitor (NaPI-II); *Plathymenia foliolosa* trypsin inhibitor (PFTI); Tobacco trypsin protease inhibitor (TTPI); Buckwheat serine protease inhibitor (BSPI); *Solanum americanum* proteinase inhibitor (SAPI).

Table 6.2 Transgenic plants over-expressing protease inhibitors and target arthropod pests

Protease inhibitor	Transgenic plant	Target species	References
Cysteine-PI[a]			
OC-I	Poplar	*Chrysomela tremulae*	Leple et al. (1995)
	Rice	*Sitophilus zeamais*	Irie et al. (1996)
	Potato	*Myzus persicae*	Gatehouse et al. (1996)
	Oilseed rape	*Ceutorhynchus assimilis*	Girad et al. (1998a)
			Jouanin et al. (1998)
		Psylliodes chrysocephala	Girad et al. (1998b)
		Baris coerulescens	Bonade-Bottino et al. (1999)
	Potato	*Leptinotarsa decemlineata*	Lecardonnel et al. (1999)
			Cloutier et al. (1999)
			Cloutier et al. (2000)
			Bouchard et al. (2003a, b)
	Oilseed rape	*Myzus prsicae*	Schuler et al. (2001)
			Rahbe et al. (2003)
		Plutella xylostella	Ferry et al. (2003)
	Eggplant	*Myzus persicae*	Ribeiro et al. (2006)
		Macrosiphum euphorbiae	
OC-IΔD86	Potato	*Myzus persicae*	Cowgill et al. (2004)
		Macrosiphum euphorbiae	
OC-II	Alfalfa	*Phytodecta fornicata*	Ninković et al. (2007)
AtCYS	Poplar	*Chrysomela populi*	Delledonne et al. (2001)
PC	Potato	*Fralkliniella occidentalis*	Outchkourov et al. (2004a, b)
HvCPI-1 G→C	Potato	*Leptinotarsa decemlineata*	Alvarez-Alfageme et al. (2007)
		Spodoptera littoralis	
HvCPI-6	Maize	*Tetranychus urticae*	Carrillo et al. (2011b)
		Brevipalpus chilensis	
	Arabidopsis	*Myzus persicae*	Carrillo et al. (2011a)
Other PIs[b]			
TCDI	Tomato	*Leptinotarsa decemlineata*	Brunelle et al. (2004)
PCPI	Rice	*Chilo suppressalis*	Quilis et al. (2007)
		Spodoptera littoralis	

[a]Cysteine-protease inhibitors: Rice (OC-I, OC-IΔD86, OC-II); Arabidopsis (AtCYS) and Barley (HvCPI-1 G→C, HvCPI-6)
[b]Other protease inhibitors: Tomato cathepsin D inhibitor (TCDI); Potato carboxypeptidase inhibitor (PCPI)

(Outchkourov et al. 2004a, b) and Acari (Carrillo et al. 2011b). Likewise, transgenic potato and rice lines expressing a tomato cathepsin D inhibitor and a caboxypeptidase inhibitor, respectively, also had an impact on lepidopteran and coleopteran larvae (Brunelle et al. 2004; Quilis et al. 2007).

Inhibitors of α-amylase (AI) function in a similar manner to PIs but interfering with carbohydrate digestion. The expression in transgenic plants of α-AI and WAI genes from bean and wheat, respectively, has shown their potential as useful control agents against coleopteran (Altabella and Chrispeels 1990; Shade et al. 1994; Schroeder et al. 1995; Ishimoto et al. 1996; Morton et al. 2000; De Sousa-Majer et al. 2004; Sarmah et al. 2004; Ignacimuthu and Prakash 2006; Solleti et al. 2008;

Table 6.3 Transgenic plants over-expressing α-amylase inhibitors and target arthropod pests

Amylase inhibitor	Transgenic plant	Target species	References
αAI-1	Tobacco	*Tenebrio molitor*	Altabella and Chrispeels (1990)
	Pea	*Callosobruchus maculatus*	Shade et al. (1994)
		Callosobruchus chinensis	
		Bruchus pisorum	Schroeder et al. (1995)
			Morton et al. (2000)
			De Sousa-Majer et al. (2004)
	Azuki bean	*Callosobruchus maculatus*	Ishimoto et al. (1996)
		Callosobruchus chinensis	
		Callosobruchus analis	
	Chickpea	*Callosobruchus maculatus*	Sarmah et al. (2004)
			Ignacimuthu and Prakash (2006)
	Coffea	*Hypotheneumus hampei*	Barbosa et al. (2010)
αAI-2	Pea	*Bruchus pisorum*	Morton et al. (2000)
WAI	Potato	*Lacanobia oleracea*	Gatehouse et al. (1997)
WMAI-1	Tobacco	*Agrotis ipsilon*	Carbonero et al. (1999)

α amylase inhibitors of bean (αAI-1 and αA-2) and wheat (WAI and WMAI-1)

Barbosa et al. 2010) and lepidopteran pests (Gatehouse et al. 1997; Carbonero et al. 1999). Most of these transgenes have been expressed under the control of seed-specific promoters conferring resistance to stored product pests (see Table 6.3 and references herein).

In general terms, insects feeding on plants expressing foreign PIs or AI endured reduced growth and development and, in some cases, increased mortality (Lecardonnel et al. 1999; Marchetti et al. 2000; Morton et al. 2000; Alfonso-Rubí et al. 2003) and reduced reproductive performance (Rahbe et al. 2003; Ribeiro et al. 2006; Carrillo et al. 2011b); whereas antisense genes that block the endogenous wound-induced expression of PIs make the plants more susceptible to the attack by insects (Royo et al. 1999). Despite these results, none of the transgenic plants produced in these studies achieved complete resistance against insect damage, and rarely produce high levels of mortality. Because of the complexity of digestive systems in the target pests, the limited spectrum of activity of most inhibitors, besides inadequate expression levels, a number of transgenic plants failed to show any significant levels of resistance. Furthermore, it has been demonstrated that many insects are physiologically adapted to circumvent the effects of exposure to specific plant hydrolyse inhibitors by up-regulation of inhibitor insensitive enzymes to compensate for the activity inhibited and/or by the proteolysis of PIs by non-target digestive proteases (Markwick et al. 1998; Rivard et al. 2004; Alvarez-Alfageme et al. 2007).

To obtain commercially viable resistant transgenic crops expressing hydrolytic enzyme inhibitors, other strategies have appeared to improve their usefulness and efficacy. The most promising results obtained so far have been obtained with genes encoding multiple inhibitory activities. The functional co-expression of different insecticidal proteins in plants is technically challenging if each one is expressed separately. However, nature provides at least two distinct ways of achieving expression of different inhibitory activities from a single encoding gene; bifunctional inhibitors and multidomain inhibitors that generate active inhibitors by post-translational fragmentation of a single gene product (Huntington 2006; Nissen et al. 2009). Likewise, engineered multidomain cystatin expressed in potato yielded resistance against *Frankliniella occidentalis* in greenhouse trials (Outchkourov et al. 2004b).

Alternatively, pyramiding (stacking) of genes encoding PIs and AI with other resistance genes have been developed as a method to prevent pest resistance and to improve pest control. Plants co-expressing a combination of hydrolytic enzyme inhibitors or PIs with Bt toxins had enhanced toxicity for lepidopteran species when compared to plants that expressed the individual genes (Zhao et al. 1998; Maqbool et al. 2001; Falco and Silva 2003; Rivard et al. 2004; Abdeen et al. 2005; Han et al. 2008; Arvinth et al. 2010; Dunse et al. 2010; Senthilkumar et al. 2010). Potentiation of the insecticidal activity of hydrolytic enzyme inhibitors has also been obtained by combining them with transgenically expressed lectins (Boulter et al. 1990; Gatehouse et al. 1997; Golmirizaie et al. 1997; Li et al. 2005) and thionins (Charity et al. 2005). However, the mechanisms by which such potentiation occurs are unknown and other combinations have failed to show any additive effects (Santos et al. 1997; Gatehouse et al. 1997; Winterer and Bergelson 2001). Many promising field trials have been carried out with these transgenic plants and in most cases the results have not been sufficiently convincing to lead to serious attempts at commercialization. Probably an exception are rice lines expressing Cry1Ac plus the cowpea trypsin inhibitor CpTI, which after being extensively tested, are now expected to obtain approval biosafety certificates for their release/explotation as commercial resistant plants in China (Chen et al. 2010). If transgenic plants expressing foreign hydrolase inhibitors for protection against pests are to be deployed as a component of Integrated Pest Management Systems, possible adverse direct or indirect effects of the transgene on natural enemies should also be considered (Table 6.4).

Additional candidate genes for insect resistance are lectins (Table 6.4). Plant lectins are a heterogeneous group of sugar binding proteins that have been shown to control mainly sap-sucking pest, particularly homopteran and Acari species. The transgenic expression of mannose specific snowdrop lectin gene (GNA), extensively studied, has conferred protection against several homopteran, coleopteran, lepidopteran and Acari (Stoger et al. 1999; Wu et al. 2002; McCafferty et al. 2008). Similarly, leaf lectins from *Allium* species (garlic agglutinin ASAL and onion agglutinin ACA, mainly), have shown to have antifeedant properties against plant hoppers and aphids (Tang et al. 2001; Dutta et al. 2005; Hossain et al. 2006). The ability of mannose-specific snowdrop lectin GNA as a carrier protein to deliver insecticidal peptides to the lepidopteran larvae haemolymph (Fitches et al. 2002, 2004) opened a new protective approach of transforming plants with hybrid proteins. Zhu-Salzman et al. (2003) covalently linked a soyacystatin to a legume lectin using a linker region from the potato multicystatin. This fusion protein presented a novel binding ability, localised at the anterior of the insect guts and prevented proteolytic degradation of the cystatin, enhancing the anti-insect

Table 6.4 Transgenic plants over-expressing a combination of genes (gene pyramiding) and target arthropod pests

Gene pyramiding	Transgenic plant	Target pest species	References
SKTI + SBBI	Sugarcane	*Diatraea saccharalis*	Falco and Silva (2003)
OC-IΔD86 + CpTI	*Arabidopsis*	*Leptinotarsa decemlineata*	Rivard et al. (2004)
PPI-II + CPI	Tomato	*Heliothis obsoleta* *Liriomyza trifolii*	Abdeen et al. (2005)
PPI-I + PPI-II	Cotton	*Helicoverpa punctigera* *Helicoverpa armigera*	Dunse et al. (2010)
NaPI-II + PPI-II	Cotton	*Helicoverpa punctigera*	Dunse et al. (2010)
CeCPI + Sporamin	Tobacco	*Helicoverpa armigera*	Senthilkumar et al. (2010)
CpTI + P-lec	Tobacco	*Heliothis virescens*	Boulter et al. (1990)
CpTI + GNA	Sweet potato	*Euscepes postfasciatus*	Golmirizaie et al. (1997)
SBTI + GNA	Rice	*Nilaparvata lugens* *Cnaphalocrocis medinalis*	Li et al. (2005)
NaPI-II + β-HTH		*Helicoverpa armigera*	Charity et al. (2005)
WAI + GNA	Potato	*Lacanobia voleracea*	Gatehouse et al. (1997)
CpTI + Cry1Ac	*Arabidopsis*	*Spodoptera exigua* *Helicoverpa zea* *Pseudoplusia includens* *Heliothis virescens*	Santos et al. (1997)
CpTI + Cry1Ac	Tobacco	*Helicoverpa armigera*	Zhao et al. (1998)
PPI-II + Cry1Ac	Oilseed rape	*Plutella xylostella*	Winterer and Bergelson (2001)
Cry1Ac + Cry2A + GNA	Rice	*Naphalocrocis medinalis* *Scirpophaga incertulas* *Nilaparvata lugens*	Maqbool et al. (2001)
CpTI + Cry1Ac	Rice	*Cnaphalocrocis medinalis*	Han et al. (2008)
Aprotinin + Cry1Ab	Sugarcane	*Chilo infuscatellus*	Arvinth et al. (2010)
Multiple combinations	Rice	multiple insects	Chen et al. (2010)

SKTI Soybean Kunitz trypsin inhibitor, *SBBI* Soybean Bowman-Birk inhibitor, *OC-IΔD86* Oryzacystatin I, *CpTI* Cowpea trypsin inhibitor, *PPI-I and II* potato protease inhibitors, *CPI* cowpea trypsin inhibitor, *NaPI-II Nicotiana alata* multidomain trypsin and chymotrypsin inhibitor, *CeCPI* taro cystatin, *Sporamin* sweet potato trypsin inhibitor, *P-lec* pea lectin, *GNA* snowdrop lectin, *SBTI* soybean trypsin inhibitor, *β-HTH* barley beta-hordothionin, *WAI* wheat amylase inhibitor, *Cry1Ac, Cry2A and Cry1Ab* Bt toxins, *Aprotinin* bovine pancreatic trypsin inhibitor

activity synergistically. Mehlo et al. (2005) showed that Bt covalently fused to the non-toxic B-chain of ricin (RB) provided wider receptor sites within target species. Bioassays against *Chilo suppressalis* and *Spodoptera littoralis* feeding on BtRB transgenic maize and rice demonstrated that the fusion protein extended the range of Cry1Ac toxicity to these two species. Furthermore, plants expressing the fusion protein gained protection against the hemipteran *Cicadulina mbila,* which is resistant to Bt toxins. The addition of different protein sequences in a unique hybrid protein has introduced extra functionality to resistant pest proteins and has potential applications in crop protection.

Another group of proteins extensively used to combat pest using biotechnological approaches are the Biotin-Binding Proteins (BBPs). When biotin, an essential vitamin for vertebrates and insects, is bound to these proteins it cannot develop its action as enzyme cofactor of important metabolic pathways. Deficiencies in biotin have negative effects on growth and development of organisms. BBPs are effective insecticides across a broad range of insect orders. In particular, avidin and streptavidin have been reported as severe growth reduction agents in at least 40 insect species across five insect orders (Christeller et al. 2010 and references herein). BBPs resulted toxic when produced in cytoplasm, because biotin is synthesized in this compartment in plants. Storage vacuole- and apoplast-targeting of avidin/streptavidin have been used to express these proteins in transgenic plants (Murray et al. 2010). Moreover, BBPs have been pyramided with Bt genes and increased the efficacy to control pest using transgenic plants (Zhu et al. 2005; Cooper et al. 2006).

Multiple candidate genes could also be included in this section as genes with a potential use of defence proteins for phytophagous arthropod resistance. Among them, have to be mentioned genes regulated via the octadecanoid-wound-signalling pathway as polyphenol oxidases (Constabel et al. 1995), vegetative storage proteins (Liu et al. 2005) and plant proteases (van der Hoorn 2008; Ankala et al. 2009) have to be mentioned. However, apart from the enhanced resistance obtained by polyphenol oxidase over-expression in transgenic poplar (Wang and Constabel 2004), and the growth reduction of lepidopteran larvae feeding in maize callus expressing a cysteine-protease (Pechan et al. 2000), the effects of these potential protective proteins have not still been directly demonstrated directly using transgenic plants.

6.2 Novel Protective Genes and Approaches for Pest Control

Besides insecticidal δ-toxins produced by *B. thuringiensis*, other bacteria such as *Photorhabdus* and *Xenorhabdus*, symbiotic with entomopathogenic nematodes, contain genes encoding large insecticidal toxin complexes (ffrench-Constant et al. 2007). The bacteria live in the nematode gut and when the insect is infected by the nematode, the bacteria are released into the insect hemocele and cause a lethal septicaemia in insects, being subsequently used as nutrients for the nematode. *Arabidopsis* plants transformed with the toxin A from *P. luminescens* resulted highly toxic to the tobacco hornworm *Manduca sexta* and to the southern corn rootworm *Diabrotica undecimpunctata* (Liu et al. 2003). The mode of action is unknown, although the toxins show significant sequence similarity to the recently described neurotoxin beta-leptinotarsin-h isolated from the blood of the Colorado potato beetle (ffrench-Constant et al. 2007). Protein toxins with high activity against a wide range of plant feeding insects are naturally produced by multiple strains of *Photorhabdus* spp. and their close relative and *Xenorhabdus* spp. Maybe in a future these toxins could be used as alternatives to Bt genes.

A promising approach for pest control could come from the use of transcription factors (TFs), regulators of gene expression not yet been fully exploited. TF members of ERF, WRKY, bZIP and MYB families have been strongly implicated in defence against plant pathogens (Century et al. 2008), but little is known on their putative role in defence against plant pest. An example to support this approach are the transgenic lines of *Arabidopsis* constitutively expressing a conserved MYB TF of phenylpropanoid biosynthesis that resulted in solid-purple leaves and had significantly increased resistance to leaf feeding by first instar fall armyworms (*Spodoptera frugiperda*). The reduction in feeding by *S. frugiperda* was significantly positively correlated with reduction in the weight of survivors, but both were negatively correlated with the concentration of anthocyanins (Johnson and Dowd 2004). These results indicate that a single gene regulator can activate a defensive pathway sufficiently to produce increased resistance to insects but that this activation confers a cost in plant productivity. However, it seems reasonable to expect that TFs will be a significant component of the next round of agricultural biotechnology products.

It is also important to mention as a new biotechnological approach for pest control based on the expression of key genes expression for insect development or biochemical metabolism through RNA interference (RNAi) using gene fragments from target pest.

As it has been described in this chapter the factors involve in plant defence against herbivores are multiple and most of them are additive and change among plant-arthropod systems. Several genes and pathways that are induced by herbivore attack are also involved in other biological processes (Walling 2000). The difficulty to differentiate the specific defence strategies from different systems and the study of the connexion of these pathways with other biological processes make plant defence against insect a very complex phenomenon to be study by traditional approaches. Nowadays new technologies in genomics, proteomics, metabolomics, lipidomics and bioinformatics allow us to understand these issues more efficiently.

The study of genes, proteins, metabolites and other molecular compounds candidates to take part in plant defence response will accelerate the identification of specific targets that are critical for plant defence in specific plant-insect systems.

Several genes and proteins have been found to be involved in plant defence in the last few years in different systems using microarray technology or proteomics approaches, to select candidate genes/proteins involved in defence. In general, the data obtained indicate the involvement of reactive oxygen species and calcium in early signalling and salycilic and jasmonic acids in the regulation of defence responses. Transcripts or proteins related to senescence, biosynthesis of anti-herbivore proteins, as well as several TFs used to be over-expressed (Kusnierczyk et al. 2008; Delp et al. 2009). Additionally, studies on defence-involved secondary metabolites revealed deposition of callose at the pest feeding sites and variation in the levels of glucosinolates, terpenes, phenolic compounds or nitrogen containing secondary products after pest attack (Grubb and Abel 2006; Kusnierczyk et al. 2007). Deciphering of the signals regulating herbivore responsive gene expression will afford many opportunities to manipulate the response. A wider knowledge of these interactions can be exploited in the rational design of transgenic plants with increased herbivore resistance.

Finally, besides the use of transgenic plants expressing the insecticidal and/or acaricidal transgenes described, most of the above mentioned protective molecules are stable enough to be applied as a spray or powder as part of IPM system. The large-scale production by fermentation or preparation of powder or extracts from transgenic plants will be followed by their application as chemicals.

Acknowledgements We thank to Dr. Gonzalez-Melendi for critically reading the manuscript. Financial support from the Ministerio de Ciencia e Innovación (project BFU2008-01166) and from the Universidad Politecnica de Madrid/Comunidad de Madrid (project CCG10-UPM/AGR-5,242) are gratefully acknowledged.

References

Abdeen A, Virgos A, Olivella E, Villanueva J, Gabarra R, Prat S (2005) Multiple insect resistance in transgenic tomato plants over expressing two families of plant proteinase inhibitors. Plant Mol Biol 57:184–202

Alfonso-Rubí J, Ortego F, Castañera P, Carbonero P, Díaz I (2003) Transgenic expression of trypsin inhibitor CMe from barley in *indica* and *japonica* rice, confers resistance to the rice weevil *Sitophilus oryzae*. Transgenic Res 12:23–31

Altabella T, Chrispeels MJ (1990) Tobacco plants transformed with the bean *αai* gene express an inhibitor of insect α-amylase in their seeds. Plant Physiol 93:805–810

Altpeter F, Diaz I, McAuslane H, Gaddour K, Carbonero P, Vasil IK (1999) Increased insect resistance in transgenic wheat stably expressing trypsin inhibitor CMe. Mol Breed 5:53–63

Alvarez-Alfageme F, Martinez M, Pascual-Ruiz S, Castañera P, Diaz I, Ortego F (2007) Effects of potato plants expressing a barley cystatin on the predatory bug *Podisus maculiventris* via herbivorous prey feeding on the plant. Transgenic Res 16:1–13

Ankala A, Luthe DS, Williams WP, Wilkinson JR (2009) Integration of ethylene and jasmonic acid signaling pathways in the expression of maize defense protein Mir1-CP. Mol Plant Microbe Interact 22:1555–1564

Arvinth S, Arun S, Selvakesavan RK, Srikanth J, Mukunthan N, Kumar PA, Premachandran MN, Subramonian N (2010) Genetic transformation and pyramiding of aprotinin-expressing sugarcane with Cry1Ab for shoot borer (*Chilo infuscatellus*) resistance. Plant Cell Rep 29:383–395

Barbosa A, Alburquerque EV, Silva MC, Souza DS, Oliveira-Neto OB, Valencia A, Rocha TL, Grossi-deSa (2010) α-amylase inhibitor-1 gene from *Phaseolus vulgaris* expressed in *Coffea arabica* plants inhibits α-amylases from the coffee berry borer pest. BMC Biotechnol 10:44

Bell HA, Fitches EC, Down RE, Ford L, Marris GC, Edwards JP, Gatehouse JA, Gatehouse AMR (2001) Effect of dietary cowpea trypsin inhibitor (CpTI) on the growth and development of the tomato moth *Lacanobia oleracea* (Lepidoptera: Noctuidae) and on the success of the gregarious ectoparasitoid *Eulophus pennicornis* (Hymenoptera: Eulophidae). Pest Manag Sci 57:57–65

Bonade-Bottino M, Lerin J, Zaccomer B, Jouanin J (1999) Physiological adaptation explains the insensitivity of *Baris coerulescens* to transgenic oilseed rape expressing oryzacystatin I. Insect Biochem Mol Biol 29:131–138

Bouchard E, Cloutier C, Michaud D (2003a) Oryzacystatin I expressed in transgenic potato induces digestive compensation in an insect natural predator via its herbivorous prey feeding on the plant. Mol Ecol 12:2439–2446

Bouchard E, Michaud D, Cloutier C (2003b) Molecular interaction between an insect predator and its herbivore prey on transgenic potato expressing a cysteine proteinase inhibitor from rice. Mol Ecol 12:2429–2437

Boulter D, Edwards GA, Gatehouse AMR, Gatehouse JA, Hilder VA (1990) Additive protective effects of different plant-derived insect resistance genes in transgenic tobacco plants. Crop Prot 9:351–354

Brunelle F, Cloutier C, Michaud D (2004) Colorado potato beetles compensate for tomato cathepsin D inhibitor expressed in transgenic potato. Arch Insect Biochem Physiol 55:103–113

Carbonero P, Diaz I, Vicente-Carbajosa J, Alfonso-Rubi J, Gaddour K, Lara P (1999) Cereal α-amylase/trypsin inhibitors and transgenic insect resistance. In: Scarascia Mugnozza GT, Porceddu E, Pagnotta MA (eds) Genetics and breeding for crop quality and resistance. Kluwer Academic Publishers, The Netherlands, pp 147–158

Carriere Y, Crowder DW, Tabashnik BE (2010) Evolutionary ecology of insect adaptation to Bt crops. Evol Appl 3:561–573

Carrillo L, Martinez M, Alvarez-Alfageme F, Castañera P, Smagghe G, Diaz I, Ortego F (2011a) A barley cysteine-proteinase inhibitor reduces the performance of two aphid species in artificial diets and transgenic Arabidopsis plants. Transgenic Res 20:305–319

Carrillo L, Martínez M, Ramessar K, Cambra I, Castañera P, Ortego F, Diaz I (2011b) Expression of a barley cystatin gene in maize enhances resistance against phytophagous mites by altering their cysteine-proteases. Plant Cell Rep 30:101–112

Castagnoli M, Caccia R, Liguori M, Simoni S, Marinari S, Soressi GP (2003) Tomato transgenic lines and *Tetranychus urticae*: changes in plant suitability and susceptibility. Exp Appl Acarol 31:177–189

Century K, Reuber TL, Ratcliffe OJ (2008) Regulating the regulators: the future prospect for transcription-factor-based agricultural biotechnology products. Plant Physiol 147:20–29

Charity JA, Anderson MA, Bittisnich DJ, Whitecross M, Higgins TJV (1999) Transgenic tobacco and peas expressing a proteinase inhibitor from *Nicotiana alata* have increased insect resistance. Mol Breed 5:357–365

Charity JA, Hughes P, Anderson MA, Bittisnich DJ, Whitecross M, Higgins TJV (2005) Pest and disease protection conferred by expression of barley beta-hordothionin and *Nicotiana alata* proteinase inhibitor genes in transgenic tobacco. Funct Plant Biol 32:35–44

Chen M, Shelton A, Ye G (2010) Insect-resistant genetically modified rice in China: from research to commercialization. Annu Rev Entomol 56:81–101

Christeller JT, Marwicke NP, Burguess EPJ, Malone LA (2010) The use of biotin-binding proteins for insect control. J Econ Entomol 103:497–508

Christou P, Capell T, Kohli A, Gatehouse JA, Gatehouse AMR (2006) Recent developments and future prospect in insect pest control in transgenic crops. Trends Plant Sci 11:302–308

Christy LA, Arvinth S, Saravanakumar M, Kanchana M, Mukunthan N, Srikanth J, Thomas G, Subramonian N (2009) Engineering sugarcane cultivars with bovine pancreatic trypsin inhibitor (aprotinin) gene for protection against top borer (*Scirpophaga excerptalis* Walker). Plant Cell Rep 28:175–184

Cipriani G, Michaud D, Brunelle F, Golmirzaie A, Zhang DP (1999) Expression of soybean proteinase inhibitor in sweetpotato. CIP Program Rep 98:271–277

Cloutier C, Fournier M, Jean C, Yelle S, Michaud D (1999) Growth compensation and faster development of Colorado potato beetle (Coleoptera: Chrymelidae) feeding on potato foliage expressing oryzacystatin I. Arch Insect Biochem Physiol 40:69–79

Cloutier C, Jean C, Fournier M, Yelle S, Michaud D (2000) Adult Colorado potato beetles, *Leptinotarsa decemlineata* compensate for nutritional stress on oryzacystatin I-transgenic potato plants by hypertrophic behaviour and over-production of insensitive protease. Arch Insect Biochem Physiol 44:69–81

Confalonieri M, Allegro G, Balestrazzi A, Fogher C, Delledonne M (1998) Regeneration of *Populus nigra* transgenic plants expressing a Kunitz proteinase inhibitor (KTi3) gene. Mol Breed 4:137–145

Constabel C, Bergey C, Ryan C (1995) Systemin activates synthesis of wound-inducible tomato leaf polyphenol oxidase via the octadecanoid defense signaling pathway. Proc Natl Acad Sci USA 92:407–411

Cooper SG, Douches DS, Grafius EJ (2006) Insecticidal activity of avidin combined with genetically engineered and traditional host plant resistance against Colorado potato beetle (Coleoptera: Chrysomelidae). J Econ Entomol 99:527–536

Cowgill SD, Danks C, Atkinson HJ (2004) Multitrophic interactions involving genetically modified potatoes, nontarget aphids, natural enemies and hyperparasitoids. Mol Ecol 13:639–647

Da Silveira V, Machado MG, Postali JR, Rodriguez ML (2009) Regulatory effects of an inhibitor from *Plathymenia foliolosa* seeds on the larval development of *Anagasta kuehniella* (Lepidoptera). Comp Biochem Physiol A 152:255–261

De Leo F, Gallerani R (2002) The mustard trypsin inhibitor 2 affects the fertility of *Spodoptera littoralis* larvae fed on transgenic plants. Insect Biochem Mol Biol 32:489–496

De Leo F, Bonadé-Bottino MA, Ceci LR, Gallerani R, Jouanin L (1998) Opposite effects on *Spodoptera littoralis* larvae of high expression level of a trypsin proteinase inhibitor in transgenic plants. Plant Physiol 118:997–1004

De Leo F, Bonadé-Bottino MA, Ceci LR, Gallerani R, Jouanin L (2001) Effects of a mustard trypsin inhibitor expressed in different plants on three lepidopteran pests. Insect Biochem Mol Biol 31:593–602

De Sousa-Majer MJ, Turner NC, Hardie DC, Morton RL, Lamont B, Higgins TJV (2004) Response to water deficit and high temperature of transgenic peas (*Pisum sativum* L.) containing a seed-specific alpha-amylase inhibitor and the subsequent effects on pea weevil (*Bruchus pisorum* L.) survival. J Exp Bot 55:497–505

Delledonne M, Allegro G, Belenghi B, Balestrazzi A, Picco F, Levine A, Zelasco S, Calligari P, Confalonieri M (2001) Transformation of white poplar (*Populus alba* L.) with a novel *Arabidopsis thaliana* cysteine proteinase inhibitor and analysis of insect pest resistance. Mol Breed 7:35–42

Delp G, Gradin T, Ahman I, Jonsson LMV (2009) Microarray analysis of the interaction between the aphid *Rhopalosiphum padi* and the host plants reveals both differences and similarities between susceptible and partially resistant barley lines. Mol Genet Genomics 281:233–248

Ding LC, Hu CY, Yeh KW, Wang PJ (1998) Development of insect-resistant transgenic cauliflower plants expressing the trypsin inhibitor gene isolated from local sweet potato. Plant Cell Rep 17:854–860

Duan X, Li X, Xue Q, Abo-El-Saad M, Xu D, Wu R (1996) Transgenic rice plants harboring an introduced potato proteinase inhibitor II gene are insect resistant. Nat Biotechnol 14:494–498

Dunse KM, Stevens JA, Lay FT, Gaspar YM, Heath RL, Anderson MA (2010) Coexpression of potato type I and II proteinase inhibitors gives cotton plants protection against insect damage in the field. Proc Natl Acad Sci USA 107:15011–15015

Dutta I, Saha P, Majumder P, Sarkar A, Chakraborti D, Banerjee S, Das S (2005) The efficacy of a novel insecticidal protein, *Allium sativum* leaf lectin (ASAL), against homopteran insects monitored in transgenic tobacco. Plant Biotechnol J 3:601–611

Falco MC, Silva MC (2003) Expression of soybean proteinase inhibitors in transgenic sugarcane plants: effects on natural defense against *Diatraea saccharalis*. Plant Physiol Biochem 41:761–766

Fang HJ, Li DL, Wang GL, Li YH (1997) An insect resistant transgenic cabbage plant with the cowpea trypsin inhibitor (CpTi) gene. Acta Bot Sin 39:940–945

Ferry N, Raemaekers RJM, Majerus MEN, Jouanin L, Port G, Gatehouse JA, Gatehouse AMR (2003) Impact of oilseed rape expressing the insecticidal cysteine protease inhibitor oryzacystatin on the beneficial predator *Harmonia axyridis* (Multicoloured Asian ladybeetle). Mol Ecol 12:493–504

Ferry N, Edwards MG, Gatehouse J, Capell T, Christou P, Gatehouse AMR (2006) Transgenic plants for insect pest control: a forward looking scientific perspective. Transgenic Res 15:13–19

ffrench-Constant RH, Dowling A, Waterfield NR (2007) Insecticidal toxins from *Photorhabdus* bacteria and their potential use in agriculture. Toxicon 49:436–451

Fitches E, Audsley N, Gatehouse JA, Edwards JP (2002) Fusion proteins containing neuropeptides as novel insect control agents: snowdrop lectin delivers fused allatostatin to insect haemolymph following oral ingestion. Insect Biochem Mol Biol 32:1653–1661

Fitches E, Edwards MG, Mee C, Grishin E, Gatehouse AMR, Gatehouse JA (2004) Fusion proteins containing insect-specific toxins as pest control agents: snowdrop lectin delivers fused insecticidal spider venom toxin to insect haemolymph following oral ingestion. J Insect Physiol 50:61–71

Gatehouse JA (2008) Biotechnological prospects for engineering insect-resistant plants. Plant Physiol 146:881–887

Gatehouse AMR, Shi Y, Powell KS, Brough C, Hilder VA, Hamilton WDO, Newell CA, Merryweather A, Boulter D, Gatehouse JA (1993) Approaches to insect resistance using transgenic plants. Philos Trans R Soc Lond Ser B 342:279–286

Gatehouse AMR, Down RE, Powell KS, Sauvion N, Rahbe Y, Newell CA, Merryweather A, Hamilton WDO, Gatehouse JA (1996) Transgenic potato plants with enhanced resistance to the peach-potato aphid Myzus persicae. Entomol Exp Appl 79:295–307

Gatehouse AMR, Davison GM, Newell CA, Merryweather A, Hamilton WDO, Burgues EPJ, Gilbert RJC, Gatehouse JA (1997) Transgenic potato plants with enhanced resistance to the tomato moth, Lacanobia oleracea: growth room trials. Mol Breed 3:49–63

Gatehouse AMR, Norton E, Davison GM, Babbé SM, Newell CA, Gatehouse JA (1999) Digestive proteolytic activity in larvae of tomato moth, Lacanobia oleracea; effects of plant protease inhibitors in vitro and in vivo. J Insect Physiol 45:545–558

Girad C, Bonade-Bottino M, Pham-Delegue MH, Jouanin L (1998a) Two strains of cabbage seed weevil (Coleoptera: Curculionidae) exhibit differential susceptibility to a transgenic oilseed rape expressing oryzacystatin I. J Insect Physiol 44:569–577

Girad C, Le Metayer M, Zaccomer B, Baellet E, Williams I, Bonade-Bottino M, Pham-Delegue M-H, Jouanin L (1998b) Growth stimulation of beetle larvae reared on a transgenic oilseed rape expressing a cysteine proteinase inhibitor. J Insect Physiol 44:263–270

Golmirizaie A, Zhang DP, Nopo L, Newell CA, Vera A, Cisneros F (1997) Enhanced resistance to West Indian sweet potato weevil (Euscepes postfaciatus) in transgenic 'Jewel' sweet potato with cowpea trypsin inhibitor and snowdrop lectin. HortSci 32:435

Graham J, Gordon SC, McNicol RJ (1997) The effect of the CpTi gene in strawberry against attack by vine weevil (Otiorhynchus sulcatus F Coleoptera: Curculionidae). Ann Appl Biol 131:133–139

Grubb CD, Abel S (2006) Glucosinolate metabolism and its control. Trends Plant Sci 11:89–100

Han L, Wu K, Peng Y, Wang F, Guo Y (2008) Efficacy of transgenic rice expressing Cry1Ac and CpTI against the rice leaffolder, Cnaphalocrocis medinalis (Guene'e). J Invertebr Pathol 96:71–79

Hao Y, Ao G (1997) Transgenic cabbage plants harbouring cowpea trypsin inhibitor (CPTI) gene showed improved resistance to two major insect pests, Pieris rapae nd Heliothis armigera. FASEB J 1:A868

Heath RL, McDonald G, Christeller JT, Lee M, Bateman K, West J, Van Heeswijck R, Anderson MA (1997) Proteinase inhibitors from Nicotiana alata enhance plant resistance to insect pests. J Insect Physiol 43:833–842

Hesler LS, Li Z, Cheesbrough TM, Riedell WE (2005) Nymphiposition and population growth of Rhopalosiphum padi L. (Homoptera: Aphididae) on conventional wheat cultivars and trans-genic wheat isolines. J Entomol 40:186–196

Hilder VA, Boulter D (1999) Genetic engineering of crop plants for insect resistance - a critical review. Crop Prot 18:177–191

Hilder VA, Gatehouse AMR, Sheerman SE, Barker RF, Boulter D (1987) A novel mechanism of insect resistance engineered into tobacco. Nature 330:160–163

Hilder VA, Gatehouse AMR, Boulter D (1989) Potential for exploiting plant genes to genetically engineer insect resistance, exemplified by the cowpea trypsin inhibitor gene. Pest Sci 27:165–171

Hoffman MP, Zalom FG, Wilson LT, Smilanick JM, Malyj LD, Kiser J, Hilder VA, Barnes WM (1992) Field evaluation of transgenic tobacco containing genes encoding Bacillus thuringiensis δ-endotoxin or cowpea trypsin inhibitor: efficacy against Helicoverpa zea (Lepidoptera: Noctuidae). J Econ Entomol 85:2516–2522

Hossain MA, Maiti MK, Basu A, Sen S, Ghose AK, Sen SK (2006) Transgenic expression of onion leaf lectin gene in Indian mustard offers protection against aphid colonization. Crop Sci 46:2022–2032

Huntington JA (2006) Shape-shifting serpins- advantages of a mobile mechanism. Trends Biochem Sci 31:427–435

Ignacimuthu S, Prakash S (2006) *Agrobacterium*-mediated transformation of chickpea with α-amylase inhibitor gene for insect resistance. J Biosci 31:339–345

Irie K, Hosoyama H, Takecuchi T, Iwabuchi K, Watanabe H, Abe M, Abe K, Arai S (1996) Transgenic rice established to express corn cystatin exhibits strong inhibitory activity against insect gut proteinases. Plant Mol Biol 30:149–157

Ishimoto M, Sato T, Chrispeels MJ, Kitamura K (1996) Bruchid resistance of transgenic azuki bean expressing seed α-amylase inhibitor of common bean. Entomol Exp Appl 79:309–315

James C (2010) Global status of commercialized biotech/GM crops, ISAAA Brief 42. ISAAA, Itahca

Johnson ET, Dowd PF (2004) Differentially enhanced insect resistance, at a cost, in *Arabidopsis thaliana* constitutively expressing a transcription factor of defensive metabolites. Agric Food Chem 52:5135–5138

Johnson R, Narvaez J, An G, Ryan C (1989) Expression of proteinase inhibitors I and II in transgenic tobacco plants: effects on natural defense against *Manduca sexta* larvae. Proc Natl Acad Sci USA 86:9871–9875

Jongsma MA, Bakker PL, Peters J, Bosch D, Stiekema WJ (1995) Adaptation of *Spodoptera exigua* larvae to plant proteinase inhibitors by induction of gut proteinase activity insensitive to inhibition. Proc Natl Acad Sci USA 92:8041–8045

Jouanin L, Bonadé-Bottino M, Girard C, Morrot G, Giband M (1998) Transgenic plants for insect resistance. Plant Sci 131:1–11

Khadeeva NV, Kochieva EZ, Yu M, Tcheredntchenkop MY, Yu E, Yakovieva EY, Sydoruk KV, Bogush VG, Dunaevsky YE, Belozersky MA (2009) Use of buckwheat seed protease inhibitor gene for improvement of tobacco and potato plant resistance to biotic stress. Biochem (Moscow) 74:260–267

Klopfenstein NB, Allen KK, Avila FJ, Heuchlin SA, Martinez J, Carman RC, Hall ER, McNabb HS (1997) Proteinase inhibitor II gene in transgenic poplar: chemical and biological assays. Biomass Bioenerg 12:299–311

Kusnierczyk A, Winge P, Midelfart H, Armbruster WS, Rossiter JT, Bones AM (2007) Transcriptional responses of Arabidopsis thaliana ecotypes with different glucosinolate profiles after attack by polyphagous Myzus persicae and oligophagous Brevicoryne brassicae. J Exp Bot 58:2537–2552

Kusnierczyk A, Winge P, Jostard TS, Troczynska J, Rossiter JT, Bones AM (2008) Towards global understanding of plant defence against aphids – timing and dynamics of early Arabidopsis defence responses to cabbage aphid (*Brevicoryne brassicae*) attack. Plant Cell Environ 31:1097–1111

Lara P, Ortego F, Gonzalez-Hidalgo E, Castañera P, Carbonero P, Diaz I (2000) Adaptation of *Spodoptera exigua* (Lepidoptera: Noctuidae) to barley trypsin inhibitor BTI-CMe expressed in transgenic tobacco. Transgenic Res 9:169–178

Lawo NC, Wackers FL, Romeis J (2009) Indian Bt cotton varieties do not affect the performance of cotton aphids. PLoS One 4:e4804

Lawrence PK, Koundal KR (2001) Plant protease inhibitors in control of phytophagous insects. Electron J Biotechnol 5:93–109

Lecardonnel A, Chauvin L, Jouanin L, Beaujean A, Prevost G, Sangwan-Norreel B (1999) Effects of rice cystatin I expression in transgenic potato on Colorado potato beetle larvae. Plant Sci 140:71–79

Lee SI, Lee SH, Koo JC, Chun HJ, Lim CO, Mun JH, Song YH, Cho MJ (1999) Soybean Kunitz trypsin inhibitor (SKTI) confers resistance to the brown planthopper (*Nilaparvata lugens* Stal) in transgenic rice. Mol Breed 5:1–9

Leple JC, Bonade-Bottino M, Augustin S, Pilate G, Le Tan VD, Deplanque A, Cornu D, Jouanin L (1995) Toxicity to *Chrysomela tremulae* (Coleoptera: Chrysomelidae) of transgenic poplars expressing a cysteine proteinase inhibitor. Mol Breed 1:319–328

Li Y, Romeis J (2010) Bt maize expressing Cry3Bb1 does not harm the spider mite, *Tetranychus urticae*, or its ladybird beetle predator, *Stethorus punctillum*. Biol Control 56:157–164

Li Y, Zhu Z, Chen ZX, Wu X, Wang W, Li SJ (1998) Obtaining transgenic cotton plants with cowpea trypsin inhibitor gene. Acta Gossip Sin 10:237–243

Li G, Xu X, Xing H, Zhu H, Fan Q (2005) Insect resistance to *Nilaparvata lugens* and *Cnaphalocrocis medinalis* in transgenic indica rice and the inheritance of *gna + sbti* transgenes. Pest Manag Sci 61:390–396

Liu D, Burton S, Glancy T, Li ZS, Hampton R, Meade T, Merlo DJ (2003) Insect resistance conferres by 283-kDa *Photorhabdus luminescens* protein TcdA in *Arabidopsis thaliana*. Nat Biotechnol 21:1222–1228

Liu YL, Ahn JE, Datta S, Salzman RA, Moon J, Huyghues-Despointes B, Pittendrigh B, Murdock LL, Koiwa H, Zhy-Salzman K (2005) Arabidopsis vegetative storage protein is an anti-insect acid phosphatase. Plant Physiol 139:1545–1556

Luo M, Wang A, Li H, Xia KF, Cai Y, Xu ZF (2009) Overexpression of a weed (*Solanum americanum*) proteinase inhibitor in transgenic tobacco results in increased glandular trychome density and enhanced resistance to Helicoverpa armigera and Spodoptera litura. Int J Mol Sci 10:1896–1910

Maheswaran G, Pridmore I., Franz P, Anderson MA (2007) A proteinase inhibitor from *Nicotiana alata* inhibits the normal development of light-brown apple moth, *Epiphyas postvittana* in transgenic apple plants. Plant Cell Rep 26:773–782

Maqbool SB, Riazuddin S, Loc NT, Gatehouse ANR, Gatehouse JA, Christou P (2001) Expression of multiple insecticidal genes confers broad resistance a range of different rice pest. Mol Breed 7:85–93

Marchetti S, Delledonne M, Fogher C, Chiaba C, Chiesa F, Savazzini F, Giordano A (2000) Soybean Kunitz, C-II and PI-IV inhibitor genes confer different levels of insect resistance to tobacco and potato transgenic plants. Theor Appl Genet 101:519–526

Markwick NP, Laing WA, Christeller JT, McHenry JZ, Newton MR (1998) Overproduction of digestive enzymes compensates for inhibitory effects of protease and α-amylase inhibitors fed to three species of leafrollers (Lepidoptera: Tortricidae). J Econ Entomol 91:1265–1276

McCafferty HRK, Moore PH, Zhu Y (2008) Papaya transformed with the *Galanthus nivalis* GNA gene produces a biologically active lectin with spider mite control activity. Plant Sci 175:385–393

McManus MT, White DWR, McGregor PG (1994) Accumulation of a chymotrypsin inhibitor in transgenic tobacco can affect the growth of insect pests. Transgenic Res 3:50–58

McManus MT, Burgess EPJ, Philip B, Watson LM, Laing WA, Voisey CR, White DWR (1999) Expression of the soybean (Kunitz) trypsin inhibitor in transgenic tobacco: effects on larval development of *Spodoptera litura*. Transgenic Res 8:383–395

Mehlo L, Gahakwa D, Nghia PT, Loc NT, Capell T, Gatehouse JA, Gatehouse AMR, Christou P (2005) An alternative strategy for sustainable pest resistance in genetically enhanced crops. Proc Natl Acad Sci USA 102:7812–7816

Mochizuki A, Nishizawa Y, Onodera H, Tabei Y, Toki S, Habu Y, Ugaki M, Ohashi Y (1999) Transgenic rice plants expressing a trypsin inhibitor are resistant against rice stem borers, *Chilo suppressalis*. Entomol Exp Appl 93:173–178

Morton R, Schroeder HE, Bateman KS, Chrispeels MJ, Armstrong E, Hs JV (2000) Bean α-amylase inhibitor 1 in transgenic peas (*Pisum sativum*) provides complete protection from pea weevil (*Bruchus pisorum*) under field conditions. Proc Natl Acad Sci USA 97:3820–3825

Murray C, Markwick NP, Kaji R, Poulton J, Martin H, Christeller TJ (2010) Expression of various biotin-binding proteins in transgenic tobacco confers resistance to potato tuber moth, *Phthorimaea operculella* (Zeller) (fam. Gelechiidae). Transgenic Res 19:1041–1051

Nandi AK, Basu D, Das S, Sen SK (1999) High level expression of soybean trypsin inhibitor gene in transgenic tobacco plants failed to confer resistance against damage caused by *Helicoverpa armigera*. J Biosci 24:445–452

Ninković S, Miljuš-Đukić J, Radović S, Maksimović V, Lazarević J, Vinterhalter B, Nešković M, Smigocki A (2007) *Phytodecta fornicata* Brüggemann resistance mediated by oryzacystatin II proteinase inhibitor transgene. Plant Cell Tissue Org Cult 91:289–294

Nissen MS, Kumar GNM, Youn B, Knowles DB, Ks L, Ballinger WJ, Knowles NR, Kang CH (2009) Characterization of *Solamun tuberosum* multicystatin and its structural comparison with other cystatins. Plant Cell 21:861–875

Nutt KA, Allsopp PG, McGhie TK, Shepherd KM, Joyce PA, Taylor GO, McQuatter RB, Smith GR, Ogarth DM (1999) In: Proceedings of the 1999 Conference of the Australian Society of Sugarcane Technologists, Townsville, Brisbane, Australia, 27–30 April 1999, pp171–176

Outchkourov NS, Kogel WJ, Schuuman-de Bruin A, Abrahamson M, Jongsma MA (2004a) Specific cysteine protease inhibitors act as deterrents of western flower thrips, *Fanklinella occidentalis* (Pergande), in transgenic potato. Plant Biotechnol J 2:439–448

Outchkourov NS, Kogel WJ, Wiegers GL, Abrahamson M, Jongsma MA (2004b) Engineered multidomain cysteine protease inhibitors yield resistance against western flower thrips (*Frankliniella occidentalis*) in greenhouse trials. Plant Biotechnol J 2:449–458

Pechan T, Ye L, Chang YM, Mitra A, Lin L, Davis FM, Williams WP, Luther SD (2000) A unique 33-kD cysteine proteinase accumulates in response to larval feeding in maize genotypes resistant to fall armyworm and lepidopteran. Plant Cell 12:1031–1040

Quilis J, Meynard D, Vila L, Avilés FX, Guiderdoni E, San Segundo B (2007) A potato carboxypeptidase inhibitor gene provides pathogen resistance in transgenic rice. Plant Biotechnol J 5:537–553

Rahbe Y, Derason C, Bonade-Bottino M, Girard C, Nardon C, Jouanin L (2003) Effects of the cysteine protease inhibitor oryzacystatin (OC-I) on different aphids and reduced performance of *Myzus persicae* on OC-I expressing transgenic oilseed rape. Plant Sci 164:441–450

Ribeiro APO, Pereira EJC, Galvan TL, Picanzo MC, Picoli EAT, da Silva DJH, Fari MG, Otoni WC (2006) Effect of eggplant transformed with oryzacystatin gene on *Myzus persicae* and *Macrosiphum euphorbiae*. J Appl Entomol 130:84–90

Rivard D, Cloutier C, Michaud D (2004) Colorado potato beetles show differential digestive compensatory responses to host plants expressing distinct sets of defense proteins. Arch Insect Biochem Physiol 55:114–123

Royo J, León J, Vancanneyt G, Albar JP, Rosal S, Ortego F, Castañera P, Sánchez-Serrano JJ (1999) Antisense-mediated depletion of a potato lipoxygenase reduces wound induction of proteinase inhibitors and increases weight gain of insects pests. Proc Natl Acad Sci USA 96:1146–1151

Sane VA, Nath P, Sane PV (1997) Development of insect-resistant transgenic plants using plant genes: expression of cowpea trypsin inhibitor in transgenic tobacco plants. Curr Sci 72:741–747

Santos MO, Adang MJ, All JN, Boerma HR, Parrott WA (1997) Testing transgenes for insect resistance using *Arabidopsis*. Mol Breed 3:183–194

Sarmah BK, Moore A, Tate W, Molvig L, Morton RL, Rees DP, Chiaiese P, Chrispeels MJ, Tabe LM, Higgins TJ (2004) Transgenic chickpea seeds expressing high levels of a bean α-amylase inhibitor. Mol Breed 14:73–82

Schluter U, Benchabane M, Munger A, Kiggundu A, Vortster J, Goulet MC, Cloutier C, Michaud D (2010) Recombinant protease inhibitors for herbivore pest control: a multitrophic perspective. J Exp Bot 61:4169–4183

Schroeder HE, Gollasch S, Moore A, Tabe LM, Craig S, Hardie C, Chrispeels MJ, Spencer D, Higgins TJ (1995) Bean α-amylase inhibitor confers resistance to the pea weevil (*Bruchus pisorum*) in transgenic peas (*Pisum sativum* L). Plant Physiol 107:1233–1239

Schuler TH, Denholm I, Jouanin L, Clark SJ, Clarak AJ, Poppy GM (2001) Population-scale laboratory studies of the effect of transgenic plants on nontarget insects. Mol Ecol 10:1845–1853

Senthilkumar R, Cheng CP, Yeh KW (2010) Genetically pyramiding protease-inhibitor genes for dual broad-spectrum resistance against insect and phytopathogens in transgenic tobacco. Plant Biotechnol J 8:65–75

Shade RE, Schroeder HE, Pueyo JJ, Tabe LM, Murdock LL, Higgins TJV, Chrispeels MJ (1994) Transgenic peas expressing the α-amylase inhibitor of the common bean are resistant to bruchid beetles. Biotechnol 12:793–796

Shulke RH, Murdock LL (1983) Lipoxygenase, trypsin inhibitor and lectin from soybeans: effects on larval growth of *Manduca sexta* (Lepidoptera: Sphingidae). Environ Entomol 12:787–791

Solleti SK, Bakshi S, Purkavastha J, Panda SK, Sahoo L (2008) Transgenic cowpea (*Vigna unguiculata*) seeds expressing a bean alpha-amylase inhibitor 1 confer resistance to storage pests, bruchid beetles. Plant Cell Rep 27:1841–1850

Srinvasan T, Kumar KRR, Kirti PB (2009) Constitutive expression of a trypsin protease inhibitor confers multiple stress tolerance in transgenic tobacco. Plant Cell Physiol 650:541–553

Stoger E, Williams S, Christou P, Down RE, Gatehouse JA (1999) Expression of the insecticidal lectin from snowdrop (*Galanthus nivalis* agglutinin; GNA) in transgenic wheat plants: effects on predation by the grain aphid *Stobion avenae*. Mol Breed 5:65–73

Tang K, Zhao E, Sun X, Wan B, Qi H, Lu K (2001) Production of transgenic rice homozygous lines with enhanced resistance to the rice brown plant hopper. Acta Biotechnol 21:117–128

Van der Hoorn RAL (2008) Plant proteases: from phenotypes to molecular mechanism. Ann Rev Plant Biol 59:191–223

Vila L, Quilis J, Meynard D, Breitler JC, Marfa V, Murillo I, Vassal JM, Messeguer J, Guiderdoni E, San Segundo B (2005) Expression of the maize proteinase inhibitor (Mpi) gene in rice plants enhances resistance against the striped stem borer (*Chilo suppressalis*): effects on larval growth and insect gut proteinases. Plant Biotechnol J 3:187–202

Walling LL (2000) The myriad plant responses to herbivores. J Plant Growth Regul 19:195–216

Wang J, Constabel CP (2004) Polyphenol oxidase overexpression in transgenic *Populus* enhances resistance to herbivory by forest tent caterpillar (*Malacosoma disstria*). Planta 220:87–96

Winterer J, Bergelson J (2001) Diamondback moth compensatory consumption of protease inhibitor-transformed plants. Mol Ecol 10:1069–1074

Wu Y, Llewellyn D, Mathews A, Dennis ES (1997) Adaptation of *Helicoverpa armigera* (Lepidoptera: Noctuidae) to a proteinase inhibitor expressed in transgenic tobacco. Mol Breed 3:371–380

Wu A, Sun X, Pang Y, Tang K (2002) Homozygous transgenic rice lines expressing GNA with enhanced resistance to the rice sap-sucking pest *Laodelphax striatellus*. Plant Breed 121:9395

Xu D, Xue Q, McElroy D, Mawal Y, Hilder VA, Wu R (1996) Constitutive expression of a cowpea trypsin inhibitor gene, CpTi, in transgenic rice plants confers resistance to two major rice insect pests. Mol Breed 2:167–173

Yeh KW, Lin MI, Tuan SJ, Chen YM, Lin CY, Kao SS (1997) Sweet potato (*Ipomoea batatas*) trypsin inhibitors expressed in transgenic tobacco plants confer resistance against *Spodoptera litura*. Plant Cell Rep 16:696–699

Zhao JZ, Fan YL, Zhao RM, Fan XL (1998) Insecticidal activity of transgenic tobacco co-expressing Bt and CpTI genes on *Helicoverpa armigera* and its role in delaying pest resistance. Rice Biotechnol Quart 34:9–10

Zhu YC, Adameczyk JJ Jr, West S (2005) Avidin, a potential biopesticide and synegsist to *Bacillus thuringiensis* toxins against field crop insects. J Econ Entomol 98:1566–1571

Zhu-Salzman K, Ahn JE, Salzman RA, Koiwa H, Shade RE, Balfe S (2003) Fusion of a soybean cysteine protease inhibitor and legume lectin enhances anti-insect activity synergistically. Agric For Entomol 5:317–323

Chapter 7
Use of RNAi for Control of Insect Crop Pests

Luc Swevers and Guy Smagghe

7.1 Introduction

RNA interference (RNAi) refers to double-stranded RNA (dsRNA)-mediated gene silencing. Since its discovery, it has developed as a powerful tool in functional genomics, and to date it is widely used in insect genetic research. It is certain that the discovery of RNAi has augmented our understanding of ~20–30 nucleotide non-coding small RNAs as critical regulators of gene expression and genome stability. Besides, gene silencing through RNAi has revolutionized the study of gene function, particularly in non-model and non-genome sequenced insect species, which is the case for most agricultural pest insects. Without doubt, it contains great potential for diverse applications in fundamental and applied research, for instance in gene therapy in medicine and disease control. More recent, a new hot point is to find a feasible way to use RNAi as an alternative method for practical application of crop protection to combat pest insects.

In this chapter, we give space for a brief introduction of RNAi discovery together with the principles of the molecular mechanisms behind, discuss the biological functions of RNAi in insects, and focus then on the development of this new technology for insect pest control. Emphasis will go to methods of dsRNA delivery, pathways of dsRNA uptake by insects, especially the engineering of plants to specifically suppress herbivorous insect gene expression, and its gene-specific and insect-selective characters, confirming potential use

L. Swevers (✉)
Insect Molecular Genetics and Biotechnology, Institute of Biology, National Center for Scientific Research "Demokritos", Athens, Greece
e-mail: swevers@bio.demokritos.gr

G. Smagghe
Department of Crop Protection, Faculty of Bioscience Engineering, Ghent University, Coupure Links 653, 9000 Ghent, Belgium
e-mail: guy.smagghe@ugent.be

G. Smagghe and I. Diaz (eds.), *Arthropod-Plant Interactions: Novel Insights and Approaches for IPM*, Progress in Biological Control 14,
DOI 10.1007/978-94-007-3873-7_7, © Springer Science+Business Media B.V. 2012

in safe environmentally-friendly control programs. Finally, we conclude with future perspectives as challenges to identify new insecticide targets, manage emerging insecticide resistance, and control important pest insects.

7.2   Principles for Use of RNAi in Insect Pest Control

As depicted in Fig. 7.1, in RNAi, dsRNA is delivered to the cells where it is processed by Dicer enzymes to 'small interfering RNAs' or siRNAs that are recruited by Argonaute proteins which use them as 'guides' to recognize and destroy homologous mRNAs. According to this pathway, if dsRNAs are introduced that are directed against mRNAs that encode proteins essential for cellular function, this will result in cellular toxicity and cell death. Because RNAi only works efficiently in case of full complementarity between trigger and target, the method is considered to be very specific to single (or a limited number of) species without causing toxicity to other organisms.

From a functional perspective, the RNAi machinery can be roughly divided into two parts, i.e. (1) the intracellular core machinery, consisting of Dicer enzymes, RNA-binding factors and Argonaute 'slicers', and (2) the 'systemic' machinery: factors that amplify the dsRNA signal and allow it to spread to other tissues within the animal or even to the next generation (Siomi and Siomi 2009).

Regarding the core machinery, it is noted that in somatic tissues in fruitflies (*Drosophila melanogaster*), and presumably in insects in general, two RNA silencing pathways exist, characterized by siRNAs and microRNAs (miRNAs), that seem to be at least partially separated with respect to biogenesis and function (Tomari et al. 2007). While the miRNA pathway primarily uses endogenous products transcribed from the cell's genome with dsRNA structure to regulate developmental processes, the siRNA pathway is thought to primarily function as a defense response against exogenous dsRNAs, as for instance generated from viruses. *Drosophila* mutants in the core siRNA machinery (Dicer-2, the RNA-binding protein R2D2, and Argonaute-2) indeed show increased sensitivity to infection by RNA viruses (Wang et al. 2006).

Besides the core factors that are dedicated to the sensing of the dsRNA trigger and effecting the destruction of the target RNAs, efficient systemic RNAi is thought to require a machinery to amplify the initial dsRNA trigger and to subsequently export it to the other tissues in the organism (Huvenne and Smagghe 2010) (Fig. 7.1). The first component relies on the function of RNA-dependent RNA polymerase (RdRp) enzymes, which are most extensively characterized in plants and the nematode *C. elegans*, and which generate 'secondary siRNAs' that can greatly sustain and amplify the RNAi response (Carthew and Sontheimer 2009). However, RdRp genes do not exist in insect genomes, and it is generally thought that a robust amplification response does not exist in this group of animals (Gordon and Waterhouse 2007). The second component involves the secretion and uptake of RNA signals from the hemolymph or even the environment and therefore relies on the existence of dsRNA transporters and/or the endocytosis machinery

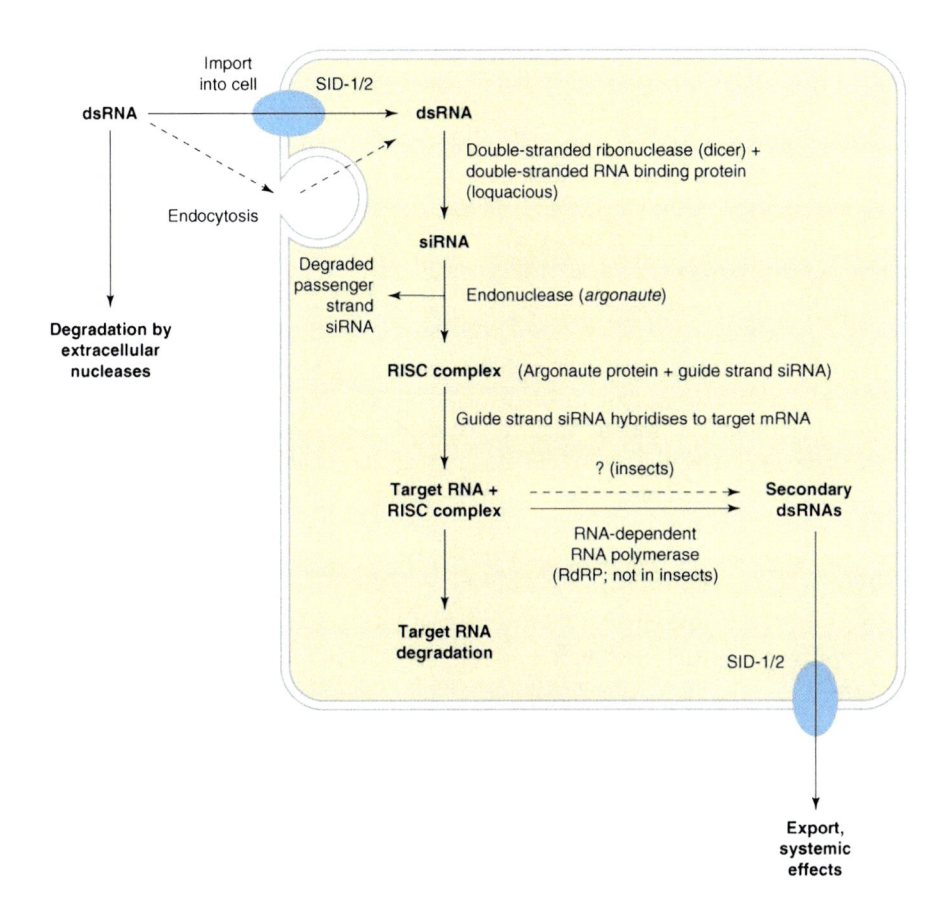

Fig. 7.1 Principle of RNAi in the cells of insects, showing the functional stages involved in local and systemic gene silencing by dsRNA. Exogenous dsRNA is imported into cells, processed by Dicer into small interfering RNA (siRNA) and assembled with the Argonaute protein into the RNA-induced silencing complex (RISC). The RISC complex targets and degrades specific mRNAs based on the siRNA sequence. Systemic RNAi effects are mediated through the production of new dsRNAs by RNA-dependent RNA polymerase (RdRP), which uses the target RNA as a template and is primed by siRNA strands. The secondary dsRNAs can be exported from the cell to spread the RNAi effect to other cells. Gene names in italics have been identified in fruitflies (*Drosophila melanogaster*). The transport proteins SID-1 and SID-2 have been identified in nematodes of *Caenorhabditis elegans*, as has the RdRP enzyme. Transport mechanisms might differ between different organisms (Reprinted from Price and Gatehouse (2008). With permission from Elsevier)

(Gordon and Waterhouse 2007; Sabin et al. 2010; Huvenne and Smagghe 2010). In the nematode *C. elegans* the spread of the systemic RNAi signal requires the function of the putative transporter protein SID-1 (Winston et al. 2002). SID-1 homologs are detected in genomes of insects belonging to Lepidoptera, Coleoptera, Hymenoptera, Hemiptera but not Diptera (Gordon and Waterhouse 2007). However, phylogenetic analysis and functional studies do not support a role for SID-1 homologs in the red flour beetle *Tribolium castaneum* (Tomoyasu et al. 2008). At this point, the pathway for spreading the RNA silencing signal remains a matter of speculation for insects, although it is generally believed that the endocytosis pathway is involved (Saleh et al. 2006; Ulvila et al. 2006).

From a perspective to control pest insects, it is evident that the point of entry for dsRNAs – whether added exogenously to the food or produced as hairpin RNAs in transgenic plants – will be the insect midgut. It is well described that the insect gut is divided into three regions; foregut, midgut and hindgut. Of these the first two are continuations of the 'outside' of the insect and are chitin-lined, so that their surfaces do not present areas of exposed cells; although receptors and transporters are present to allow processes such as taste recognition in the mouth cavity and water transfer in the hindgut to occur. The midgut region is the only part of the gut that contains surfaces of exposed cells, and it is the main site of exchange between the circulatory system (hemolymph) and the gut contents. The midgut itself is responsible for nutrient absorption, whereas excretion and water balance take place primarily in the Malpighian tubules attached to the hindgut, which carry out a function similar to that of the kidney in higher animals. RNAi-effects occurring in insects as a result of oral delivery of dsRNA are presumably mediated by the midgut surfaces through exposure of cells of the midgut epithelium and the Malpighian tubules to dsRNA in the gut contents. Conditions in the gut vary considerably between insect orders. Gut pH is an important factor in insect digestion and can vary from predominantly acidic (typical for Coleoptera) to strongly alkaline (even up to pH 10.5 in some species of Lepidoptera). In addition, within a single insect the pH changes along the gut and with distance from the gut epithelium. The stability of ingested dsRNA in the insect gut could be affected both by chemical hydrolysis (which increases with increasing pH) and by enzymes present in the gut contents (reviewed in Price and Gatehouse 2008; Hakim et al. 2010).

To achieve efficient pest control, it is sufficient that dsRNAs and/or siRNAs are taken up by the midgut epithelium where they will exert their toxic effects. Efficient spread to other tissues may not be necessary in case disruption of the midgut epithelium by itself will be lethal. Thus, the factors that determine the efficiency of RNAi, such as expression levels of the basic RNAi machinery, capacity for functional uptake from the extracellular medium of dsRNAs/siRNAs, and absence/presence of dsRNA-degrading enzymes, are especially important with respect to the cells of midgut epithelium (see Sect. 4).

7.3 Successes of RNAi in Insects of Different Orders and Other Arthropods

7.3.1 Coleopteran Insects: Sensitive to RNAi

While the RNAi machinery is conserved and could in principle be applied successfully to all insects or arthropods, it was noted rapidly that the red flour beetle (*T. castaneum*) is very sensitive to the technique. Reverse genetics by administration of dsRNA is systemic in nature and can be applied at all life stages,

Fig. 7.2 Phenotypes observed after injection in the red flour beetle (*Tribolium castaneum*) of dsRNA of different nuclear receptors. Control malE or nuclear receptor-dsRNA was injected into 1-day-old final instar larvae. Pictures shown are various abnormal phenotypes observed during larval and pupal stages. (**a** and **b**) The larvae died after EcR-dsRNA injection; (**c** and **d**) abnormal phenotypes observed in pupae developed from seven-up (SVP)-dsRNA injected larvae; (**e** and **f**) abnormal phenotypes observed in pupae developed from TcHR39-dsRNA injected larvae; (**g**) increased pigmentation observed in larvae injected with TcHR38-dsRNA; (**h** and **i**) abnormal pupae developed from TcHR38-d sRNA injected larvae. Scale bar: 1 mm (Reprinted from Tan and Palli (2008). With permission from Elsevier)

contributing to the rise of *T. castaneum* as the most prominent non-drosophilid insect model system (Tomoyasu et al. 2008). In addition, by use of RNAi in *T. castaneum*, Tan and Palli (2008) demonstrated that nuclear receptors can be attractive targets for developing new insecticides with strong effects in larval growth, metamorphosis and offspring production (Fig. 7.2). Together, the successes of RNAi in *Tribolium* may suggest that coleopteran insects are rather sensitive to RNAi, compared to other insect orders.

The potential for the use of this technique for control of coleopteran pests was corroborated in a high profile paper in Nature Biotechnology that described the successful use of this technology to control the western corn rootworm, *Diabrotica virgifera* (Baum et al. 2007; Gordon and Waterhouse 2007). As depicted in Fig. 7.3, transgenic F1-corn plants, expressing a V-ATPase A dsRNA, are protected from feeding damage. The latter article provided the basic outline for the development of an RNAi-based strategy to control insect pests:

(1) Sequences from cDNA libraries or directly obtained by high-throughput sequencing are chosen as targets for RNAi-based pest control. DsRNAs should target essential genes that are knocked down efficiently. Furthermore, gene knockdown should rapidly result in toxic effects. Examples of efficient genes as the V-ATPase proton pump, are given in Table 7.1.

western corn rootworm, *Diabrotica virgifera*

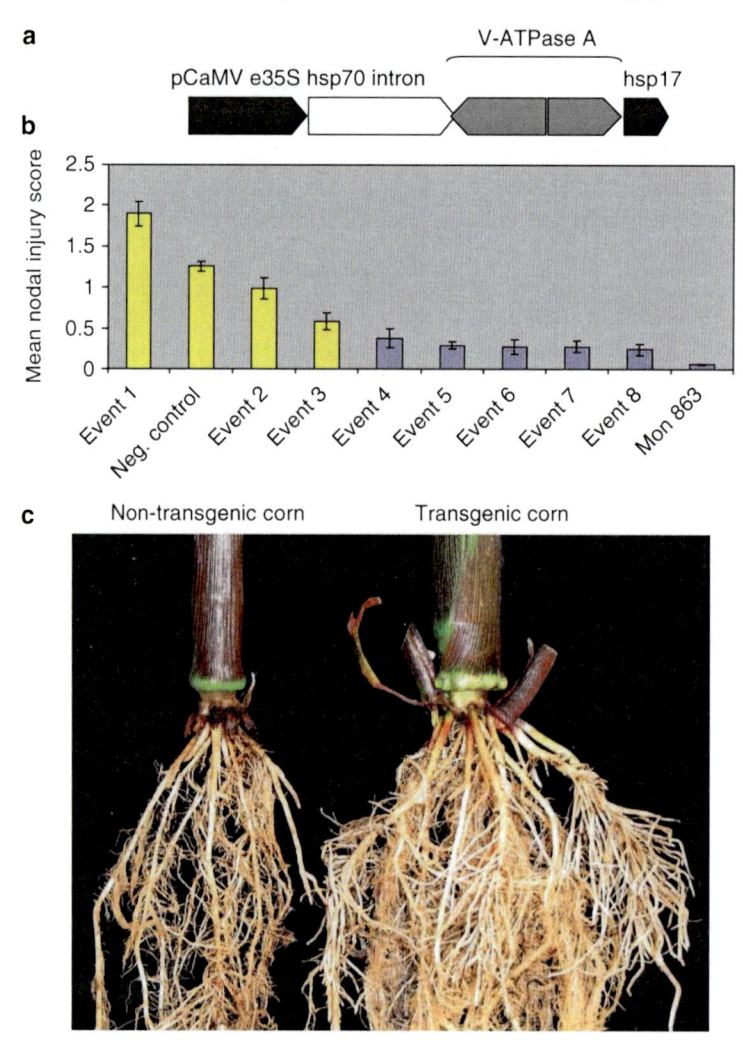

Fig. 7.3 F1-corn plants, expressing a V-ATPase A dsRNA, are protected from feeding damage by the western corn rootworm (*Diabrotica virgifera virgifera*). (**a**) Map of the expression cassette. (**b**) Mean root damage ratings for eight F1 populations, the parental inbred line (negative control)

Table 7.1 Toxicity LC_{50} values as determined in the western corn rootworm feeding bioassay employing dsRNA of different target genes (Reprinted from Baum et al. (2007). With permission from Macmillan)

Plasmid	Nearest ortholog[a]	Annotation	LC_{50} (ng dsRNA/cm^2)
pMON101053	F38E11_5	Putative COPI coatomer, β' subunit	0.57
pMON102873	CG6223	Putative COPI coatomer, β subunit	0.73
pMON78428	CG8055	Putative ESCRT III_Snf7 ortholog	1.20
pMON78416	CG11276	Putative ribosomal protein S4 ortholog	1.30
pIC16005	CG2934-PA	Putative v-ATPase D subunit 1 ortholog	1.72
pIC17504	CG3762	Putative v-ATPase A subunit 2 ortholog	1.82
pMON78412	CG3416	Putative mov34 ortholog (proteosome)	1.98
pMON98445	F37C12_9	Putative ribosomal protein rps-14 ortholog	2.60
pMON101120	M03F4.2	Putative actin ortholog	2.86
pMON96168	CG2331	Putative apple ATPase ortholog	4.16
pMON102861	CG12770	Putative ESCRT I_Vps28 ortholog	4.47
pMON78424	CG6141	Putative ribosomal protein L9 ortholog	5.20
pMON78425	CG2746	Putative ribosomal protein L19 ortholog	5.20
pIC17503	CG1913-PA	Putative alpha tubulin ortholog	5.20
pMON78440	CG3180	Putative RNA polymerase II ortholog	7.80
pMON102865	CG14542	Putative ESCRT III_vps2 ortholog	11.96
pMON30694	CG9277	Putative beta tubulin ortholog	51.98

[a]CG orthologs are from *D. melanogaster*, all others are from *C. elegans*

(2) Target dsRNAs are tested in larval feeding assays. For the western corn rootworm, an artificial agar diet is available on which dsRNAs can be applied on the surface at concentrations from <1 to >100 ng/cm^2 (Table 7.1). The procedure allows the calculation of median lethal effective concentrations (LC_{50}) and provides a preliminary screening test for the potency of dsRNAs to cause mortality in insect populations. The procedure also allows for evaluation of certain parameters of dsRNAs, such as length and which sections of the mRNAs are targeted.

(3) DsRNAs that perform well in larval feeding assays, typically with LC_{50} of less than 1 ng/cm^2, provide the basis for construction of RNA hairpin expression constructs to be introduced in transgenic plants. The expression construct is introduced into an *Agrobacterium* transformation vector to yield transgenic crops. Transgenic crops are tested in growth chamber assays to evaluate feeding damage by crop insect pests.

Based on these results, it was predicted that transgene-encoded dsRNAs could complement *Bacillus thuringiensis* (*Bt*) toxins in insect management programs

Fig. 7.3 (continued) and the corn rootworm-protected Cry3Bb event MON863. (**c**) The plant on *left* is a non-transgenic control with average root damage, whereas the plant on the *right* shows the average root protection seen when the transgene is expressed. F1 plants expressing a V-ATPase A dsRNA are protected from corn rootworm-feeding damage (Reprinted from Baum et al. (2007). With permission from Macmillan; The photos of larval and adult western corn rootworm are from the web page of The College of Agricultural and Life sciences, University of Idaho (http://www.cals.uidaho.edu/)

(Gordon and Waterhouse 2007). DsRNAs or RNA hairpins corresponding to essential genes have a different mode of action than *Bt* toxins and can be introduced as a protection trait with independent mode of action in transgenic crops. Transgenic crops with such "pyramided" insect-protection traits are generally superior to single-trait crops with respect to development of resistance by insect pests. *Bt* toxins and dsRNAs can complement also with respect to duration of effect; while *Bt* toxins act relatively quickly, RNAi effects take a longer time to develop (Zhu et al. 2011).

While in several countries transgenic crops often struggle for public acceptance, an alternative strategy could consist of the application of either heat-killed bacteria containing dsRNA or dsRNA purified from the bacteria in the field (Zhu et al. 2011), similar to the application of *Bt* toxin produced in bacteria. DsRNA from bacteria can be produced cheaply and has similar efficiency than *in vitro* synthesized dsRNA using commercial molecular biology kits (Zhu et al. 2011).

Other coleopteran species have shown a similar sensitivity to RNAi by feeding, such as the southern corn rootworm, *Diabrotica undecimpunctata*, and the Colorado potato beetle, *Leptinotarsa decemlineata* (Baum et al. 2007; Zhu et al. 2011). Feeding dsRNA to *Tribolium* larvae was also successful (Whyard et al. 2009). RNAi targeting endosymbiotic bacteria was also achieved by injection in several cereal weevil (*Sitophilus*) species (Vallier et al. 2009). However, the technique did not prove to be successful to the cotton boll weevil, *Anthonomus grandis*, for unknown reasons (Baum et al. 2007; Price and Gatehouse 2008).

Major concerns regarding the use of RNAi in insect pest control include the specificity of action and off-target effects (Auer and Frederick 2009). Comparative studies have established a correlation between the degree of identity between dsRNA and target gene and the degree of silencing (Baum et al. 2007). Thus, the specificity of the RNAi effect, whether it targets a single species or a group of species of a wider taxonomic group, could be achieved by careful selection of the dsRNA that is produced: unique sequences may target efficiently only a single species while more conserved sequences could achieve more broad specificity (Runo et al. 2011).

Emergence of resistance is always a major issue when new control strategies for insect pests are developed. At the level of the target sequence, it seems theoretically difficult how a stretch of mutations can develop along the entire length of a nucleic acid sequence such that base-pairing is impaired with the applied dsRNAs. However, the RNAi machinery itself seems to be dispensable for development (Shabalina and Koonin 2008) and therefore could accumulate mutations to resist efficient action of dsRNAs. It is also not known for insect species that are sensitive to RNAi at what frequency individuals exist in natural populations that are more resistant to RNAi. Under selection pressure, such resistant alleles could spread rapidly through the population and render deficient the RNAi approach for pest control.

7.3.2 *Lepidopteran Insects: Are These Resistant for RNAi?*

After the discovery of the RNAi process, initial experiments indicated that injection of dsRNA did not result in efficient gene silencing in lepidopteran insects.

Starting from 2007, however, an increasing number of reports were published that described successful RNAi experiments in Lepidoptera (see review by Terenius et al. 2011). To evaluate the occurrence of successful RNAi in lepidopteran insects and to determine which factors (species, tissue, life stage, gene identity, dsRNA properties) contribute to the efficiency of the process, a database (http://insectacentral.org/RNAi) was constructed of more than 200 RNAi experiments involving lepidopteran insects, including many unpublished data. Analysis of the data from the database allowed to formulate only a few general conclusions regarding RNAi efficiency in Lepidoptera: (1) silencing by injection of dsRNA seems to be efficient in the family of Saturniidae, (2) physiological processes related to the innate immune response are more efficiently silenced than other processes, and (3) epidermal tissue seems to be rather refractory to the RNAi process. Besides the above general conclusions, it was very difficult to discriminate a pattern for deciding which strategy would work best in lepidopteran insects for successful RNAi (Terenius et al. 2011).

Regarding feeding of dsRNA, it was observed that the technique worked only effectively if high or very high concentrations were taken up by the insect (10–100 µg per mg insect) (Terenius et al. 2011), casting doubt whether this technique could be effectively used to control pest insects. Insect pests for which silencing of essential genes and toxicity effects could be observed in several recent publications, include the beet armyworm, *Spodoptera exigua* (Tian et al. 2009; Chen et al. 2010; Tang et al. 2010), the fall armyworm, *Spodoptera frugiperda* (Rodríguez-Cabrera et al. 2010), the diamondback moth, *Plutella xylostella* (Bautista et al. 2009; Gong et al. 2011), the tobacco hornworm, *Manduca sexta* (Whyard et al. 2009; Cancino-Rodezno et al. 2010), the light brown apple moth, *Epiphyas postvittana* (Turner et al. 2006), the cotton bollworm, *Helicoverpa armigera* (Mao et al. 2007), the European corn borer, *Ostrinia nubilalis* (Khajuria et al. 2010), and the sugarcane borer, *Diatraea saccharalis* (Yang et al. 2010). However, for most lepidopteran species silencing effects observed after feeding showed great variability: for instance, in only 14% and 17% of experiments was high silencing observed in *Helicoverpa armigera* and *Spodoptera frugiperda*, respectively (http://insectacentral.org/RNAi; Terenius et al. 2011; own unpublished results). Similarly, in *Epiphyas postvittana*, silencing effects by feeding of dsRNA were reported as either low, or did not occur (Terenius et al. 2011).

Some studies stress the importance of the stage at which dsRNAs should be administered to achieve silencing effects: it was for instance reported that RNAi is mostly effective after the molt and following a starvation period in *Spodoptera frugiperda* (Rodríguez-Cabrera et al. 2010). However, such recommendations are not very practical in the field and cannot be applied in insect pest control.

While most studies were carried out by application of dsRNA in the artificial diet or by oral droplet feeding with a carbohydrate solution containing dsRNA, in one study the dsRNA source consisted of bacteria that produce dsRNA (Tian et al. 2009) and in another study transgenic plants expressing hairpin RNAs were employed (Mao et al. 2007). In the latter study, the plants produced hairpin RNAs that interfered with the expression of the *CYP6AE14* gene, which encodes the cytochrome P450 enzyme that inactivates the toxic plant compound gossypol.

Silencing of inducible genes is considered easier to achieve than the knockdown or knockout of genes that are constitutively expressed (Terenius et al. 2011) and could therefore explain the success of this approach in this instance, despite the perceived lower sensitivity of lepidopteran insects to RNAi.

Because high amounts of dsRNA in the food are needed to observe gene silencing and toxic effects in lepidopteran larvae, it remains to be seen whether these represent effects by a true RNAi process. Involvement of the RNAi machinery could be verified by the detection of siRNAs by Northern blot or deep sequencing methods. Alternatively, it remains to be investigated whether dsRNAs could initiate an innate immune response through interaction with (unidentified) pattern recognition receptors, such as Toll-like receptors through which dsRNAs act in vertebrates (Gantier and Williams 2009).

7.3.3   Hemipteran Insects: Delivery Issues for RNAi by Feeding

In hemipteran insects, mouthparts are formed into beak-like structures that are used to pierce plant tissues and suck the sap from the phloem or the xylem. Besides causing injury to plants and concomitant opportunities for infections by secondary insects or fungal pathogens, hemipteran insects are also vectors of many plant viruses.

Injection experiments have shown that hemipteran insects are sensitive to systemic RNAi (Hughes and Kaufman 2000). Genome projects were initiated and/or completed for important hemipteran pest insects such as the pea aphid *Acyrthosiphon pisum* (International Aphid Genomics Consortium 2010) and the whitefly *Bemisia tabaci* (Leshkowitz et al. 2006), while massive parallel transcriptome analyses were performed for the small brown planthopper, *Laodelphax striatellus*, a notorious rice pest (Zhang et al. 2010a). These hemipteran pests also show gene silencing upon injection of dsRNA or siRNA in larvae and adults (Mutti et al. 2006; Jaubert-Possamai et al. 2007; Ghanim et al. 2007; Liu et al. 2010). Knockdown of *c002* mRNA, corresponding to the most abundant transcript in the salivary glands of the pea aphid, resulted in a dramatic reduction in lifespan, apparently by disruption of feeding efforts through an unknown mechanism (Mutti et al. 2008). On the other hand, knockdown of a calcium binding protein or a cathepsin L of the gut did not result in a clear phenotype in this species (Jaubert-Possamai et al. 2007). In the whitefly, RNAi resulted in a lethal phenotype for one out of four target genes tested, although in all cases silencing was observed (Ghanim et al. 2007). Gene silencing through feeding of dsRNA was successfully attempted in two studies that involved the pea aphid. In the first study, during which dsRNA was administered in the artificial diet at 1–5 µg/µl, knockdown of the target gene encoding an aquaporin resulted in the expected increase in hemolymph osmotic pressure, but no changes in mortality or growth were detected (Shakesby et al. 2009). In the second study, incorporation of dsRNA targeting *V-ATPase* transcripts at various concentrations in the artificial diet resulted in increased mortality (LC_{50} = 3.4 µg/ml; Whyard et al. 2009).

The above results illustrate that RNAi through feeding can be used to control hemipteran pests. A major challenge, however, may concern the delivery of dsRNA to the insect, considering its mode of feeding through sap-sucking from the phloem or xylem. Systemic RNAi signals (as triggered by RNA hairpin transgenes) can spread over long distances in plants through the phloem and may consist of ~25 nucleotide ssRNAs complexed to small RNA-binding proteins (Xie and Guo 2006). It remains to be established whether these can trigger an RNAi response in the (phloem) sap-sucking hemipteran. On the other hand, it is totally unclear whether dsRNAs somehow can be introduced into and transported by the xylem vascular system of the plants to provide protection against xylem feeders.

7.3.4 Other Insect Orders: Indications of Potential Use for RNAi

Insects of the orders Diptera (mosquitoes and flies), Dictyoptera (cockroaches), Isoptera (termites), Hymenoptera (bees) and Orthoptera (locusts and crickets) are sensitive to gene silencing by dsRNA or siRNA injection (He et al. 2006; Zhou et al. 2006; Gempe et al. 2009; Rogers et al. 2009; Bellés 2010; Caljon et al. 2010; Torres et al. 2011). In termites and honeybees, dsRNA was also effective through feeding (Zhou et al. 2008; Hunter et al. 2010). In locusts, which comprise some of the most damaging agricultural pests in the world, it was shown that silencing of target genes involved in the molting process induced considerable toxicity and mortality (Wei et al. 2007; Zhang et al. 2010b). Whether this could lead to new control strategies through exogenous application of dsRNA on crops or transgenic plants expressing appropriate RNA hairpins remains to be determined. Recently, RNAi approaches were also used to identify the resistance mechanism of locusts against the insecticide malathion (Zhang et al. 2011).

7.3.5 Other Arthropods in Modern Agriculture That Can Be Targeted: for Instance Spider Mites

Besides insects, nothing is known regarding sensitivity to RNAi in other groups of arthropods that are agricultural pests, with the exception of the two-spotted spider mite, *Tetranychus urticae* (Acari, Chelicerata), which is a major agricultural pest because of its extreme polyphagy and short development time. Resources to fight this pest include the availability of the sequence of its genome (http://bioinformatics.psb.ugent.be/genomes) and the existence of protocols for gene silencing through injection of dsRNA or siRNA (Khila and Grbić 2007). The spider mite genome contains genes encoding the complete RNAi machinery, including five orthologs of RdRP genes, which are absent in insects but implicated in systemic gene silencing in *C. elegans* and plants (Gordon and Waterhouse 2007). Thus, genomic data suggest that RNAi may be an efficient process in the spider mite *T. urticae* as well as in ticks, its sister group of the Acari, where also RdRP genes were identified (Kurscheid et al. 2009; Swevers, Smagghe, unpublished results). It was nevertheless observed that RNAi by injection in females resulted in low penetrance of phenotypes in embryos (paternal RNAi) and that siRNA acted more efficiently than dsRNA (Khila and Grbić 2007). It also remains to be tested whether RNAi can be induced in the spider mite by feeding.

7.4 Factors That Could Determine Efficiency of RNAi

From the discussion above it can be inferred that major differences occur among insects regarding their sensitivity to the RNAi process. Even among coleopteran insects that in general are considered sensitive to RNAi, it is observed that some species may not respond well to the administration of dsRNA, such as the boll weevil *Anthonomus grandis* (Baum et al. 2007). For the application of the RNAi technique in insect pest control, it is therefore imperative to identify these species with sufficient sensitivity to allow the technique to work. This raises the question regarding the factors that can contribute to the sensitivity/resistance of insects to environmental RNAi. Understanding the obstacles that interfere with successful RNAi can also lead to new strategies to circumvent them and wider applications of this technology. As discussed in Sect. 2, from an insect control perspective, these factors count most with respect to the midgut, while they are likely of minor importance for other tissues. Below follows a general discussion regarding factors that could stimulate or interfere with the process of RNAi.

7.4.1 *Expression of the Basic RNAi Machinery*

It is now recognized that the RNAi machinery corresponds to an ancient mechanism of defense against invading nucleic acids such as viruses and transposable elements. In *Drosophila*, one branch of the RNAi machinery, involved in the generation of siRNAs, is considered part of the innate immune response (Sabin et al. 2010). This apparatus consists of the Dicer-2 enzyme and its cofactor, the RNA-binding protein R2D2, which cleaves the exogenous long dsRNAs into siRNAs and presents them subsequently for loading into the Ago-2 protein, which contains the 'slicer' activity of the 'RNA-induced silencing complex' (RISC) (Moazed 2009). It can be anticipated that differences in expression exist among different tissues in one insect and in tissues among different insect species, and that these differences are correlated with performance of the RNAi machinery. For instance, the silkmoth *Bombyx mori* is rather refractory against injection of dsRNA to trigger gene silencing in most tissues (Terenius et al. 2011). When the expression of RNAi genes was checked in different tissues and stages of the Daizo strain of *Bombyx*, it was noted that the co-factor R2D2 was not expressed in any tissue (Swevers et al. 2011; own unpublished data). The Daizo strain is very refractory to RNAi and this could be explained by the low performance of the RNAi machinery involved in processing of exogenous dsRNAs, due to the absence of expression of R2D2.

It is also observed that the antiviral immunity genes *Dicer-2*, *R2D2* and *Ago-2* are among the fastest evolving of all *Drosophila* genes (Obbard et al. 2006). Thus, the efficiency of the RNAi apparatus in different species and its involvement in the defense against invading nucleic acids (dsRNAs) is predicted to be determined by the evolutionary history of the species or the taxon.

7.4.2 Uptake of dsRNA from the Extracellular Medium

An obvious barrier to RNAi constitutes the uptake of dsRNA from the hemolymph for systemic RNAi, and from the gut content for environmental RNAi. However, experiments using tissue culture cells have shown that uptake of dsRNA *per se* does not necessarily trigger an RNAi response. In *Trichoplusia ni* (cabbage looper)-derived Hi5 cells and silkmoth (*Bombyx mori*)-derived Bm5 cells, fluorescently labeled dsRNA that is added to the culture medium can be internalized by the tissue culture cells, but the dsRNA does not trigger gene silencing (Fig. 7.4; own unpublished results; Swevers et al. 2011; Terenius et al. 2011). Thus, it seems that, for effective RNAi, internalization of dsRNA should be coupled to efficient transfer to the RNAi machinery. It is possible that the accessibility of the RNAi machinery is coupled to the general activation of the immune response, *i.e*, if the innate immune response is activated by other immunogenic triggers, it may somehow boost the sensitivity of the RNAi apparatus.

So far, no studies have been reported that investigate the uptake of fluorescently labeled dsRNA from the gut content and its possible spread to other tissues. Such studies could uncover possible resistance mechanisms against RNAi (despite the reservations mentioned above).

7.4.3 DsRNA- or siRNA-Degrading Enzymes

That dsRNA catabolism can influence the efficiency of RNAi was already apparent from work in *C. elegans*, where the refractoriness of the nervous system was explained by the expression of the *eri-1* nuclease (Kennedy et al. 2004). Of relevance is the presence of nucleic acid degrading enzymes in the gut content that are capable of degrading dsRNA (Arimatsu et al. 2007). In some insects, dsRNA-degrading activity in the gut is lowest during the molt and can be further reduced by starvation (Rodríguez-Cabrera et al. 2010), but it is uncertain whether

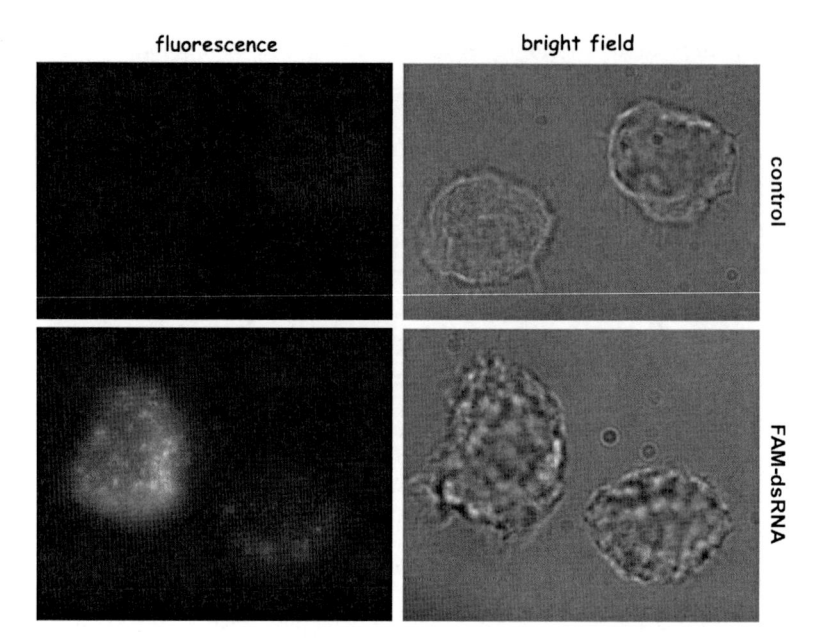

Fig. 7.4 Fluorescence microscope images of lepidopteran *Bombyx mori*-derived Bm5 cells after soaking in high concentrations (about 100 μg/ml) of fluorescein (FAM)-labeled dsRNA for 24 h. Control cells were left untreated. Uptake of FAM-labeled-dsRNA by individual cells showed considerable variation that resulted in the observation of intense signals in some cells and the absence of fluorescence in other cells. Shown are selected treated cells in which strong internalization of FAM-labeled dsRNA is detected. The experiment suggests that lepidopteran cells are able to take up dsRNA molecules (Own unpublished results)

this knowledge can be applied practically in the field. So far, there are no systematic studies that investigate the stability of dsRNA in the gut or after uptake in the animal that would uncover additional resistance mechanisms against RNAi.

7.4.4 Involvement of Immune Response

At least for some insects, such as *Drosophila*, the RNAi machinery of the siRNA pathway evolved as a defense against RNA viruses (Wang et al. 2006; Saleh et al. 2009). However, viral infection, by an unknown mechanism, also induces the JAK-STAT signaling pathway in insect cells which results in the establishment of an antiviral state (Kemp and Imler 2009). How both pathways interact is not known. Thus, it is possible that other features of viral infection somehow 'sensitize' the host cells and up-regulate the factors of the RNAi machinery dedicated to defense against exogenous dsRNA. From an insect control perspective, it may be worthwhile to search for 'co-activators' of the RNAi response that will enhance its activity and efficiency.

7.5 Conclusions and Future Perspectives for RNAi in Pest Insects

In conclusion, the problems on human and environmental safety and insecticide resistance towards neurotoxic and broad-spectrum insecticides and the limitation of the crystal toxins of *Bt* demand for new approaches for pest control. To address this issue, RNAi as a sequence-specific gene silencing tool provides a great potential in crop protection. This technology can be particularly valuable to those insects that are insensitive to current *Bt* crops or/and where *Bt* does not work very well, such as sucking pests, aphids, whiteflies, planthoppers, and ticks and mites. We believe that a better understanding of the molecular mechanisms of plant and insect RNAi and plant-insect interactions will greatly help to further development of novel technologies for crop pest control. There is no doubt that with continuous efforts toward increasing gene silencing efficacy and specificity, RNAi will find its application in a wide range of agriculture, especially crop protection. We see three major lines of interest for further importance of RNAi:

1. identification of new insecticide targets;
2. insecticide resistance management;
3. management to control pest insects

In the search for new target identification, it is evident that the insect genomics research has been providing increasingly huge amounts of data: up to-date (13 June 2011) 87 insect genomes are either finished or in progress and most of them are available in the NCBI Entrez Genome Project database (http://www.ncbi.nlm.nih.gov/genomes/leuks.cgi), and this number will continue to increase at accelerating pace. Indeed, Robinson et al. (2011) recently announced the launch of the "i5k" initiative to sequence the genomes of 5,000 species of insects and other arthropods during the next 5 years. Besides such genome projects, initiatives on data-mining and RNAi screen analyses are greatly appreciated. Two recent examples are FLIGHT (http://flight.icr.ac.uk/) and the Lepidoptera database (http://insectacentral.org/RNAi; Terenius et al. 2011). FLIGHT is an online resource compiling data from high-throuput *Drosophila in vivo* and *in vitro* RNAi screens (Sims et al. 2010). It includes details of RNAi reagents, predicted off-target effects, RNAi screen hits, scores and phenotypes, including images from high-content screens. Users can integrate RNAi screen data with microarray gene expression as well as genomic annotations and genetic/physical interaction datasets to provide a single interface for RNAi screen analysis and data-mining in *Drosophila*. The online Lepidoptera database comprises data from more than 200 experiments including all to date published as well as many unpublished experiments. Despite a large variation in the data, trends that are found are that RNAi is particularly successful in the family Saturniidae and in genes involved in immunity. The database also points to a need to further investigate the mechanism of RNAi in Lepidoptera and its possible connection to the innate immune response. Here we believe therefore that an efficient international RNAi platform is invaluable.

In addition, it is evident that the modern technology of second-generation sequencing will also significantly contribute forming a key in the success for the RNAi approach. Illumina's RNA-sequencing and digital gene expression tag profile (DGE-tag) technologies are useful to screen optimal RNAi targets from different insects and their tissues, especially species that are not whole genome sequenced which is the case for most part of the insect pests important in agriculture. So far several have been analyzed; good examples are the brown plant hopper *Nilaparvata lugens* (Xue et al. 2010), the whiteflies *Bemisia tabaci* (Wang et al. 2010) and *Trialeurodes vaporariorum* (Karatolos et al. 2011), the potato psyllid *Bactericera cockerelli* (Hail et al. 2010), the Colorado potato beetle *Leptinotarsa decemlineata* (Zhu et al. 2011; own unpublished data), the Asian corn borer *Ostrinia furnalalis* (Wang et al. 2011) and the exotic invasive insect pest emerald ash borer *Agrilus planipennis* (Mittapalli et al. 2010). In the latter studies, it was of interest that the authors were able to broaden the target selection for RNAi from just insect midgut-specific genes to targets in the whole insect body. We believe these tools will help to identify new sensitive insecticide targets for designing RNAi-based technology against insect damage.

In addition to careful target selection, a convenient delivery system also plays a crucial role in the success of RNAi as discussed above. In the year of 2007, two papers in Nature Biotechnology reported the use of plant-mediated RNAi to suppress the growth of two agricultural pests, the cotton bollworm *H. armigera* (Mao et al. 2007) and the western corn rootworm *D. virgifera* (Baum et al. 2007). These breakthroughs brought lights into further development of RNAi technology for crop protection. More recent Rocha et al. (2011) reported on the systemic character of RNAi in insects, that phagocytic uptake of dsRNA offers a potential route for systemic spread of RNAi. Although these results were obtained in cells of the model insect of *Drosophila*, we believe that the current finding can provide a robust push in the effectiveness of RNAi in pest insects.

It is also of interest to mention here that RNAi can contribute in the emphasis of a joint action of potential insecticidal compounds. As reported by Moar et al. (2010) the next generation of pyramided insect-protection traits may employ *Bt* genes and efficacious dsRNA target genes. These different insecticidal compounds act in an independent complementary manner in a single set of pests. *Bt* toxin acts fast but less persistent, while dsRNA has its best effects after >10 days and reduced the risk of insecticide resistance, providing a successful alternative to high dose approach and reduction of refuge requirements.

Furthermore, the fascinating data of Zhu et al. (2010) shed new light on the understanding of the molecular basis and evolution of insecticide resistance as they realized in a significant reduction of the resistance to the pyrethroid insecticide deltamethrin by RNAi knockdown in the expression of CYP6BQ9. The brain-specific insect P450 enzyme, *CYP6BQ9*, is involved in high levels of deltamethrin resistance in the QTC279 strain of *T. castaneum* by an increased metabolic detoxification. We believe these data are true indicatives of the power of RNAi in the management of insecticide resistance in agriculture.

Finally, because RNAi selectively inhibits target genes with a high specificity and fidelity, it is naturally a strategy of choice to be used in controlling pest insects in the frame of integrated pest management (IPM) in combination with beneficial organisms as pollinators and natural enemies for biological control. Here risk assessment for hazards is essential for confirming the compatible use in the field with beneficial organisms. This can follow the tier testing approach as for classical plant protection products and agents as formulated by international organizations as IOBC and EPPO. Besides, a careful selection of target genes and constructs can help in the high specificity of RNAi. Here, the accumulation of genome data (for instance of honeybees, parasitic wasps, bumblebees) will therefore also serve for assessment of environmental safety. As a consequence, we believe that together with all stakeholders, RNAi can develop as a new technology for environmental-friendly insect pest control strategy and form outstanding challenges in modern agriculture.

Acknowledgements   The authors acknowledge support for their research by the Fund for Scientific Research-Flanders (FWO-Vlaanderen), the Flemish agency for Innovation by Science and Technology (IWT-Vlaanderen), the Special Research Funds of Ghent University, in Belgium, and the General Secretariat for Research and Technology, Hellenic Republic Ministry of National Education and Religious Affairs, in Greece. Luc Swevers acknowledges the support with a scholarship for foreign researcher by the Special Research Fund of Ghent University.

References

Arimatsu Y, Kotanib E, Sugimurab Y, Furusawa T (2007) Molecular characterization of a cDNA encoding extracellular dsRNase and its expression in the silkworm, *Bombyx mori*. Insect Biochem Mol Biol 37:176–183

Auer C, Frederick R (2009) Crop improvement using small RNAs: applications and predictive ecological risk assessments. Trends Biotechnol 27:644–651

Baum JA, Bogaert T, Clinton W, Heck GR, Feldmann P, Ilagan O, Johnson S, Plaetinck G, Munyikawa T, Pleau M, Vaughn T, Roberts J (2007) Control of coleopteran insect pests through RNA interference. Nat Biotechnol 25:1322–1326

Bautista MAM, Miyata T, Miura K, Tanaka T (2009) RNA interference-mediated knockdown of a cytochrome P450, CYP6BG1, from the diamondback moth, *Plutella xylostella*, reduces larval resistance to permethrin. Insect Biochem Mol Biol 39:38–46

Bellés X (2010) Beyond *Drosophila*: RNAi *in vivo* and functional genomics in insects. Annu Rev Entomol 55:111–128

Caljon G, De Ridder K, De Baetselier P, Coosemans M, Van den Abbeele J (2010) Identification of a tsetse fly salivary protein with dual inhibitory action on human platelet aggregation. PLoS One 5:e9671

Cancino-Rodezno A, Alexander C, Villaseñor R, Pacheco S, Porta H, Pauchet Y, Soberón M, Gill SS, Bravo A (2010) The mitogen-activated protein kinase p38 is involved in insect defense against Cry toxins from *Bacillus thuringiensis*. Insect Biochem Mol Biol 40:58–63

Carthew RW, Sontheimer EJ (2009) Origins and mechanisms of miRNAs and siRNAs. Cell 136:642–655

Chen J, Tang B, Chen H, Yao Q, Huang X, Chen J, Zhang D, Zhang W (2010) Different functions of the insect soluble and membrane-bound trehalase genes in chitin biosynthesis revealed by RNA interference. PLoS One 5:e10133

Gantier MP, Williams BRG (2009) siRNA delivery not Toll-free. Nat Biotechnol 27:911–912

Gempe T, Hasselmann M, Schiøtt M, Hause G, Otte M, Beye M (2009) Sex determination in honeybees: two separate mechanisms induce and maintain the female pathway. PLoS Biol 7:e1000222

Ghanim M, Kontsedalov S, Czosneck H (2007) Tissue-specific gene silencing by RNA interference in the whitefly *Bemisia tabaci* (Gennadius). Insect Biochem Mol Biol 37:732–738

Gong LA, Yang XQ, Zhang BL, Zhong G, Hu M (2011) Silencing of Rieske iron-sulfur protein using chemically synthesised siRNA as a potential biopesticide against *Plutella xylostella*. Pest Manag Sci 67(5):514–520

Gordon KHJ, Waterhouse PM (2007) RNAi for insect-proof plants. Nat Biotechnol 25:1231–1232

Hail D, Hunter WB, Dowd SE, Bextine BR (2010) Expressed sequence tag (EST) survey of life stages of the potato psyllid, *Bactericera cockerelli*, using 454 pyrosequencing. Southwest Entomol 35(3):463–466

Hakim RS, Baldwin K, Smagghe G (2010) Regulation of midgut growth, development, and metamorphosis. Annu Rev Entomol 55:593–608

He Z-B, Cao Y-Q, Yin Y-P, Wang Z-K, Chen B, Peng G-X, Xia Y-X (2006) Role of *hunchback* in segment patterning of *Locusta migratoria manilensis* revealed by parental RNAi. Dev Growth Differ 48:439–445

Hughes CL, Kaufman TC (2000) RNAi analysis of deformed, proboscipedia and sex combs reduced in the milkweed bug *Oncopeltus fasciatus*: novel roles for Hox genes in the hemipteran head. Development 127:3683–3694

Hunter W, Ellis J, van Engelsdorp D, Hayes J, Westervelt D, Glick E, Williams M, Sela I, Maori E, Pettis J, Cox-Foster D, Paldi N (2010) Large-scale field application of RNAi technology reducing Israeli acute paralysis virus disease in honey bees (*Apis mellifera*, Hymenoptera: Apidae). PLoS Pathog 6:e1001160

Huvenne H, Smagghe G (2010) Mechanisms of dsRNA uptake in insects and potentials of RNAi for pest control: a review. J Insect Physiol 56:227–235

International Aphid Genomics Consortium (2010) Genome sequence of the pea aphid *Acyrthosiphon pisum*. PLoS Biol 8:e1000313

Jaubert-Possamai S, Le Trionnaire G, Bonhomme J, Christophides GK, Rispe C, Tagu D (2007) Gene knockdown by RNAi in the pea aphid *Acyrthosiphon pisum*. BMC Biotechnol 7:63

Karatolos N, Pauchet Y, Wilkinson P, Chauhan R, Denholm I, Gorman K, Nelson DR, Bass C, ffrench-Constant RH, Williamson MS (2011) Pyrosequencing the transcriptome of the greenhouse whitefly, Trialeurodes vaporariorum reveals multiple transcripts encoding insecticide targets and detoxifying enzymes. BMC Genomics 12:56. doi:10.1186/1471-2164-12-56

Kemp C, Imler J-L (2009) Antiviral immunity in *Drosophila*. Curr Opin Immunol 21:3–9

Kennedy S, Wang D, Ruvkun G (2004) A conserved siRNA-degrading RNase negatively regulates RNA interference in *C. elegans*. Nature 427:645–649

Khajuria C, Buschman LL, Chen M-S, Muthukrishnan S, Zhu KY (2010) A gut-specific chitinase gene essential for regulation of chitin content of peritrophic matrix and growth of *Ostrinia nubilalis* larvae. Insect Biochem Mol Biol 40:621–629

Khila A, Grbić M (2007) Gene silencing in the spider mite *Tetranychus urticae*: dsRNA and siRNA parental silencing of the Distal-less gene. Dev Genes Evol 217:241–251

Kurscheid S, Lew-Tabor AE, Valle MR, Bruyeres AG, Doogan VJ, Munderloh UG, Guerrero FD, Barrero RA, Bellgard MI (2009) Evidence of a tick RNAi pathway by comparative genomics and reverse genetics screen of targets with known loss-of-function phenotypes in *Drosophila*. BMC Mol Biol 10:26

Leshkowitz D, Gazit S, Reuveni E, Ghanim M, Czosnek H, McKenzie C, Shatters RL Jr, Brown JK (2006) Whitefly (*Bemisia tabaci*) genome project: analysis of sequenced clones from egg, instar, and adult (viruliferous and non-viruliferous) cDNA libraries. BMC Genomics 7:79

Liu S, Ding Z, Zhang C, Yang B, Liu Z (2010) Gene knockdown by intro-thoracic injection of double-stranded RNA in the brown planthopper, *Nilaparvata lugens*. Insect Biochem Mol Biol 40:666–671

Mao Y-B, Cai W-J, Wang J-W, Hong G-J, Tao X-Y, Wang L-J, Huang Y-P, Chen X-Y (2007) Silencing a cotton bollworm P450 monooxygenase gene by plant-mediated RNAi impairs larval tolerance of gossypol. Nat Biotechnol 25:1307–1313

Mittapalli O, Bai XD, Mamidala P, Rajarapu SP, Bonello P, Herms DA (2010) Tissue-specific transcriptomics of the exotic invasive insect pest emerald ash borer (*Agrilus planipennis*). PLoS One 5(10):e13708. doi:10.1371/journal.pone.0013708

Moar WJ, Clark T, Segers G, Ramaseshadri P, Hibbard B, Head G (2010) dsRNA: the next generation of pyramided insect protection traits. Abstract book 58th annual meeting of the Entomological Society of America, 12–15 Dec 2010, San Diego, CA, p 108

Moazed D (2009) Small RNAs in transcriptional gene silencing and genome defence. Nature 457:413–420

Mutti NS, Park Y, Reese JC, Reeck GR (2006) RNAi knockdown of a salivary transcript leading to lethality in the pea aphid, Acyrthosiphon pisum. J Insect Sci 6(38):1–7

Mutti NS, Louis J, Pappan LK, Pappan K, Begum K, Chen M-S, Park Y, Dittmer N, Marshall J, Reese JC, Reeck GR (2008) A protein from the salivary glands of the pea aphid, Acyrthosiphon pisum, is essential in feeding on a host plant. Proc Natl Acad Sci U S A 105:9965–9969

Obbard DJ, Jiggins FM, Halligan DL, Little TJ (2006) Natural selection drives extremely rapid evolution in antiviral RNAi genes. Curr Biol 16:580–585

Price DRG, Gatehouse JA (2008) RNAi-mediated crop protection against insects. Trends Biotechnol 26:393–400

Robinson GE, Hackett KJ, Purcell-Miramontes M, Brown SJ, Evans JD, Goldsmith MR, Lawson D, Okamuro J, Robertson HM, Schneider DJ (2011) Creating a buzz about insect genomes. Science 331:1386

Rocha JJE, Korolchuk VI, Robinson IM, O'Kane CJ (2011) A phagocytic route for uptake of double-stranded RNA in RNAi. PLoS One 6(4):e19087. doi:10.1371/journal.pone.0019087

Rodríguez-Cabrera L, Trujillo-Bacallao D, Borrás-Hidalgo O, Wright DJ, Ayra-Pardo C (2010) RNAi-mediated knockdown of a *Spodoptera frugiperda* trypsin-like serine-protease gene reduces susceptibility to a *Bacillus thuringiensis* Cry1Ca1 protoxin. Environ Microbiol 12 (11):2894–2903

Rogers DW, Baldini F, Battaglia F, Panico M, Dell A, Morris HR, Catteruccia F (2009) Transglutaminase-mediated semen coagulation controls sperm storage in the malaria mosquito. PLoS Biol 7:e1000272

Runo S, Alakonya A, Machuka J, Sinha N (2011) RNA interference as a resistance mechanism against crop parasites in Africa: a 'Trojan horse' approach. Pest Manag Sci 67(2):129–136

Sabin LR, Hanna SL, Cherry S (2010) Innate antiviral immunity in *Drosophila*. Curr Opin Immunol 22:4–9

Saleh MC, van Rij RP, Hekele A, Gillis A, Foley E, O'Farrell PH, Andino R (2006) The endocytic pathway mediates cell entry of dsRNA to induce RNAi silencing. Nat Cell Biol 8:793–802

Saleh MC, Tassetto M, van Rij RP, Goic B, Gausson V, Berry B, Jacquier C, Antoniewski C, Andino R (2009) Antiviral immunity in *Drosophila* requires systemic RNA interference spread. Nature 458:346–350

Shabalina SA, Koonin EV (2008) Origins and evolution of eukaryotic RNA interference. Trends Ecol Evol 23:578–587

Shakesby AJ, Wallace IS, Isaacs HV, Pritchard J, Roberts DM, Douglas AE (2009) A water-specific aquaporin involved in aphid osmoregulation. Insect Biochem Mol Biol 39:1–10

Sims D, Bursteinas B, Jain E, Gao QO, Baum B, Zvelebil M (2010) The FLIGHT *Drosophila* RNAi database 2010 update. Fly 4(4):344–348

Siomi H, Siomi MC (2009) On the road to reading the RNA-interference code. Nature 457:396–404

Swevers L, Liu J, Huvenne H, Smagghe G (2011) Search for limiting factors in the RNAi pathway in silkmoth tissues and the silkmoth-derived Bm5 cell line: the RNA-binding proteins R2D2 and Translin. PLoS One 6(5):e20250. doi:10.1371/journal.pone.0020250

Tan A, Palli SR (2008) Identification and characterization of nuclear receptors from the red flour beetle, *Tribolium castaneum*. Insect Biochem Mol Biol 38:430–439

Tang B, Wang S, Zhang F (2010) Two storage hexamerins from the beet armyworm Spodoptera exigua: cloning, characterization and the effect of gene silencing on survival. BMC Mol Biol 11:65

Terenius O, Papanicolaou A, Garbutt JS, Eleftherianos I, Huvenne H, Sriramana K, Albrechtsen M, An C, Aymeric J-L, Barthel A, Bebas P, Bitra K, Bravo A, Chevalier F, Collinge DP, Crava CM, de Maagd RA, Duvic B, Erlandson M, Faye I, Felföldi G, Fujiwara H, Futahashi R, Gandhe AS, Gatehouse HS, Gatehouse LN, Giebultowicz J, Gómez I, Grimmelikhuijzen CJ, Groot AT, Hauser F, Heckel DG, Hegedus DD, Hrycaj S, Huang L, Hull J, Iatrou K, Iga M, Kanost MR, Kotwica J, Li C, Li J, Liu J, Lundmark M, Matsumoto S, Meyering-Vos M, Millichap PJ, Monteiro A, Mrinal N, Niimi T, Nowara D, Ohnishi A, Oostra V, Ozaki K, Papakonstantinou M, Popadic A, Rajam MV, Saenko S, Simpson RM, Soberón M, Strand MR, Tomita S, Toprak U, Wang P, Wee CW, Whyard S, Zhang W, Nagaraju J, ffrench-Constant RH, Herrero S, Gordon K, Swevers L, Smagghe G (2011) RNA interference in Lepidoptera: an overview of successful and unsuccessful studies and implications for experimental design. J Insect Physiol 57:231–245

Tian H, Peng H, Yao Q, Chen H, Xie Q, Tang B, Zhang W (2009) Developmental control of a Lepidopteran pest Spodoptera exigua by ingestion of bacteria expressing dsRNA of a non-midgut gene. PLoS One 4:e6225

Tomari Y, Du T, Zamore PD (2007) Sorting of Drosophila small silencing RNAs. Cell 130:299–308

Tomoyasu Y, Miller SC, Tomita S, Schoppmeier M, Grossmann D, Bucher G (2008) Exploring systemic RNA interference in insects: a genome-wide survey for RNAi genes in Tribolium. Genome Biol 9:R10

Torres L, Almazán C, Ayllón N, Galindo RC, Rosario-Cruz R, Quiroz-Romero H, de la Fuente J (2011) Functional genomics of the horn fly, Haematobia irritans (Linnaeus, 1758). BMC Genomics 12:105. doi:10.1186/1471-2164-12-105

Turner CT, Davy MW, MacDiarmid RM, Plummer KM, Birch NP, Newcomb RD (2006) RNA interference in the light brown apple moth, Epiphyas postvittana (Walker) induced by double-stranded RNA feeding. Insect Mol Biol 15:383–391

Ulvila J, Parikka M, Kleino A, Sormunen R, Ezekowitz RA, Kocks C, Ramet M (2006) Double-stranded RNA is internalized by scavenger receptor-mediated endocytosis in Drosophila S2 cells. J Biol Chem 281:14370–14375

Vallier A, Vincent-Monégat C, Laurençon A, Heddi A (2009) RNAi in the cereal weevil Sitophilus spp Systemic gene knockdown in the bacteriome tissue. BMC Biotechnol 9:44

Wang XH, Aliyari R, Li WX, Li HW, Kim K, Carthew R, Atkinson P, Ding SW (2006) RNA interference directs innate immunity against viruses in adult Drosophila. Science 312:452–454

Wang XW, Luan JB, Li JM, Bao YY, Zhang CX, Liu SS (2010) De novo characterization of a whitefly transcriptome and analysis of its gene expression during development. BMC Genomics 11:400. doi:10.1186/1471-2164-11-400

Wang Y, Zhang H, Li H, Miao X (2011) Second-generation sequencing supply an effective way to screen RNAi targets in large scale for potential application in pest insect control. PLoS One 6 (4):e18644. doi:10.1371/journal.pone.0018644

Wei Z, Yin Y, Zhang B, Wang Z, Peng G, Cao Y, Xia Y (2007) Cloning of a novel protease required for the molting of Locusta migratoria manilensis. Dev Growth Differ 49:611–621

Whyard S, Singh AD, Wong S (2009) Ingested double-stranded RNAs can act as species-specific insecticides. Insect Biochem Mol Biol 39:824–832

Winston WM, Molodowitch C, Hunter CP (2002) Systemic RNAi in C. elegans requires the putative transmembrane protein SID-1. Science 295:2456–2459

Xie Q, Guo H-S (2006) Systemic antiviral silencing in plants. Virus Res 118:1–6

Xue J, Bao Y-Y, B-l L, Cheng Y-B, Peng Z-Y, Liu H, Xu H-J, Zhu Z-R, Lou Y-G, Cheng J-A, Zhang C-X (2010) Transcriptome analysis of the brown planthopper Nilaparvata lugens. PLoS One 5(12):e14233. doi:10.1371/journal.pone.0014233

Yang Y, Zhu YC, Ottea J, Husseneder C, Leonard BR, Abel C, Huang F (2010) Molecular characterization and RNA interference of three midgut aminopeptidase N isozymes from *Bacillus thuringiensis*-susceptible and -resistant strains of sugarcane borer, *Diatraea saccharalis*. Insect Biochem Mol Biol 40:592–603

Zhang F, Guo H, Zheng H, Zhou T, Zhou Y, Wang S, Fang R, Qian W, Chen X (2010a) Massively parallel pyrosequencing-based transcriptome analyses of small brown planthopper (*Laodelphax striatellus*), a vector insect transmitting rice stripe virus (RSV). BMC Genomics 11:303

Zhang J, Liu X, Zhang J, Li D, Sun Y, Guo Y, Ma E, Zhu KY (2010b) Silencing of two alternative splicing-derived mRNA variants of chitin synthase 1 gene by RNAi is lethal to the oriental migratory locust, *Locusta migratoria manilensis* (Meyen). Insect Biochem Mol Biol 40:824–833

Zhang J, Zhang J, Yang M, Jia Q, Guo Y, Ma E, Zhu KY (2011) Genomics-based approaches to screening carboxylesterase-like genes potentially involved in malathion resistance in oriental migratory locust (*Locusta migratoria manilensis*). Pest Manag Sci 67:183–190

Zhou X, Qi FM, Scharf ME (2006) Social exploitation of hexamerin: RNAi reveals a major caste-regulatory factor in termites. Proc Natl Acad Sci U S A 103:4499–4504

Zhou X, Wheeler MM, Qi FM, Scharf ME (2008) RNA interference in the termite Reticulitermes flavipes through ingestion of double-stranded RNA. Insect Biochem Mol Biol 38:805–815

Zhu F, Parthasarathy R, Bai H, Woithe K, Kaussmann M, Nauen R, Harrison DA, Palli SR (2010) A brain-specific cytochrome P450 responsible for the majority of deltamethrin resistance in the QTC279 strain of *Tribolium castaneum*. Proc Natl Acad Sci U S A 107(19):8557–8562

Zhu F, Xu J, Palli R, Ferguson J, Palli SR (2011) Ingested RNA interference for managing the populations of the Colorado potato beetle, *Leptinotarsa decemlineata*. Pest Manag Sci 67(2):175–182

Chapter 8
Regulatory Approvals of GM Plants (Insect Resistant) in European Agriculture: Perspectives from Industry

Jaime Costa and Concepcion Novillo

8.1 Genetically Modified (GM) Crops and Agricultural Sustainability

It is well known that historical progress in agriculture, including the genetic improvement of domesticated crops over the past 10,000 years, has been linked with the development of our civilization (Diamond 1998; García Olmedo 1998; Ridley 2010). Following this initial phase of empirical trial and error approach, plant breeding was refined in a more structured way, looking for sources of desirable traits and combining them into new varieties. This approach started three centuries ago and has intensified in the last 80 years (Cubero 2003; García Olmedo 2009). By 1993, thanks to modern biotechnology, it was possible to add genes from non-related species to single plant cells, which were allowed the regeneration of viable plants with stable genetic modifications (GM). This new tool complemented and advanced previous methods of plant breeding (García Olmedo 1998), offering better precision, knowledge and predictability of the results from the introduced changes when compared to conventional breeding. In addition to these benefits, GM technology offers the opportunity to produce high yields of quality produce more sustainably. Improving agricultural sustainability is a critical objective for the future, in which growers must find ways of increasing food, feed and fibre production for a burgeoning global population, but with much reduced demand on the environment and reduced consumption of natural resources, especially fossil fuels.

Experience has already shown that GM technology is contributing significantly to agricultural sustainability. For example, the use of insect-resistant GM varieties of maize and cotton in the period 1996–2008 has enabled a reduction in insecticide applications equivalent, respectively, to 29.89 and 140.6 million kg of active

J. Costa (✉) • C. Novillo
Monsanto Agricultura España S.L., Avda. de Burgos 17, 28036 Madrid, Spain
e-mail: jaime.costa@monsanto.com

ingredients (Brookes and Barfoot 2010), contributing to higher yields and more affordable food prices (Brookes et al. 2010).

Despite this and similar evidence of the potential for GM crops to improve agricultural sustainability, however, the European Union (EU) has been reluctant to approve GM crops for cultivation in Member States. From the time of the first plantings of Roundup Ready® soybean in the US in 1996, until 2009, a total of 318 import or cultivation approvals were granted in 25 countries throughout the world (James 2009). However, if only commercial planting approvals are considered, there are 113 cultivation approvals for biotech crops either grown or previously grown in North America and only 6 cultivation approvals for Europe, including two Romanian approvals for Roundup Ready Soybeans that were discontinued upon Romania entering the EU.

As a result, the number of hectares of biotech crops now planted in North America vs. the EU is dramatically skewed. Statistics for 2009, for example, show that a total of 71.7 million hectares of biotech crops were planted in North America compared with only 94,000 ha for Europe. This represents greater than a 750-fold difference in the biotech hectares between the two continents. This is due to a variety of reasons including the magnitude of difference in the number of cultivation approvals, the lack of official support for Roundup Ready soybean cultivation in Romania upon entering the EU and the result of basing the regulatory approvals on a science-based risk and safety assessment process (as in North America) versus a more politically-driven process based on the precautionary principle (as in Europe), which is addressed later in this chapter. Moreover, whereas only one GM crop, the insect-protected maize MON 810, is currently cultivated to any extent in Europe, in North America, a diversity of GM crop varieties containing multiple agronomic traits are available commercially, including traits for herbicide tolerance and insect resistance (Lepidopteran and/or Coleopteran) in maize (*Zea mays*) and cotton (*Gossypium hirsutum*). Following this example, similar stacks with the same crops and traits have been approved for commercial planting in South America (Brazil and Argentina), Africa (South Africa) and Asia (Philippines) and Australia.

8.2   Regulatory Oversight of GM Crops and Their Products in the EU

The development of GM plants triggered an EU regulatory process through which GM plants need to pass prior to being commercialized or consumed. According to the European Commission, by 2006 the EU regulatory system was "one of the strictest in the world".

The cultivation of GM plants in the European Union has been regulated since 1990, prior to the regulation of novel foods (1997) or feeds (2003). It follows a

---

® Roundup Ready is a registered trademark of Monsanto.

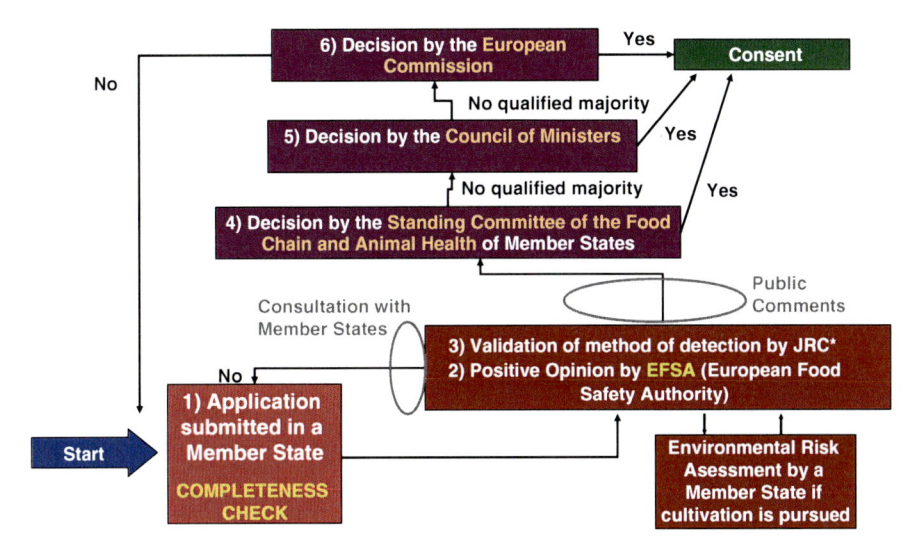

Fig. 8.1 Centralized approval system in the EU for GM food and feed (may include cultivation) according to EC Regulation 1829/2003. Public consultations are indicated as *grey ellipses*. *JRC is the official EC Joint Research Center

"case-by-case" and "step-by-step" approach including the regulatory oversight along the entire path of the product development leading to commercial release authorization:

- Directive 2001/18/EC on the deliberate release into the environment of GMOs (governing confined testing, field trial releases and placing on the market), as transposed in Member State legislation.
- Regulation (EC) no 1829/2003 on genetically modified food and feed (allowing use for food and feed of products derived from GM plants, which may also include cultivation)
- Member State or European Catalog approval for varieties derived from a GM plant and which have fulfilled the requirements of homogeneity, stability and distinctiveness

The "one door, one key" process facilitated by EC Regulation 1829/2003, which can include approvals for import and cultivation for GM plants, has become the preferred route for approval of GM plants in Europe and the use for feed or food of their fractions since April of 2004 and is described in Fig. 8.1.

To support the safety of the GM product, the company or institution developing the GM product must deliver an application fulfilling EFSA Guidelines (http://www.efsa.europa.eu/en/science/gmo/gmo_guidance.html) with a technical information package including:

- Information on the parental plant, and the genetic modification
- characterization of the inserted DNA sequences
- characterization and safety of the new proteins, including expression levels in the plant

- genetic stability of the insert
- information on any toxic, allergenic or other harmful effects arising from the GM food/feed
- proposal for a specific method of detection (primers overlapping inserted and native DNA sequences)
- plant characterization under field conditions
- effects of GM modification on plant composition
- effects of GM modification on nutritive value
- potential interactions of the GM plant with the biotic/abiotic environments
- Environmental Risk Assessment on direct and indirect effects, immediate or delayed
- Proposal of Monitoring Plan, considering case-specific monitoring of anticipated effects and general surveillance for unanticipated effects.

The goal of the environmental risk assessment required by EC Directive 2001/18 is to systematically collect information to support decision making on:

- Identification of characteristics of the GMO which may cause adverse effects
- Evaluation of the potential consequences of each adverse effect, if it occurs
- Evaluation of the likelihood of the occurrence of each identified adverse effect
- Estimation of the risk posed by each identified characteristic of the GMO (considering both the hazard and exposure)
- Proposal of management strategies for risks from the GMO
- Determination of the overall risk of the GMO.

Each of these six points must be completed to consider:

- The likelihood of the GM plant becoming more persistent or invasive than the parental plants in natural habitats
- Any selective advantage or disadvantage conferred to the GM plant
- Potential for gene transfer to other plants under conditions of planting and selective advantage or disadvantage conferred
- Potential immediate and/or delayed environmental impact resulting from the interactions between the GM plant and target organisms
- Potential immediate and/or delayed environmental impact on non-target organisms, including competitors, herbivores, potential symbionts, parasites and pathogens
- Possible immediate and/or delayed effects on human health resulting from the interactions of the GM plant with people in the vicinity of the release
- Possible immediate and/or delayed effects on animal health and consequences from consumption of the GMO for the food chain
- Possible immediate and/or delayed effects on biogeochemical processes resulting from the interaction of the GM plant and other organisms
- Possible immediate and/or delayed, direct and indirect environmental impacts of the specific cultivation, management and harvesting techniques used for the GM plant where these are different from those used for conventional plants.

This environmental risk assessment is also required for regulated field trials, normally under official supervision to ensure that the regulated GM crop is appropriately isolated and not used for food or feed. A well-constructed risk assessment should follow a logical progression or "tiered approach", where the available information is gathered to determine if additional data is needed to reach satisfactory conclusions (Tencalla et al. 2009).

For the approval process for placing GM crops and their products on the market according to EC Regulation 1829/2003, the cornerstone is a favorable Scientific Opinion by the European Food Safety Authority (EFSA), with the target of completion within 6 months after the submission has been considered complete, although the clock is often stopped when the evaluators raise additional questions, and the safety assessment may be considerably delayed by Member States involvement in the environmental risk assessment, as well as by other interventions, including by EU officials. The final Scientific Opinion by EFSA may consider comments made by Member States and potentially any other relevant points raised by individuals.

If the approval process were exclusively science-based, as in North America, a positive EFSA Opinion on the safety of a GM crop would inform all subsequent decision-making. However, in the EU such an Opinion merely opens the way for an approval decision to be voted on a complex comitology procedure (see Fig. 8.1 with the process followed until 2010). While, under the comitology procedure, approvals for pesticides, additives, food contact materials and other EU regulated products are almost exclusively granted by qualified majority vote in the Standing Committee of the Food Chain and Animal Health in accordance with a positive EFSA Scientific Opinion, the situation for the approval of GM crops and their products is different. There has been only one example in 20 years, for a non food GM carnation approved for cultivation in 1998, where Member States agreed unanimously with the positive evaluation by the Netherlands according to Directive 90/220 (http://www.gmo-compass.org/eng/gmo/db/91.docu.html), and only one example of an approval for insect protected GM crop being granted by qualified majority – for the cultivation of MON 810 maize in 1998.

8.3 A Malfunctioning Approval Process?

The adoption of GM plants protected against insect pests is now supported by 15 years of extensive cultivation. During this period, the area planted with insect-resistant crops reached 50 million ha in 2009 with a wide range of adopting countries including USA, India, China, Brazil, Burkina-Faso, Spain and others with no adverse effects to human health or organisms in the natural environment (James 2009). This is supported by thorough meta-analysis where the effects of Bt crops on non-target arthropods were found to be lower than those from the application of approved insecticides (Romeis et al. 2006; Marvier et al. 2007; Wolfenbarger et al. 2008). Confirmatory results by many independent experts on the effects of GM crops protected from insects have found that the impacts on non-target organisms are either non significant or within the levels now found with conventional practices (Kessler and Economidis 2001; Bartsch et al. 2009; Castañera et al. 2010).

It is not surprising, therefore, that the number of commercial GM approvals world-wide in 2008 reached 21 for insect resistance and 5 for virus resistance. Against this background, however, it is surprising that only one, MON 810, was approved for EU cultivation at that time (Stein and Rodriguez-Cerezo 2009), a situation which persists today. Two similar insect-protected GM maize crops have received positive Opinions in recent years from EFSA, which concluded that these GM crops are "unlikely to have adverse effects on human and animal health or the environment in the context of their proposed uses."

Clearly, factors other than the safety assessment itself are being brought to bear on the EU approval process for GM crops.

One of the reasons most often given to repeated delays in the regulatory process, and the non-approval of GM plants in the EU is the precautionary principle. This principle states that when there is reason to believe that a technology or activity may result in harm, and there is scientific uncertainty regarding the nature and extent of that harm, then measures to anticipate and prevent harm are necessary and justifiable (Raffensperger and Barret 2001). The principle should be applied in a manner that is proportional, non-discriminatory, and consistent in such a way that is not blocking agricultural progress. One may question why this principle is only applicable to plant improvements or crop protection using genetic engineering and not to conventional methods of plant breeding or crop protection based on the repeated use of insecticides. The reality is that this principle is embedded in Directive 2001/18/EC and Regulation EC No. 1829/2003, which now regulates the environmental release of GM plants in the European Union. NGOs, the Institutions of the EU and Member States have all invoked the principle and in doing so, have challenged the scientific Opinions, and indeed the credibility, of EFSA. Consequently, the timing for advancement of pending applications, in particular for the decision-making process, is often delayed for reasons not related to product safety. The complexity of the Environmental Risk Assessment (http://www.efsa.europa.eu/en/efsajournal/pub/1879.htm), unjustified local bans and cumbersome coexistence rules have given European farmers a very limited access to GM crops, even though these same GM crops are often approved for import and consumption but not approved for planting.

The environmental risk assessments in agriculture must consider that crop production itself has modified the natural environment and is related to changing agro-ecosystems (García Olmedo 2009). Crop production, regardless of GM involvement, is subjected to variations within time, location, and environmental factors including weather or soil conditions and agronomic practices such as crop rotations, tillage, irrigation, fertilizer or pesticide applications. As explained in Fig. 8.2, the goal of agriculture should not be to preserve the populations of the different organisms related to the agro-ecosystem but to improve the sustainability of food production within the baseline ranges considered acceptable under conventional practices.

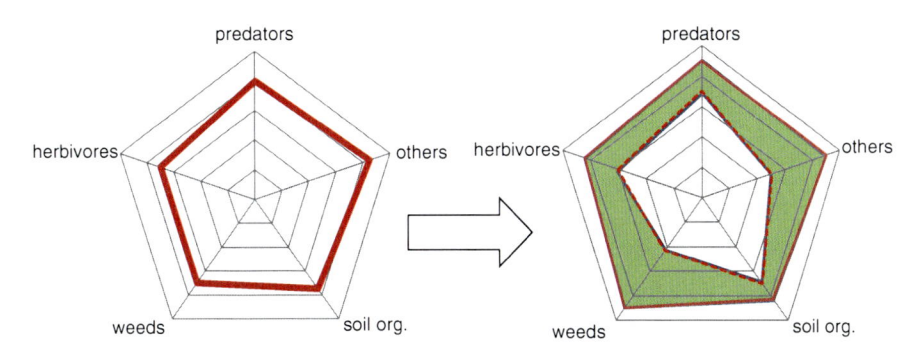

Fig. 8.2 In the environment risk evaluation of a GM crop, the objective should not be to maintain today's conditions, but to improve sustainability of food production within the accepted baseline ranges

When a crop variety includes a genetic modification, the changes in the genotype may not be larger than with other breeding technologies. As an example, the introduction of the insect-protection MON 810 trait in maize to confer resistance to European corn borer *Ostrinia nubilalis* has been found to have less changes in gene expression than other variations through conventional breeding (Coll et al. 2008; Batista et al. 2008). In fact, environmental factors may cause more variation in the different transcript/protein/metabolite profiles than the different genotypes (Barros et al. 2010).

Reaching a positive scientific Opinion from EFSA on cultivation of new GM crops is a necessary condition, but not sufficient, for final approval. The European Commission is then expected to propose the approval to the Member States through a comitology procedure (see Fig. 8.1) but the proposal may take several years in spite of demands by farmers that the process is streamlined to facilitate competitiveness (http://www.asajanet.org/asaja/inicio/procesar.do?id=28631&accion=noticia). While it is likely that a new comitology procedure will soon be established in the EU to reach decisions, in spite of the global concern for rising food prices (http://www.fao.org/news/story/en/item/50519/icode/), the process for new cultivation approvals in the EU is stalled in late 2011.

Regulation needs to be kept in perspective. Consider the regulation of exotic invasive species versus GM plants. As the result of global trade and tourism, over 100 exotic plant species have been introduced in Spain in the last 30 years (Del Monte and Aguado 2003). Unlike GM crops, some of these exotic introductions have become weeds (*Abutilon teophrasti)* and others are harmful for pets (*Opuntia tunicata)* or pose clear environmental and agricultural hazards. Although new regulations are being proposed for control of exotic invasive species, these regulations are slow in coming and do not reflect the same level of urgency from regulators and law makers as do GM crops. On the contrary, based on the safety record and benefits of GM crops, the regulatory hurdles in place to monitor GM variants of domesticated species are excessive and the cost to address them is now considered prohibitive for many organizations, especially for small and medium enterprises and especially in the EU. Recently, prominent scientists have denounced the excessive regulation of GM plants as this has created a barrier for further improvement of crop plants (García Olmedo 2009; Potrykus 2010). For example, rice (*Oryza sativa* L.) engineered to produce pro-vitamin A, also known as "golden rice" was first described by Ye et al. (2000) as a means to provide a solution to vitamin A deficiency in developing countries. Potrykus (2010) describes the excessive regulation involving golden rice from the time of their production in 1999 to when he expects deregulation to occur in 2012. He blamed the regulatory barriers as the reason why golden rice is not available to the millions of people that suffer from vitamin A deficiency in developing countries and stated "I therefore hold the regulation of genetic engineering responsible for the death and blindness of thousands of children and young mothers" (Potrykus 2010).

Discriminatory regulations on coexistence and environmental liability, GM labeling of identical products, or even proposals for display on separate shelves have been proposed. Since there are neither health nor safety issues to be expected for GM products that pass the rigorous safety assessment and approval process, such additional regulations are excessive and are not helping farmers or consumers in the EU or in developing countries that want to adopt this technology (Christou and Capell 2009). Last but not least, the European Commission proposal to entirely to revise GM legislation in 2012 will undoubtedly lead to further delays and uncertainty in the approval process.

8.4 Risk Evaluation and Social Perception

In a perfect world, regulatory oversight should be proportionate to potential risks of a given technology because the social perception of people and their elected political representatives who write the laws and regulations would recognize technology has the potential to improve our quality of life. In a perfect world, safety would be achieved thanks to public agencies who complete independent risk evaluation (consequences of hazard and exposure) for all products of concern. If a risk assessment shows that the risk is acceptable or manageable relative to accepted baseline practices, it should lead to an efficient positive regulatory decision enabling commercialization, possibly including risk management conditions. Transparent communication on the regulatory process and the principles and conclusions of the risk assessment by credible agencies is important for stakeholders. These and other points have been considered in the safety evaluations on GM food and plants, probably the most strict ever required in agriculture, and which have been summarized in Fig. 8.3.

Despite 15 years of worldwide consumption of GM food without a single health issue, in 2010, 8% of Europeans spontaneously associate food risks with GMOs, behind other problems such as chemical products (19%), bacteria food poisoning (12%), diet-related diseases (10%), obesity (9%), lack of freshness (9%) and food

Risk assessment ⟹ **Risk management** ⟹ **Risk Communication**

- Selection of gene donors and recipient plant
- Safety of new protein
- Screens to select transformants
- Insert characterization
- Effects on plant composition
- Effects on nutritive value
- Field comparisons with conventional varieties over > 4 years
- Tier I hazards for non target organisms
- Tier II risk (hazard x exposure) under field conditions
- Peer reviewed articles with methods and data
- Scientific Opinions on GM food or plant safety by relevant Authorities (EPA, FDA, USDA, EFSA, etc.).

- Approval decision by relevant Authorities (USDA, European Commission, Member States, etc) and use conditions
- Product labels including clear use recommendations
- Product stewardship to insure correct commercialization and use by distributors and farmers
- Monitoring plans and general surveillance.

- Publication of safety conclusions in the approval decisions
- Food safety or environmental alerts related with GM products.

Fig. 8.3 Safety considerations in the cultivation of GM plants

additives (9%) (http://www.efsa.europa.eu/en/riskcommunication/riskperception. htm). Public perceptions about GM crops are undoubtedly one of the drivers behind political resistance to approving GM crops for cultivation.

Some cases where GM grains or GM food have been rejected were because of the absence of tolerance levels for GM products, safely approved and consumed in the country of origin, but awaiting approval in the European Union (http://ec. europa.eu/food/food/rapidalert/rasff_publications_en.htm). The lack of established low level tolerances for GM food safely consumed in other countries, has seriously disrupted some soybean commodity product imports into the EU in 2009, in particular for the EU feed and livestock production sectors (Stein and Rodríguez-Cerezo 2009).

Social perception related with GM crops might have decreased the safety of maize when attack of corn borers increases the presence of fumonisins. Fumonisin B_1 is a mycotoxin produced by infections of *Fusarium nivale* in maize – facilitated by the attack of corn borers- and has been considered carcinogenic by the European Food Safety Authority (http://www.efsa.europa.eu/EFSA/efsa_locale-1178620753812_1178620762453.htm). In December 2006, the European Commission published EC Regulation 1881/2006, setting tolerances for mycotoxins and a maximum level of 2,000 micrograms/kg for fumonisins $B_1 + B_2$ in maize. The use of GM varieties of maize resistant to corn borers, has been shown to reduce the levels of fumonisins in maize grain both in USA and Europe (Munkwold et al. 1997; Serra et al. 2008). This was a real risk because 62 batches of products derived from maize were reported in the Rapid Alert System (http://ec.europa.eu/food/food/rapidalert/index_en.htm) as not allowed to be sold in the European Union between 2003 and 2008 (Escobar and Quintana 2008) because of the presence of mycotoxin contamination. From the 62 batches of maize products, 39 were from conventional maize, 19 were from organic maize and 0 from transgenic maize. While the cultivation of GM maize protected against corn borers was being restricted in several European countries, for reasons not related with safety according to the European Food Safety Authority, the answer of the European Commission to manage the mycotoxin issue was not to require an upgrade of the growing conditions for organic maize, by far the most problematic given the low surface (0.1% of total maize in Spain) grown in Europe, but with justifications including "to avoid market perturbations" doubled the tolerance for fumonisin in maize up to 4,000 micrograms/kg according to the new EC Regulation 1126/2007 published in September 2007.

Public opinion is being frequently measured in many countries and the European Commission has sponsored "Eurobarometer" studies on biotechnology since 1996. These studies show varying degrees of acceptance of GM crops and GM food among Member States of the EU [lowest in Greece with 12% and highest in Spain with 74% in 2005 (http://ec.europa.eu/research/press/2006/pdf/pr1906_eb_64_3_final_report-may2006_en.pdf)]. In 2010 the question on tolerance was changed to "The development of GM food should be encouraged" and the positive answers ranged from 8% in Cyprus to 36% in Czech Republic (http://ec.europa.eu/public_opinion/archives/ebs/ebs_341_en.pdf). Another recent survey on food safety in the European Union has found that 63% of Europeans believe that their diet is today less healthy compared to 10 years ago (http://www.efsa.europa.eu/en/riskcommunication/riskperception.htm), which is not supported by life expectation data, and suggests that better risk communication efforts are needed.

In separate surveys in the European Union, biotechnology continues to be among the top five environmental concerns for some EU people. The percentage of those concerned about GM plants, however, has decreased from 30% in 2002 to 20% in 2007, and concern for climatic change is clearly on the rise. The top concern in 2002 was nuclear power, but this option was not reported in the 2007 survey.

8.5 Impacts of Labeling, Traceability and Segregation

At the end of last century, European citizens were surveyed on behalf of the European Commission and 90% of them answered "yes" to the question "Do you think that food coming from GM plants should be labeled?". This opinion was used to support the need for labeling and traceability according to EC Regulations numbers 258/97, 1139/98 and associated regulations, 1829/2003 and 1830/2003, requiring identification of live GM seeds, fractions containing traces of the new DNA or the new protein, or fractions from GM plants where traces of new DNA or the new protein cannot be found (refined oils, starch, alcohol, etc.).

In other (more pragmatic) surveys around the same time in the USA, citizens were asked about their satisfaction with the information on food labels: the majority of American people (77%) were satisfied with the current food labels, and only 2% of the consumers named anything related to GM as an item they would like to see added to a food label (http://www.foodinsight.org/Press-Release/Detail.aspx? topic=IFIC_Survey_Americans_Acceptance_of_Food_Biotechnology_Matches_ Growers_Increased_Adoption_of_Biotech_Crops).

The safety assessment process of most GM crops includes considerable testing to confirm that there is substantial equivalence for composition when compared with non-GM crops. The question is, if there is no significant difference between GM and non-GM crops and both are proven to be equally safe, why label the foods that are made from GM plants? However, labeling of GM food does occur in the European Union because the approval system is technology or process based, rather than product based. Consider this example. Foods with traces of GM ingredients that are safe are required to be labeled but foods containing up to 20 ppm of wheat gluten protein, which could be harmful to celiac patients, are permitted to be labeled as "gluten free" (EC Regulation no 41/2009) (http://eur-lex.europa.eu/LexUriServ/LexUriServ. do?uri=OJ:L:2009:016:0003:0005:EN:PDF).

Furthermore, additional labeling of foods with GM ingredients often adds to the price of these foods because of increased costs of separate storage, transport and traceability, while in European studies most people are neither really interested in, nor very alert to, the presence of GM ingredients or products (Moses 2010). However, in developing countries, needless labeling is not always possible, and a food expense that they should not have to pay.

The reason why labeling is expensive is because it requires traceability of GM products and the segregation of grains. Segregation of organic non-GM materials can be afforded by farmers and traders thanks to some higher prices for the certified materials. But this is not justified for the GM-value chain by any safety considerations, and adds a processing cost leading to higher food prices. The non scientific threshold requiring GM labeling ranges from 0.9% in the European Union to 1% in Brazil and 5% in Japan. In the EU, the traceability and labeling applies to food or feed derived from the GM plants, but has not so far been required for meat, milk and eggs from animals fed with GM products, which are identical to meat, milk and eggs from animals fed with conventional products (Beever and Kemp 2000).

One of the objectives of GM labeling is to offer informed choice to the consumers. This has hardly been met in the places where labeling is compulsory as it happens in the European Union. Food manufacturers tend to choose non-GM grain sources to defend their brands and image from the threats of activist groups. The financing of these anti-GM groups in Europe is heavily supported by public funds and contributes to deny access of subsistence farmers to agricultural bio-technology (Apel 2010).

Another consequence of GM labeling is that a farmer, growing a GM crop able to cross pollinate with a conventional or organic crop of the same species, might force the neighbor's crop to be labeled as GM if the adventitious presence of the genetic modification is above the threshold established in the regulations of each country.

This shows another paradox; expensive coexistence rules have been implemented to satisfy a capricious tolerance unrelated with human or environmental safety.

In principle, farmers should be able to cultivate the types of agricultural crops they choose be it GM crops, conventional or organic crops. The European Commission adopted a 2003 Guideline (2003/556/EC) which specifies that farmers introducing a new regional production system need to implement farm management measures (including cleaning of equipment, buffer zones, isolation distances and communication between neighboring farmers) allowing the neighbors a harvested product meeting the above mentioned 0.9% threshold.

Cross pollination is a "two-way street", taking place between GM and non-GM crops of the same species interchangeably, yet planting restrictions for farmers growing GM crops have been defined in some countries as obligations for farmers growing GM crops to allow coexistence. Pollen mediated gene flow is recognized as one source of impurities, which under most conditions can be managed with isolation distances of 20–50 m or 10–20 m buffers planted with conventional maize (Husken et al. 2007). Adventitious presence can also happen because of GM traces in seed, volunteers from the previous GM crop or some mixing with GM in the processes of harvesting, transport, storage and processing.

Most coexistence regulations implemented so far in the European Union concentrate the attention on minimum distances from conventional or organic crops. Good agricultural practices requested in Portugal (Carvalho et al. 2008) are successfully avoiding coexistence issues, but the more strict these regulations, the less likely is the coexistence as farmers interested in growing GM crops see the potential benefits reduced or gone (Skevas et al. 2010).

8.6 Conclusions from GM Monitoring and Surveillance in the EU

When the first cultivation approvals for GM maize protected against corn borers were approved by the European Commission in 1997 (Bt event 176) the discussion among stakeholders about potential undesirable effects such as development of resistance in the target pest (Not considered a problem for the environment in France by the Haute Conseil des biotechnologies (2009) in "Avis sur les réponses de l'AESA aux questions posées par les Etats members au sujet de la culture et de la consummation du maïs Mon 810, Dossier EFSA-GMO-RX-MON 810") or adverse effects on non target species was not settled. As Spain is importing large amounts of GM maize approved by the EC for the livestock industry, the Spanish Authorities took the decision in 1998 to allow the cultivation of some varieties derived from Bt176 conditioned to 5 years Monitoring Plans including:

• Communication of sales data for each variety by June 15th of each year.
• Monitoring of insecticide efficacy

- Monitoring of potential resistance development
- Further studies of effects on NTO entomofauna and soil microorganisms
- Studies on digestive flora (related to antibiotic resistance marker in Bt176)
- Information to farmers on planting of refuges, control of volunteer plants and buffer rows of conventional maize
- Notification procedures and emergency measures in case of resistance development or detection of adverse effects.

No adverse effects were reported from this monitoring in the 1998–2005 period where Bt176 maize was grown in Spain. Similar monitoring plans (but without the studies related to antibiotic resistance marker) were required for a growing number (over 100 in 2010) of varieties derived from MON 810 which have been grown in Spain since 2003.

In parallel, the European Commission introduced in Directive 2001/18/EC the obligation of Monitoring Plans and limited to 10 years the validity of each approval. The environmental safety information developed for the MON 810 Monitoring Plans and other publications in the scientific literature were thoroughly revised by EFSA. EFSA concluded "the likelihood of adverse effects on non-target organisms or on ecological functions very low, especially if appropriate management measures are adopted to mitigate exposure" and "the information available for maize MON 810 addresses the scientific comments raised by Member States and that maize MON 810 is as safe as its conventional counterpart with respect to potential effects on human and animal health. The EFSA GMO Panel also concluded that maize MON 810 is unlikely to have any adverse effect on the environment in the context of its intended uses, especially if appropriate management measures are put in place in order to mitigate possible exposure of non-target Lepidoptera" (http://www.efsa.europa.eu/EFSA/efsa_locale-1178620753812_1211902628240.htm).

Worldwide monitoring of insect resistance to genetically modified crops has shown some populations of Lepidoptera insects have succeeded in developing resistance after 10 years of cultivation (Tabashnik and Carrière 2009), but the area planted in the USA (http://www.ers.usda.gov/Data/BiotechCrops/), India (www.isaaa.org) is still growing. Integrated pest management (IPM) has been defined as the combination of all appropriate pest management techniques such as enhancing natural enemies, planting pest-resistant crops, adopting cultural management, and using pesticides judiciously, among others (http://www.ipmnet.org/ipmdefinitions/index.pdf). Experience so far indicates Bt crops can contribute to IPM better than with chemical insecticides (Romeis et al. 2006).

8.7 Patents and Liability

Because of the extremely thorough safety studies and risk evaluations, and the need for approval in different countries (US, EU, Japan, etc.) the development of GM plants has become a time consuming (10–12 years) and expensive (over 100 million

US$) project. The large investment required explains why most GM plants are used to improve large acreage crops, and requires some temporary commercial exclusivity in the hope of recovering the invested resources.

Patents are a world recognized tool to stimulate innovation. They offer no physical property rights but only an exclusivity of commercial rights during 20 years as an exchange for its publication for innovations which require inventive activity (not just discoveries) and have industrial applications. While in the US, patents can be granted to GM plant varieties, in the European Union, according to Directive 98/44/EC, plant varieties cannot be protected, but legal protection is applicable to biotechnological inventions provided they are new, involving an inventive step (not just discoveries) and susceptible of industrial application. In this way and through cross licensing, biotechnological inventions can be found in hundreds of varieties developed by different companies and including GM.

The approval process in the European Union of a GM plants protected against pests is a long and unpredictable process which may, when added to the time required to develop the many data required by regulators, consume most if not all the 20 year period of exclusive commercial rights granted by the patent. But additionally to this, the Environmental Liability Directive 2004/35/CE further discriminates against genetic modification by including deliberate releases, transport and placing on the market of approved GMOs among the Annex III activities requiring financial security or insurance which is not required for conventional breeding techniques. As the liability in the European Union persists up to 30 years following commercialization, another penalty against innovation is incurred.

8.8   Conclusions

Following 15 years of extensive use under closely controlled conditions, GM has proven to offer a sustainable tool to protect crop varieties from arthropod damage and to increase in this way the efficiency of food, feed and fiber production (Brookes et al. 2010). Strict regulation has been implemented in the EU to ensure than new GM varieties do not represent a risk for humans or the environment, but further legislation (the EU is planning to completely revise all GM legislation in 2012), labeling, segregation, coexistence and liability are superfluous if proper risk communication has been completed. However, it is appropriate to deploy insect-protected GM plants with proper stewardship to insure long-term performance, notably the planting of refuges with unprotected varieties to delay the development of resistance in target pests.

Data from controlled studies and monitoring in the commercial phase have shown that GM varieties protected from damage of target arthropods are more compatible with Integrated Pest Management than alternative treatments with insecticides when part of the damage has already happened. Optimum crop protection without the need to manufacture, package, transport and apply chemicals and minimum damage to non-target organisms is a key tool to reduce the environmental footprint of every unit of feed, fiber or food produced.

Acknowledgements   We thank Ivo O. Brants, T.G.A. Clemence, Gary F. Hartnell, Graham P. Head and David Songstad for their valuable comments to the manuscript.

References

Apel A (2010) The costly benefits of opposing agricultural biotechnology. New Biotechnology 27:635–640

Barros E, Lezar S, Anttonen MJ, van Dijk JP, Röhling RM, Kok EJ, Engel K-H (2010) Comparison of two GM maize varieties with a near-isogenic non-GM variety using transcriptomics, proteomics and metabolomics. Plant Biotechnol J 8:436–451

Bartsch D, Schmitz G (2002) Recent experience with biosafety research and post-market environmental monitoring in risk management of plant biotechnology derived crops. In: Thomas JA, Fuchs RL (eds) Biotechnology and safety assessment. Academic, New York

Bartsch D, Buhk HJ, Engel KH, Ewen C, Flachowsky G, Gathmann A, Heinze P, Koziolek C, Leggewie G, Meisner A, Neemann G, Rees U, Scheepers A, Schmidt S, Schulte E, Sinemus K, Vaasen A (2009) BEETLE final report on long-term effects of genetically modified (GM) crops on health and the environment (including biodiversity). European Commission, Brussels

Batista R, Saibo N, Lourenço T, Oliveira MM (2008) Microarray analyses reveal that plant mutagenesis may induce more transcriptomic changes tan transgene insertion. PNAS 105:3640–3645

Beever DE, Kemp CF (2000) Safety issues associated with the DNA in animal feed derived from genetically modified crops. A review of scientific and regulatory procedures. Nutr Abstr Rev Ser B: Livest Feeds Feeding 70:175–182

Brookes G, Barfoot P (2010) Global impact of biotech crops: environmental effects, 1996–2008. AgBioForum 13:76–94

Brookes G, Yu THE, Tokgoz S, Elobeid A (2010) The production and price impact of biotech corn, canola, and soybean crops. AgBioForum 13:25–52

Carvalho PC, Quedas F, Rocha F (2008) Manual de boas práticas de coexistencia para a cultura do milho. Direcção-Geral de Agricultura e Desenvolvimento Rural (DGADR). Ministerio de Agricultura, Desenvolvimento Rural e das Pescas, Lisboa, 31 p

Castañera P, Ortego F, Hernández-Crespo P, Farinós GP, Albajes R, Eizaguirre M, López C, Lumbierres B, Pons X (2010) El maíz Bt en España: experiencia tras 12 años de cultivo. PHYTOMA España 164:25–28

Christou P, Capell T (2009) Transgenic crops and their applications for sustainable agriculture and food security. In: Ferry N, Gatehouse AMR (eds) Environmental impact of genetically modified crops. CABI, Wallingford

Coll A, Nadal A, Palaudelmàs M, Messeguer J, Melé E, Puigdomènech P, Pla M (2008) Lack of repeatable differential expression patterns between MON 810 and comparable commercial varieties of maize. Plant Mol Biol 68:105–117

Costa J, Novillo C (2009) Regulación y conocimiento sobre plantas transgénicas. In: Dorado G, Jorrín J, Tena M, Fernández-Reyes E (eds) Biotecnología. Universidad de Córdoba, Spain

Cubero JI (2003) Introducción a la mejora genética vegetal, 2nd edn. Mundi-Prensa, Madrid

Del Monte JP, Aguado PL (2003) Survey of the non native plant species in the Spanish Iberia in the period 1975–2002. Flora Mediterr 13:241–259

Diamond J (1998) Guns, germs and steel. Random House Mondadori, Barcelona

Escobar J, Quintana J (2008) Reducción de riesgos sanitarios con el cultivo de un maíz transgénico. In: Libro de Resúmenes XIII Congreso Anual en Ciencia y Tecnología de los Alimentos, Madrid, Spain pp 29–31

García Olmedo F (1998) La tercera revolución verde. Plantas con luz propia. Editorial Debate, Madrid

García Olmedo F (2009) El ingenio y el hambre. De la revolución agrícola a la transgénica. Ed. Crítica, Barcelona

Husken A, Ammann K, Messeguer J, Papa R, Robson P, Schiemann J, Squire G, Stamp P, Sweet J, Wilhelm R (2007) A major European synthesis of data on pollen and seed mediated gene flow in maize in the SIGMEA project. In: Third international conference on coexistence between genetically modified (GM) and non-GM based agricultural supply chains, vol Book of Abstracts. Sevilla, Spain

Hutchinson WD, Burkness EC, Mitchell PD, Moon RD, Leslie TW, Fleischer SJ, Abrahamson M, Hamilton KL, Steffey KL, Gray ME, Hellmich RL, Kaster LV, Hunt TE, Wright RJ, Pecinovsky K, Rabaey TL, Flood BR, Raun ES (2010) Areawide suppression of European corn borer with Bt maize reaps savings to non-Bt maize growers. Science 330:222–225

James C (2009) Global status of commercialized biotech/GM crops. ISAAA Brief 41.Ithaca, NY

James C (2010) Global status of commercialized biotech/GM crops. ISAAA Brief 42.Ithaca, NY

Kessler C, Economidis I (2001) EC-sponsored research on safety of genetically modified organisms. European Commission. Community Research, Brussels, 246 p

Marvier M, McCreedy C, Regetz J, Kareiva P (2007) A meta-analysis of effects of Bt cotton and maize on nontarget invertebrates. Science 316:1475–1477

Messeguer J, Peñas G, Ballester J, Bas M, Serra J, Salvia J, Palaudelmàs M, Melé E (2006) Pollen-mediated gene flow in maize in real situations of coexistence. Plant Biotechnol J 4:633–645

Messeguer J, Palaudelmàs M, Peñas G, Serra J, Salvia J, Ballester J, Bas M, Pla M, Nadal A, Melé E (2007) Three year study of a real situation of coexistence in maize. In: Third international conference on coexistence between genetically modified (GM) and non-GM based agricultural supply chains. Book of Abstracts, pp 93–96

Moses V (2010) Do European consumers buy GMO foods? In: A decade of EU-funded GMO research (2001–2010). Project information. European Commission, European Research Area, Food, Agriculture & Fisheries & Biotechnology, Brussels, pp 240–245

Munkwold GP, Hellmich RL, Showers WB (1997) Reduced Fusarium ear rot and symptomless infection in kernels of maize genetically engineered for European corn borer resistance. Phytopathology 87:1071–1077

Novillo C, Soto J, Costa J (1999) Resultados en España con variedades de algodón, protegidas genéticamente contra las orugas de las cápsulas. Bol San Veg Plagas 25:383–393

Novillo C, Fernández-Anero FJ, Costa J (2003) Resultados en España con variedades de maíz derivadas de la línea MON 810, protegidas genéticamente contra taladros. Bol San Veg Plagas 29:427–439

Novillo C, Ojembarrena A, Tribó F, Alcalde E, Biosca D, Aragón M, Costa J (2007) Nine years of consumer-driven coexistence for GM-crops in Spain. In: Third international conference on coexistence between genetically modified (GM) and non-GM based agricultural supply chains. Book of Abstracts, pp 31–34

Ortega JI (2006) The Spanish experience with co-existence after 8 years of cultivation of GM maize. In: Proceedings of the Co-existence of GM, conventional and organic crops, Freedom of Choice Conference, Vienna, Apr 2006

Ortego F, Pons X, Albajes R, Castañera P (2009) European commercial genetically modified plantings and field trials. In: Ferry N, Gatehouse AMR (eds) Environmental impact of genetically modified crops. CAB International, Wallingford

Plan D, Van den Eede G (2010) The EU legislation on GMOs. An overview. JRC scientific and technical reports. Institute for Health and Consumer protection. European Commission

Potrykus I (2010) Regulation must be revolutionized. Nature 466:561

Raffensperger C, Barret K (2001) In defense of the precautionary principle. Nat Biotechnol 19:811–812

Ridley M (2010) The rational optimist. How prosperity evolves. Harper Business, New York

Rodrigo-Simón A, de Maagd RA, Avilla C, Bakker PL, Molthoff J, González-Zamora JE, Ferré J (2006) Lack of detrimental effects of Bacillus thuringiensis Cry toxins on the insect predator Chrysoperla carnea: a toxicological, histopathological and biochemical analysis. Appl Environ Microbiol 72:1595–1603

Romeis J, Meissle M, Bigler F (2006) Transgenic crops expressing *Bacillus thuringiensis* toxins and biological control. Nat Biotechnol 24:63–71

Serra J, Voltas J, López A, Capellades G, Salvia J, Coll A, Esteve T, Baixas S, Repiso C, Marrupe S (2008) Les micotoxines en el cultiu del blat de moro per a gra. Dossier Tècnic del DARP. Generalitat de Catalunya 27:15–18

Skevas T, Fevereiro P, Wesseler J (2010) Coexistence regulations and agriculture production: a case study of five Bt maize producers in Portugal. Ecol Econ 69:2402–2408

Stein AJ, Rodríguez-Cerezo E (2009) The global pipeline of new GM crops. Implications of asynchronous approval for international trade. JRC scientific and technical reports. Institute for Prospective Technological studies. European Commission

Tabashnik BE, Carrière Y (2009) Insect resistance to genetically modified crops. In: Ferry N, Gatehouse AMR (eds) Environmental impact of genetically modified crops. CABI, Wallingford

Tencalla FG, Nickson TE, García-Alonso M (2009) Environmental risk assessment. In: Ferry N, Gatehouse AMR (eds) Environmental impact of genetically modified crops. CABI, Wallingford

Wolfenbarger LL, Naranjo SE, Lundgreen JG, Bitzer RJ, Watrud LS (2008) Bt crops effects on functional guilds of non-target arthropods: a meta-analysis. PLoS One 3:e2118

Ye X, Al-Babili S, Kloti A, Zhang J, Lucca P, Beyer P, Potrykus I (2000) Engineering the provitamin A (β-Carotene) biosynthetic pathway into (carotenoid-free) rice endosperm. Science 287:303–305

Index

A

Abdeen, A., 166
Acetyl cholinesterase (AChE), 91, 92
Acyrthosiphon pisum, 78
Agrotis ipsilon, 81
Alarm signal, 52–55
Albano, S., 135
Alfonso-Rubí, J., 162
Alkaloid, 42
Allene oxide synthase/cyclase (AOS/AOC), 26
Allison, J.D., 51
Al-mazra'awi, M.S., 136, 137, 142
α-Amylase inhibitors, 163–164
Altabella, T., 164
Alternative food, 20, 56
Altpeter, F., 162
Alvarez-Alfagente, F., 163
Annotation-directed improvement, 3
Anthocoris nemoralis, 51
Anthocyanin, 31
Anthonomus grandis, 188
Anticarsia gemmatalis, 131
AOS/AOC. *See* Allene oxide synthase/cyclase
    (AOS/AOC)
Aphid responsive proteome, wheat
    antioxidants and other stress responses, 115
    metabolic re-programming, 111–112
    resistance genes, 110–111
    signal transduction, 112–114
    transcription and translation changes, 112
Apis mellifera and strawberry pathogen
    advantages, 150
    Bee-Treat dispenser, 147
    marketable yield overview, 148–149
    study locations, 146–147
Apoptosis, 18
Arabidopsis thaliana, 3, 10, 107

Argonaute proteins, 178
Argout, X., 4
Arms race, 41, 55
Aromatics, 31
Arthropod counter-adaptations
    avoidance and resistance, 41–42
    induced plant volatiles suppression, 44–46
    suppression, 42–44
Arthropod-plant coevolution, 1–2
Arthropod plant-related genomic projects, 6–8
Arvinth, S., 166
Avirulence (Avr) gene, 94

B

Bacillus subtilis, 143
Bacillus thuringiensis (Bt) crops
    cotton, 99–100
    cry toxins resistance, 100–101
    economic benefits, 96–97
    maize, 98–99
    potato, 100
    toxins, 97–98
Bacillus thuringiensis (Bt) toxin, 159
Bacillus thuringiensis var kurstaki, 136
Baldwin, I.T, 40, 108
Barbosa, A., 164
Baylor College of Medicine (BCM), 6
BBPs. *See* Biotin-binding proteins (BBPs)
B-chain of ricin (RB), 166
B. cinerea, 142
BCM. *See* Baylor College of Medicine (BCM)
Beauveria bassiana GHA, 135–137
BeeTreat dispenser, 139, 140
Beijing Genomics Institute (BGI), 3
Bell, H.A., 161
Benzoate, 31

Bergelson, J., 162, 166
Bernays, E.A., 80
β-Caryophyllene, 32
Beta-leptinotarsin-h, 167
BGI. *See* Beijing Genomics Institute (BGI)
Bilu, A, 139
Binab-T-vector, 142
Bioinformatic tools, 8–9
Biological control, 92–93
Biopesticides, 92
Biotechnological approaches.
                    See Phytophagous arthropods,
                    biotechnological approaches
Biotin-binding proteins (BBPs), 167
Boguski, M., 107
Bollgard[r] cotton, 99–100
Bombus impatiens, 135–137
Bombus terrestris, 150–152
Bonade-Bottino, M., 163
Botrytis cinerea, 134, 135, 151
Bouchard, E., 163
Boulter, D., 161, 166
Boyko, E.V., 110
Broadway, R.M., 81
Bruchins, 25
Bruchis pisorium, 25
Bruessow, F., 44
Brunelle, F., 163
Bt crops. *See Bacillus thuringiensis (Bt)* crops
Bt toxin. *See Bacillus thuringiensis*
                    (Bt) toxin
Buchnera, 78
Bumble bee, 150–152
Butt, T.M., 136

C
Calmodulin-binding proteins, 113
Carbonero, P., 164
Carnivore, 20, 56
Carreck, N.L., 137
Carrillo, L., 163
Carson, R., 91
Caryophyllene, 32
Case-by-case approach, 201
Castagnoli, M., 161
CDD, 8, 9
Chamberlain, D.J., 80
Chan, A.P., 4
Charity, J.A., 162, 166
Chemical defenses. *See* Phytophagous insects
Chemotaxis, 47, 48
Chen, M., 166

Chilo suppressalis, 166
Chrispeel, M.J., 164
Christy, LA., 162
Cicadulina mbila, 166
Cipriani, G., 161
Cloutier, C., 163
2-C-methyl-Derythritol 4-phosphate (DOXP)
                    pathway, 30
Coexistence regulations, 206, 210
Cohen, A.C., 132
Comparative genomics, 8
Confalonieri, M., 161
Coronatine, 43
Cowgill, S.D., 163
Cowpea trypsin inhibitor (CPTI) gene, 160
Cross pollination, 210
cry1Ab gene, 98
Cry wolf strategy, 53–54
C_6-volatiles, 19
Cyanogenesis, 31
Cysteine-protease inhibitors, 160, 163
Cytochrome P450 genes, 96

D
Daizo strain, 189
Da Silveira, V., 162
Dedej, S., 134, 143
Defence proteins
        α-amylase inhibitors, 163–164
        BBPs, 167
        CPTI and serine-protease inhibitor genes,
                    160–162
        cysteine-protease inhibitors and
                    phytocystatins, 160, 163
        gene pyramiding, 165–166
        multiple inhibitory activities coding genes,
                    164–165
        PI gene, 160
Defenses
        constitutive, 16
        direct plant, 42–44
        indirect plant (*see* Induced
                    plant volatiles)
        induced, 16, 17
        plant, 16–18
Degenhardt, J., 95
De Leo, F., 161
Delledonne, M., 163
De Moraes, C.M., 45
De Sousa-Majer, M.J., 164
Detoxification and insect modulation, 96
De Vos, M., 109

DGE tag technologies. *See* Digital gene expression (DGE) tag technologies
Diabrotica undecimpunctata, 167, 184
Diabrotica virgifera, 181–182
Dicer enzymes, 178
Diezel, C., 43
Digestive physiology, phytophagous insects. *See* Phytophagous insects
Digital gene expression (DGE) tag technologies, 192
Dimethylallyl diphosphate (DMAPP), 29, 30
4,8-Dimethyl-1,3*(E)*,7-nonatriene (DMNT), 32, 36
Ding, L.C., 162
Diurnal rhythms, 33–35
DMAPP. *See* Dimethylallyl diphosphate (DMAPP)
DMNT. *See* 4,8-Dimethyl-1,3*(E)*,7-nonatriene (DMNT)
DOXP pathway. *See* 2-C-methyl-D-erythritol 4-phosphate (DOXP) pathway
2D-PAGE. *See* Two-dimensional polyacrylamide gel electrophoresis (2D-PAGE)
Drukker, B., 40
dsRNA, 189–190
Duan, X., 162
Dunse, K.M., 166

E
EC Regulation 1829/2003, 201
Ecto-endoperitrophic flow model, 77
Effectors, 22
Ehrlich, P.R., 2
Elad, Y., 143
Elicitors. *See* Herbivore
El-Sayed, A.M., 50
Entomovector technology. *See* Multitrophic interactions, entomovector technology
Environmental RNAi, 188, 189
Enzyme inhibitors, 104–105
Erwinia amylovora, 128, 138, 143
Escande, A.R., 135
Ethylene (ETH), 94
Eurobarometer, 208
European Food Safety Authority (EFSA), 201–203
European Union (EU), 200
Eurosta solidaginis, 45
Evolutionary dynamics. *See* Induced plant volatiles

F
FACs. *See* Fatty acid conjugates (FACs)
Falco, M.C., 166
Fang, H.J., 161
Farnesyldiphosphate synthase (FPS), 30
Fatty acid conjugates (FACs), 23
Felton, G.W., 22, 79
Ferry, N., 110, 111, 113, 114, 163
Feuillet, C., 5
Flavonoid, 31
FLIGHT, 191
Fourier-transform ion cyclotron mass spectrometry (FT-ICR-MS), 115
FPS. *See* Farnesyldiphosphate synthase (FPS)
Frankliniella occidentalis, 165
Freeman, S., 143
Frequency-dependent selection, 53, 55
FT-ICR-MS. *See* Fourier-transform ion cyclotron mass spectrometry (FT-ICR-MS)
Fumonisin B_1, 207
Functional genomics, 107 *See also* Plant–insect interactions, successes and failures
Fusarium nivale, 207

G
GABA. *See* Gamma-aminobutyrate (GABA)
GABA-TP. *See* Gamma-amino butyric acid transaminase (GABA-TP)
Galanthus nivalis agglutinin (GNA), 105
Gamma-aminobutyrate (GABA), 18
Gamma-amino butyric acid transaminase (GABA-TP), 29
Gatehouse, A.M.R., 161, 163, 164, 166
Gene defense. *See* Defence proteins
Gene families, 9–10
Gene pyramiding, 165–166
Genes co-evolution, arthropod-plant interactions
    future directions, 10
    gene families involved, 9–10
    gene family bioinformatics tools, 8–9
    pest genomes, 6–8
    plant-arthropod co-evolution, 1–2
    plant genomes
        community-defined categories, 3
        crop plant genome projects, 6
        sequencing projects, 3–5

Genetically modified (GM) plants,
　　regulatory approvals
　agricultural sustainability and, 199–200
　labeling, traceability and segregation
　　impacts, 208–210
　malfunctioning approval process
　　adoption, 203–204
　　environmental risk assessments, 204–205
　　exotic invasive species *vs.*, 205
　　golden rice, 206
　　precautionary principle, 204
　monitoring and surveillance, 210–211
　patents and liability, 211–212
　regulatory oversight and products
　　centralized approval system, 201
　　commercial release authorization,
　　　200–201
　　EFSA guidelines, 201–202
　　EFSA Scientific Opinion, 203
　　environmental risk assessment,
　　　202–203
　risk evaluation and social perception,
　　206–208
Genomes On-Line Database (GOLD), 3
Geranylgeranyl diphosphate (GGPP), 30
GGPP. *See* Geranylgeranyl diphosphate (GGPP)
Girad, C., 163
Glandular trichomes, 15
Gliocladium catenulatum, 144
Glucose oxidase, 80
Glucosinolates, 31, 42
GLVs, Se Green leaf volatiles (GLVs)
GM plants. *See* Genetically modified (GM)
　plants, regulatory approvals
GNA. *See* Galanthus nivalis agglutinin (GNA)
Goff, S.A., 5
GOLD. *See* Genomes On-Line Database
　(GOLD)
Golden rice, 206
Golmirizaie, A., 161, 166
Gossypium hirsutum, 99
Gouinguené, S., 39
Gouinguené, S.P., 36
Graham, J., 161
Green leaf volatiles (GLVs), 28–29
GreenPhylDB, 9, 10
Gross, H.R., 136

H
Halitschke, R., 40
HAMPs. *See* Herbivory associated molecular
　patterns (HAMPs)
Han, L., 166
Hao, Y., 161

Hare, J.D., 51
Heath, R.L., 162
Helicoverpa armigera, 79, 185
Heliothis nuclear polyhedrosis
　virus (HNPV), 145
Heliothis subflexa, 41–42
Heliothis virescens, 18, 160
Heliothis zea, 79, 81
Herbivore
　chewing, 18, 40
　effectors, 22
　elicitors
　　beta-glucosidase, 24
　　bruchins, 25
　　caeliferins, 23–24
　　FACs, 23
　　glucose oxidase, 43
　　isomerase, 25
　　orally secreted proteins, 24–25
　　volicitin, 23
　regurgitant, 22–25
　saliva, 24, 28
　stylet feeder, 22
Herbivore-induced plant volatiles (HIPV), 20
Herbivory associated molecular patterns
　(HAMPs), 1
Hesler, L.S., 162
Hieter, P., 107
High-quality draft, 3
Hilder, V.A., 161
HIPV. *See* Herbivore-induced plant
　volatiles (HIPV)
HNPV. *See* Heliothis nuclear polyhedrosis
　virus (HNPV)
Hoffman, M.P., 161
Hokkanen, H.M.T., 132
Honey bee. *See* Apis mellifera and strawberry
　pathogen
Honeydew, 22
Hopke, J., 24
HPL. *See* Hydroperoxide lyase (HPL)
HPL, 31
HR. *See* Hyper-sensitive response (HR)
Huang, S., 4
Human Genomic Sequencing Center, 6
Hydrolase inhibitor, 160
Hydroperoxide lyase (HPL), 29, 95
Hyper-sensitive response (HR), 16

I
Ignacimuthu, S., 164
Illumina's RNA-sequencing, 192
Improved high-quality draft, 3
Inceptines, 24

Indirect defence/indirect defenses, 95–96
    See also Induced plant volatiles
Induced plant volatiles
    defense strategies
        induced, 16
        onset, 17–18
        resource allocation, 16–17
    evolutionary dynamics
        cry wolf strategy, 53–54
        natural selection, 54–55
        odor blends, 54
        predictive modeling, 52–53
    future prospects, 55–56
    indirect defenses
        arthropod counter-adaptations
            (*see* Arthropod counter-adaptations)
        arthropod learning and consequences,
            49–51
        caeliferins, 23–24
        FACs, 23
        inceptines, 24
        onset of, 20–22
        orally secreted proteins, 24–25
        oviposition-derived cues, 25
        structural barriers, 15–16
        tritrophic interaction, 20, 21
    plant-odor recognition
        Bombyx mori, 47
        components ratio, 48–49
        Drosophila melanogaster, 46
        mixtures of, 47–48
    production variation
        diurnal rhythms, 33–35
        factors, 33, 34
        genotypes, species and different
            herbivores, 37–40
        growth conditions, 36–37
        olfactory choice assay, 37, 38
        tissue age and position, 35–36
    transgenic approaches, 31–33
    tritrophic interactions mediated, 40–41
    upstream signaling pathways
        aromatics and others, 31
        biosynthesis pathways, 26, 27
        GLVs, 28–29
        jasmonic acid, 26
        salicylic acid, 28
        terpenes, 29–30
    VOCs, 19
Induced systemic resistance (ISR), 94
Innate attraction, 47, 51
Innate immune response, 186

Insect control strategies. *See* Plant–insect
        interactions, successes and failures
Insect crop pests control, RNAi. *See* RNA
        interference (RNAi)
Insect gut physiological adaptations. *See*
        Phytophagous insects
Insecticides, 91–92
Integrated pest management (IPM), 92, 93,
        165, 193, 211, 212
InterPro, 8, 9
IPM. *See* Integrated pest management (IPM)
IPP. *See* Isopentenyl diphosphates (IPP)
Irie, K., 163
Ishimoto, M., 164
(+)-7-Iso-jasmonoyl-L-isoleucine
        (JA-Ile), 26
Isolate-and-kill strategy, 16
Isopentenyl diphosphates (IPP), 29, 30
ISR. *See* Induced systemic resistance (ISR)

J
JA. *See* Jasmonic acid (JA)
JA-Ile. *See* (+)-7-Iso-jasmonoyl-L-isoleucine
        (JA-Ile)
Jaillon, O., 5
Jansen, J.J., 116
Jansen, V.A.A, 52
Jasmonate Zim domain (JAZ) transcriptional
        repressors, 26
cis-Jasmone, 26, 47
Jasmonic acid (JA), 26–27, 94
JAZ transcriptional repressors. *See* Jasmonate
        Zim domain (JAZ) transcriptional
        repressors
J. Craig Venter Institute (JCVI), 3
JGI. *See* Joint Genome Institute (JGI)
Johnson, R., 162
Joint Genome Institute (JGI), 3
Jongsma, M.A., 162
Jouanin, L., 163
Jyoti, J.L., 136

K
Kapongo, J.P., 135, 144
Kessler, A., 40
Kessler, D., 33
Khadeeva, N.V., 162
Klopfenstein, N.B., 162
Koundal, K.R., 161
Kovach, J., 134

L

LA. *See* Linolenic acid (LA)
Labelling, 208–210
Lara, P., 162
Lawrence, P.K., 161
Learning
  associative, 50
  olfactory, 50
Lecardonnel, A., 163
Lectins, 105–106, 165–166
Lee, S.I., 161
Lepidoptera database, 191
Leple, J.C., 163
Leptinotarsa decemlineata, 100, 184
Leucine-rich repeat (LRR), 110
Liability, 211–212
Li, G., 166
Linolenic acid (LA), 42
Li, X.C., 96
Li, Y., 161
Losey, J.E., 102
LRR. *See* Leucine-rich repeat (LRR)
Luo, M., 162

M

Maccagnani, B.B.C., 128, 134–136, 143
MacWorm, 20
Maheswaran, G., 162
Mannose-specific snowdrop lectin GNA, 165
MAPK. *See* Mitogen activated protein kinase
    (MAPK)
Maqbool, S.B., 166
Marchetti, S., 161
Mason bee, 128, 133
Mass-spectrometric (MS) analysis, 109
Matsushima, R., 45
Mazumdar-Leighton, S., 81
MCA. *See* Microbiological control agents
    (MCA)
McManus, M.T., 161, 162
Mehlo, L., 166
Menzler-Hokkanen, I., 132
MEP. *See* Non-Mevalonate pathway (MEP)
MeSA. *See* Methyl salicylate (MeSA)
Metabolomics, 115–116
Metarhizium anisopliae, 136, 137
Methyl jasmonate (MeJA), 27
Methyl salicylate (MeSA), 28
Mevalonate (MVA) pathway, 29, 30
Microbiological control agents (MCA)
  dissemination
    dilutions and formulations role, 140–142

dispenser types and loading capacity,
    138–140
  efficacy and pest management agents
    barriers, 132
    conditions, 130
    examples, 130–131
    selection criteria and approaches, 131
  reliability
    pest insects, 145
    plant pathogens *vs.*, 142–144
MicroRNAs (miRNAs), 178
Ming, R., 4
miRNAs. *See* MicroRNAs (miRNAs)
Mites
  Iphiseius degenerans, 21, 38
  Panonychus ulmi, 37
  Phytoseiulus persimilis, 32, 45, 48
  Tetranychus evansi, 44–46
  Tetranychus kanzawai, 45
  Tetranychus urticae, 24, 43, 50
Mitogen activated protein kinase
    (MAPK), 113
Moar, W.J., 192
Mochizuki, A., 162
Mommaerts, V., 128, 136, 150
MON 810, 200, 211
Monarch butterfly, 102
Monilinia vaccinii-corymbosi, 143
Moran, P.J., 108
Morton, R., 164
MS analysis. *See* Mass-spectrometric (MS)
    analysis
Multitrophic interactions, entomovector
    technology
  biological control, 128–130
  definition, 127
  future perspectives, 152–153
  key components, 128
  MCA (*see* Microbiological control
    agents (MCA))
  pollinator vector
    selection, 132–133
    studies overview, 133–137
    transport and deposition, 138
  schematic view, 132, 133
  vs. strawberry pathogen
    Apis mellifera (*see Apis mellifera* and
      strawberry pathogen)
    Bombus terrestris, 150–152
    production and problems, 145–146
Murdock, L.L., 161
MVA pathway. *See* Mevalonate (MVA)
    pathway

N

Nandi, A.K., 161
Natural selection, 20, 39, 54
NCBI Entrez Genome Project database, 191
Ngugi, H.K.
N-17-hydroxylinolenoyl-L-glutamine, 23
Nicotine, 31, 33, 43
Ninković, S., 163
NMR spectroscopy. *See* Nuclear magnetic
        resonance (NMR) spectroscopy
Non-Mevalonate pathway (MEP), 30
Nuclear magnetic resonance (NMR)
        spectroscopy, 115
Nutt, K.A., 162

O

Olfactometer, 38
Olfactory choice assay, 37, 38
Olfactory receptor, 46, 47
One door, one key process, 201
One-way dispenser, 138–139
OPDA. *See cis*-Oxophytodienoic acid (OPDA)
Orally secreted proteins, 24–25
Osmia cornuta, 128
Ostrinia nubilalis, 205
Outchkourov, N.S., 163
Oviposition-derived cues, 25
cis-Oxophytodienoic acid (OPDA), 26

P

Palli, S.R., 181
Parasitism, 41
Parasitoid, 48, 50, 51
Paré, P.W, 23
Park, J.R., 97
Paterson, A.H., 5
Peng, G., 134
Perimicrovillar membrane, 77
Peritrophic membrane (PM), 77
Pest insects, 145
Phaseolus lunatus, 20
Phenylpropanoid, 31, 33
Phipps, R.H., 97
Photorhabdus, 167
Photorhabdus spp., 106
Photosynthesis, down regulation of, 15
Phytochrome-interacting factor 4 (PIF4), 112
Phytocystatins, 160, 163
Phytophagous arthropods, biotechnological
        approaches
    defence proteins and transgenic plants
        α-amylase inhibitors, 163–164

BBPs, 167
CPTI and serine-protease inhibitor
        genes, 160–162
cysteine-protease inhibitors and
        phytocystatins, 160, 163
gene pyramiding, 165–166
multiple inhibitory activities coding
        genes, 164–165
PI gene, 160
insect-resistant crops and traits, 159
pest control, 167–169
Phytophagous insects
chemical defenses
    allelochemicals, 80–81
    genes encoding, 81–82
    insecticidal proteins, 81
    salivary enzymes, 80
    transcriptome analysis, 82
digestive physiology
    cDNA screening, 78–79
    hemiptera and thysanoptera, 77–78
    limiting factors, 76
    pH and redox potential, 79–80
    PM, 77
    proteases, 78
evolution, 75–76
orders involved, 75
Phytoseiulus persimilis, 48
Phytozome, 9
PIF4. *See* Phytochrome-interacting factor 4
        (PIF4)
PIs. *See* Protease inhibitors (PIs)
Plant-arthropod co-evolution, 1–2
Plant-arthropod interactions, induced
        plant volatiles. *See* Induced
        plant volatiles
Plant defence proteins
    enzyme inhibitors, 104–105
    lectins, 105–106
Plant endogenous defenses. *See* Plant–insect
        interactions, successes and failures
Plant feeding insects. *See* Phytophagous insects
Plant genomes
    community-defined categories, 3
    crop plant genome projects, 6
    sequencing projects, 3–5
Plant hormones
    ABA, 18, 26, 28
    auxin, 31
    ethylene, 24–26
    gibberellins, 26, 30
    JA, 26
    JA/SA cross talk, 44
    SA, 26, 28

Plant–insect interactions, successes
and failures
functional genomics
aphid responsive proteome (*see* Aphid
responsive proteome, wheat)
proteome level, 109–110
transcriptome level, 107–109
future prospects, 106–107
insect control strategies
biological control, 92–93
biopesticides, 92
insecticides, 91–92
IPM, 93
insecticidal molecules sources, 106
metabolomics, 115–116
phytophagous pests, 89–90
plant defence proteins
enzyme inhibitors, 104–105
lectins, 105–106
plant endogenous defenses
detoxification and insect modulation, 96
herbivore-induced transcriptome
studies, 95
indirect defence, 95–96
induced disease resistance types, 93–94
nonspecific induced defense
pathways, 94
transgenic crops
Bt cotton, 99–100
Bt cry toxins resistance, 100–101
Bt maize, 98–99
Bt potato, 100
Bt toxins, 97–98
economic benefits, 96–97
environmental impacts, 101–103
Plant-odor recognition
Bombyx mori, 47
components ratio, 48–49
Drosophila melanogaster, 46
mixtures of, 47–48
Plant pathogens. *See* Microbiological
control agents (MCA)
Plant-predator alliance, 52, 53, 55
PLAZA, 9
PM. *See* Peritrophic membrane (PM)
Pollen mediated gene flow, 210
Pollinator vector. *See* Multitrophic interactions,
entomovector technology
Polymorphism, 46
Potrykus, I., 206
Prakash, S., 164
Predator, 48–51
Protease inhibitors (PIs), 160
for controlling pests, 104
transcripts, 108, 109

PR-protein, 28
Pyramided insect-protection traits, 184, 192
Pyrethroids, 91

Q

Quilis, J., 163

R

Rahbe, Y., 163
Rapid Alert System, 207
Raven, P.H., 2
RB. *See* B-chain of ricin (RB)
RdRp enzymes. *See* RNA-dependent RNA
polymerase (RdRp) enzymes
Reactive oxygen species (ROS), 37, 115
Regurgitation, 22–23
Rescue-the-resources response, 17
Resistance *(R)* gene mediated defense, 93–94
Resource flow, manipulation of, 17
Resources, re-allocation of, 17
Ribeiro, A.P.O., 163
Rice membrane protein library, 109
Rice proteome database, 109
Richards, S., 7
RISC. *See* RNA-induced silencing
complex (RISC)
Rivard, D., 166
RNA-dependent RNA polymerase (RdRp)
enzymes, 178
RNA hairpin, 183, 184, 187
RNAi. *See* RNA interference (RNAi)
RNA-induced silencing complex (RISC), 188
RNA interference (RNAi), 168
coleopteran insects
basic outline, 181, 183
Bt toxin, 183–184
red flour beetle, 180–181
resistance, 184
toxicity LC_{50} values, 183
western corn rootworm, 181–183
definition and application, 177
factors determining efficiency
dsRNA or siRNA degrading enzymes,
189–190
dsRNA uptake, 189, 190
expression, 188–189
immune response involvement, 190
future perspectives, 191–193
hemipteran insects, 186–187
lepidopteran insects, 184–186
other insect orders, 187
principles, 178–180
spider mites, 187–188

Robinson, G.E., 191
Rocha, J.J.E., 192
ROS. *See* Reactive oxygen species (ROS)
Roundup Ready® soybean, 200

S
Salicylic acid (SA), 16, 28, 94
Salicylic acid methyl transferase (SAMT), 33
Sallaud, C., 30
SAMT. *See* Salicylic acid methyl
        transferase (SAMT)
Sane, V.A., 161
Sanger sequencing technology, 3
Santos, M.O., 166
SAR. *See* Systemic acquired resistance (SAR)
Sarmah, B.K., 164
Schistocerca americana, 23
Schistocerca gregaria, 80
Schmelz, E.A., 35
Schmutz, J., 4
Schnable, P.S., 5
Schroeder, H.E., 164
Schuler, T.H., 163
Sclerotinia sclerotiorum, 143
Scutareanu, P., 40
Senescence, 17
Senthilkumar, R., 166
Serine-protease inhibitor genes, 160–162
Shade, R.E., 164
Shafir, S., 134, 144
Shikimate, 28, 31
Shulaev, V., 4
Shulke, R.H., 161
Signal (dis)honesty, 52–55
Silva, M.C., 166
siRNA. *See* Small interfering RNA (siRNA)
Sitobion avenae, 110
Slow-them-down strategy, 16, 17
Small interfering RNA (siRNA), 189–190
Spider mites, 187–188
Spodoptera exigua, 23, 81
Spodoptera frugiperda, 24, 168, 185
Spodoptera littoralis, 20, 166
Srinvasan, T., 162
Standard draft, 3
Step-by-step approach, 201
Strawberry. *See* Multitrophic interactions,
        entomovector technology
Suppression
    defense, 42–44
    volatiles, 44–46

Sutton, J.C., 134, 139
Systemic acquired resistance (SAR), 94
Systemic RNAi, 178, 179

T
Takabayashi, J., 36, 37
Tan, A., 181
Terpenes, 29–30
Terpene synthases (TPS), 30
Tetranychus urticae, 187
T. evansi, 45
TFs. *See* Transcription factors (TFs)
Thompson, G.A., 108
Thomson, S.V., 134
Thrips, 38, 40
Tiered approach, 203
Tooker, J.F., 45
TPS. *See* Terpene synthases (TPS)
Traceability, 208–210
Transcription factors (TFs), 168
Transgenic approaches, induced plant volatiles,
        31–33
Transgenic crops. *See Bacillus thuringiensis
        (Bt)* crops
Transgenic plants, 180, 183
Tribolium castaneum, 81, 179–181
Trichome, glandular, 15, 19, 33
4,8,12-Trimethyltrideca-1,3,7,11-tetraene
        (TMTT), 27, 37
Tritrophic interaction, 20, 21 *See also* Induced
        plant volatiles
Tumlinson, J.H, 22
Turlings, T.C.J., 36, 39
Tuskan, G.A., 4
Two-dimensional polyacrylamide gel
        electrophoresis (2D-PAGE), 109
Two-way dispenser, 138–141

U
Ubiquitin/proteasome system (UPS), 112
United States Department of Agriculture
        (USDA), 99
UPS. *See* Ubiquitin/proteasome system (UPS)
USDA. *See* United States Department of
        Agriculture (USDA)

V
Van Baalen, M., 52
Van Loon, J.J.A, 41

Velasco, R., 4
Vila, L., 162
Volatile organic compounds (VOCs),
        19, 95
Volatile production. *See* Indirect defence
Volatiles
    aromates, 19
    bergamotene, 32, 40
    caryophyllene, 32
    C_6-volatiles, 19
    diterpene, 29, 30, 32
    diurnal, 33–35
    DMNT, 32, 36
    elicitors, 22–23
    ent-kaurene, 19, 30
    farnesene, 32
    GLVs, 28–29
    headspace, 26, 28
    hexenal, 35, 48
    hexenyl acetate, 28, 37
    HIPV, 20
    isoprene, 29, 30, 32
    isoprenoids, 29
    linalool, 32, 36
    MeJA, 27
    MeSA, 28
    monoterpene, 29, 30
    nerolidol, 32
    nitriles, 37
    ocimene, 36
    oxime, 31, 37
    plant (*see* Induced plant volatiles)
    plant volatiles induction (*see* Induced
        plant volatiles)

sesquiterpene, 29, 30, 32
terpenoids, 19
TMTT, 27, 37

W
Western corn rootworm, 181–183
WideStrike, 100
Winterer, J., 162, 166
WORLD-2DPAGE, 109
Wounding response, 96
Wraight, C.L., 102
Wu, Y., 162

X
Xenorhabdus, 167
Xia, Q., 7
Xu, D., 161

Y
Yeh, K.W., 162
Ye, X., 206
YieldGard technology, 99
Yu, H., 134, 139
Yu, J., 5

Z
Zhang, F., 108
Zhao, J.Z., 166
Zhu, F., 192
Zhu-Salzman, K., 110, 165